Functional Biomaterial for Bone Regeneration

Functional Biomaterial for Bone Regeneration

Guest Editors

Marija Čandrlić
Željka Perić Kačarević
Matej Tomas
Arun Kumar Rajendran

MDPI

Basel • Beijing • Wuhan • Barcelona • Belgrade • Novi Sad • Cluj • Manchester

Guest Editors
Marija Čandrlić
Department of Integrative Dental Medicine
J.J. Strossmayer University of Osijek
Osijek
Croatia

Željka Perić Kačarević
Department of Anatomy, Histology, Embriology, Pathology Anatomy and Pathology Histology
J.J. Strossmayer University of Osijek
Osijek
Croatia

Matej Tomas
Department of Dental Medicine
J.J. Strossmayer University of Osijek
Osijek
Croatia

Arun Kumar Rajendran
School of Healthcare Science and Engineering (SHINE)
Vellore Institute of Technology
Vellore
India

Editorial Office
MDPI AG
Grosspeteranlage 5
4052 Basel, Switzerland

This is a reprint of the Special Issue, published open access by the journal *Journal of Functional Biomaterials* (ISSN 2079-4983), freely accessible at: https://www.mdpi.com/journal/jfb/special_issues/22LJI7A58B.

For citation purposes, cite each article independently as indicated on the article page online and as indicated below:

Lastname, A.A.; Lastname, B.B. Article Title. *Journal Name* **Year**, *Volume Number*, Page Range.

ISBN 978-3-7258-6055-5 (Hbk)
ISBN 978-3-7258-6056-2 (PDF)
https://doi.org/10.3390/books978-3-7258-6056-2

Contents

About the Editors

Marija Čandrlić

Marija Čandrlić is an assistant professor at the Department of Dental Medicine, Faculty of Dental Medicine and Health Osijek, J. J. Strossmayer University of Osijek, Croatia. Her scientific work focuses on the biology and application of biomaterials in regenerative dental medicine, guided bone and tissue regeneration, dental implantology, and oral surgery. She has participated in several national and institutional research projects addressing histological, immunohistochemical, and micro-CT evaluation of bone substitute materials and barrier membranes. Her current research interests include magnesium-based biomaterials, xenografts modified with hyaluronate, and autologous dentin as a natural grafting material. She is actively involved in undergraduate and postgraduate teaching and has served as Guest Editor of Special Issues dedicated to functional biomaterials and bone regeneration.

Željka Perić Kačarević

Željka Perić Kačarević is an associate professor at the Department of Anatomy, Histology, Embryology, Pathological Anatomy, and Pathological Histology, Faculty of Dental Medicine and Health Osijek, J. J. Strossmayer University of Osijek, Croatia. Her scientific interests lie in biomaterials, histology, bone tissue regeneration, and regenerative medicine, with a particular focus on host–material interactions and macrophage-mediated immune responses. She has co-authored numerous experimental and translational studies evaluating novel bone grafts, collagen membranes, and resorbable metals. Her research combines histological and molecular approaches to explore the biological mechanisms underlying bone healing. In addition to her scientific work, she contributes extensively to academic teaching and mentorship in preclinical and biomedical sciences.

Matej Tomas

Matej Tomas is an assistant professor at the Department of Dental Medicine, Faculty of Dental Medicine and Health Osijek, J. J. Strossmayer University of Osijek, Croatia. His research focuses on biomaterials, histology, bone regeneration, immunohistochemistry, and histomorphometry, as well as the use of artificial intelligence in dental medicine. He has contributed to preclinical and clinical studies investigating the biological and structural outcomes of bone substitute materials and regenerative procedures in periodontology and implantology. Besides his research activity, he is engaged in clinical practice and teaching, integrating evidence-based regenerative concepts into dental education and patient care.

Arun Kumar Rajendran

Arun Kumar Rajendran currently works as an Associate Professor at the School of Healthcare Sciences and Engineering (SHINE), Vellore Institute of Technology, Vellore, India. He has worked as an Assistant Professor of Research at the School of Chemical and Biological Engineering, the Institute of Chemical Processes, Seoul National University, Seoul, South Korea. He is a dental surgeon and holds a master's degree in Nanoscience and Technology. He obtained his PhD from Tokyo Medical and Dental University, Tokyo, Japan. His major research interests include designing and synthesising bioceramics, biopolymers, decellularised ECM-based composite scaffolds, membranes, tissue adhesives, hydrogels and peptide-based photocurable hydrogels for tissue engineering and regenerative medical applications. He also focuses on utilising mechanosignalling for controlling cellular differentiation.

Preface

This Reprint, "Functional Biomaterial for Bone Regeneration", presents current scientific and clinical perspectives on the design and application of biomaterials that promote bone regeneration. Its scope covers bioactive scaffolds, resorbable and non-resorbable membranes, grafting materials, and implant surface modifications that modulates osteogenesis and tissue integration. The aim is to highlight recent experimental, preclinical, and clinical findings that deepen understanding of how biomaterial composition and structure influence biological performance. Motivated by the continuous progress in regenerative dentistry and bone tissue engineering, this Reprint is intended for researchers, clinicians, and postgraduate students engaged in developing and applying functional biomaterials for bone regeneration.

Marija Čandrlić, Željka Perić Kačarević, Matej Tomas, and Arun Kumar Rajendran
Guest Editors

Journal of *Functional Biomaterials*

MDPI

Article

Regeneration of Critical Calvarial Bone Defects Using Bovine Xenograft, Magnesium-Enriched Bovine Xenograft and Autologous Dentin in Rats: Micro-CT, Gene Expression and Immunohistochemical Analysis

Marija Čandrlić [1,†], Ana Terezija Jerbić Radetić [2,†], Hrvoje Omrčen [3], Barbara Franović [2,4], Lara Batičić [5], Tamara Gulić [6], Tea Čaljkušić-Mance [7,8], Sanja Zoričić Cvek [2], Lucija Malešić [2,9], Željka Perić Kačarević [10,11,*] and Olga Cvijanović Peloza [2,*]

1 Department of Integrative Dental Medicine, Faculty of Dental Medicine and Health Osijek, J.J. Strossmayer University of Osijek, 31000 Osijek, Croatia; marija.candrlic@fdmz.hr
2 Department of Anatomy, Faculty of Medicine, University of Rijeka, Braće Branchetta 20/1, 51000 Rijeka, Croatia; ana.jerbic.radetic@uniri.hr (A.T.J.R.); barbara.franovic@medri.uniri.hr (B.F.); sanja.zoricic@medri.uniri.hr (S.Z.C.); lucija.malesic@gmail.com (L.M.)
3 Department of Clinical Microbiology, Teaching Institute of Public Health of Primorsko-Goranska County, Krešimirova 52a, 51000 Rijeka, Croatia; hrvoje.omrcen@zzjzpgz.hr
4 Doctoral School of Biomedicine and Health, Faculty of Medicine, University of Rijeka, Braće Branchetta 20/1, 51000 Rijeka, Croatia
5 Department of Medical Chemistry, Biochemistry and Clinical Chemistry, Faculty of Medicine, University of Rijeka, Braće Branchetta 20/1, 51000 Rijeka, Croatia; lara.baticic@uniri.hr
6 Department of Physiology, Immunology and Pathophysiology, Faculty of Medicine, University of Rijeka, Braće Branchetta 20/1, 51000 Rijeka, Croatia; tamara.gulic@uniri.hr
7 Department of Ophthalmology, Faculty of Medicine, University of Rijeka, Braće Branchetta 20/1, 51000 Rijeka, Croatia; tea.mance.caljkusic@uniri.hr
8 Clinic for Ophthalmology, Clinical Hospital Center Rijeka, Krešimirova 42, 51000 Rijeka, Croatia
9 Dental Clinic Rident, Franje Čandeka 39, 51000 Rijeka, Croatia
10 Department of Anatomy, Histology, Embriology, Pathology Anatomy and Pathology Histology, Faculty of Dental Medicine and Health Osijek, J.J. Strossmayer University of Osijek, 31000 Osijek, Croatia
11 Botiss Biomaterials GmbH, 15806 Zossen, Germany
* Correspondence: zpkacarevic@fdmz.hr (Ž.P.K.); olga.cvijanovic@medri.uniri.hr (O.C.P.)
† These authors contributed equally to this work.

Abstract: The aim of this study was to evaluate the efficacy of autologous dentin (AD), bovine xenograft (BX) and magnesium-enriched bovine xenograft (BX + Mg) in the healing of critical cranial bone defects (CCBDs) in rats. Eighty male Wistar rats were divided into four groups: BX, BX + Mg, AD and the control group (no intervention). Eight mm CCBDs were created and treated with the respective biomaterials. Healing was assessed 7, 15, 21 and 30 days after surgery by micro-computed tomography (micro-CT), real-time polymerase chain reaction (RT-PCR) and immunohistochemical analysis. Micro-CT analysis showed that AD had the highest bone volume and the least amount of residual biomaterial at day 30, indicating robust bone formation and efficient resorption. BX + Mg showed significant bone volume but had more residual biomaterial compared to AD. RT-PCR showed that the expression of osteocalcin (OC), the receptor activator of nuclear factor κB (RANK) and sclerostin (SOST), was highest in the AD group at day 21 and vascular endothelial growth factor (VEGF) at day 15, indicating increased osteogenesis and angiogenesis in the AD group. Immunohistochemical staining confirmed intense BMP-2/4 and SMAD-1/5/8 expression in the AD group, indicating osteoinductive properties. The favorable gene expression profile and biocompatibility of AD and BX + Mg make them promising candidates for clinical applications in bone tissue engineering. Further research is required to fully exploit their potential in regenerative surgery.

Keywords: bovine xenograft; autologous dentin; critical calvaria size defect; magnesium; rats

1. Introduction

The formation of bone tissue, known as osteogenesis or morphogenesis, involves the development of bone tissue through a process that involves the differentiation of osteogenic cells into osteoblasts (a concept known as osteoinduction). For bone formation to occur, the osteoblasts and newly formed bone tissue require a suitable scaffold (a concept known as osteoconduction) that provides a porous structure that allows for the inward three-dimensional growth of bone tissue from the surface. Finally, osteogenesis essentially comprises two processes: the formation (osteoformation) and the breaking down (osteoresorption) of bone tissue. Both processes occur simultaneously and synergistically as they act in a balanced manner to create and maintain the tissue homeostasis of bone tissue according to functional requirements [1,2].

The formation of bone tissue involves a complex interplay of gene regulatory networks that orchestrate the differentiation of osteogenic cells into osteoblasts. These networks, particularly the SMAD proteins, play an important role in signaling through morphogenetic proteins (BMPs), which are essential for the regulation of osteogenesis and bone remodeling processes. BMPs promote the expression of osteogenic genes through the activation of SMAD-1/5/8, thus driving bone formation and homeostasis [3–5].

The alveolar bone, located in maxilla and mandible, plays a crucial role as the main support for the teeth. Although similar in basic structure to other bone tissues, the alveolar bone undergoes rapid and continuous remodeling due to the eruption of teeth and the functional stresses of chewing [6]. This rapid remodeling process is crucial for the adaptation of tooth position and is influenced by local factors such as growth factors and cytokines as well as systemic factors such as calcitonin and estrogen, which together ensure the maintenance of bone homeostasis [7]. However, they can also contribute to an increased loss of bone volume during bone remodeling after tooth extraction [8,9]. Therefore, various biomaterials are used for bone augmentation to minimize volume loss and achieve bone regeneration following tooth extraction.

Biomaterials for bone regeneration are categorized as autogenous bone, allografts, xenografts and alloplasts [10]. In the last decade, the use of autologous dentin (AD) has also been recognized as a successful biomaterial for bone regeneration in dentistry. It is suitable for bone regeneration as its chemical composition of organic and inorganic substances is most like native bone tissue [11]. The presence of morphogenetic bone proteins in AD indicates its pronounced osteoinductive properties, which primarily distinguishes it from xenogeneic biomaterials that do not contain proteins in their composition [12]. In a recent study in which regenerated bone was histomorphometrically analyzed after the application of autologous dentin, 85% new bone formation and 25% residual dentin were found 7 months after guided bone regeneration (GBR) [13]. Pang et al. [14] histomorphometrically compared the outcome of alveolar ridge preservation using bovine xenograft (BX) and AD, and it was shown that there was no statistically significant difference in bone tissue volume.

Today, cerabone® (botiss biomaterials, Zossen, Germany) is widely used in procedures aimed at regenerating the alveolar ridge. It is derived from trabecular bovine bone, with its organic components, including osteoinducing molecules, immune cells and pathogens, removed by physical and chemical processes, leaving behind calcium hydroxyapatite [15,16].

In the available literature analyzing the osteoconductive properties of xenogeneic biomaterials in a calvarial defect model, only one paper used magnesium in combination with porcine bone [17]. Most published studies have analyzed the percentage of bone volume and remaining biomaterial at two time points [18,19], while very few studies have analyzed bone structural parameters or performed immunohistochemical analyses [18]. Previously, our group of authors published the results of an animal study on the use of four different BX in critical cranial bone defects (CCBDs) of 5 mm in size. The bone samples were analyzed by micro-CT, histology and immunohistochemistry. Overall, the xenogeneic magnesium alloy biomaterial showed key properties of osteoinduction and biodegradability during CCBD healing [20].

However, there is still a knowledge gap regarding the application of magnesium-enriched biomaterials in CCBDs, and, in general, their use in oral surgery is not yet fully understood. Considering the unexplored biological properties and unexplained gene expression of factors involved in bone remodeling during CCBD repair with AD and BX + Mg, the main aim of this study was to investigate them at different time points during the healing of 8 mm CCBDs. Therefore, the relative expression of vascular endothelial growth factor (VEGF), osteocalcin (OC), the receptor activator of nuclear factor κB (RANK) and sclerostin (SOST) were analyzed by real-time polymerase chain reaction (RT-PCR). In addition, the values of the 3D parameters were determined by micro-CT analyses of the bone samples and the expression of osteoinductive proteins and their intercellular messenger molecule were analyzed.

2. Materials and Methods

2.1. Experimental Animals and Study Design

This study was conducted in accordance with the guidelines of the Declaration of Helsinki and was approved by the Ethics Committee of the University of Rijeka and the Ministry of Agriculture (EP 302/2021).

For this study, 80 male Wistar rats aged around 2.5 months were used. The animals were reared and kept under the laboratory conditions of the Institute of Physiology, Pathophysiology and Immunology of the Faculty of Medicine in Rijeka, fed ad libitum, provided with drinking water and subjected to a daily light and dark cycle in accordance with the regulation on the protection of animals used for scientific purposes (Official Gazette 55/13). The animals were randomly divided into 4 groups of 20 animals each. Each group of animals was named after the name of the material used to promote the healing of the critical calvarial bone defect (CCBD). The animals were sacrificed at 4 different times—on the 7th, 15th, 21st and 30th day of CCBD healing. The data on the test animal groups are listed in Table 1.

Table 1. Presentation of the distribution of experimental animals according to the group and experimental duration of CCBD healing.

Group Name	Number of Animals	Time Points	Total
BX [1]	5	7th, 15th, 21st and 30th day	20
BX + Mg [2]	5	7th, 15th, 21st and 30th day	20
AD [3]	5	7th, 15th, 21st and 30th day	20
Control	5	7th, 15th, 21st and 30th day	20
Total			80

[1] Bovine xenograft, [2] bovine xenograft and magnesium, [3] autologous dentin.

2.2. Materials

Three different types of biomaterials were used to stimulate the healing of CCBD, namely the following:

1. Cerabone® (botiss GmbH, Zossen, Germany) with magnesium (BX + Mg group), produced at Biotrics Biomiplants AG (Berlin, Germany) in the form of Mg granules in which the mass fraction of the magnesium alloy (a solid solution of magnesium with yttrium, zinc, manganese and calcium) is 3%. Additional information is not available because the biomaterial itself is still the subject of research, and thus all data are confidential.
2. Cerabone® (BX group) made from trabecular bovine bone from which the organic component of the bone tissue was removed by physical and chemical processes.
3. Autologous dentin from rat teeth (AD group) ground with a dentin grinder (KometaBio Smart Dentin Grinder, Tenafly, NJ, USA) and then prepared according to a protocol previously described by our group of authors before being applied to the defect [21].

To collect dentin material for the AD group, tooth extractions were performed on donor inbred Wistar rats six days prior to the CCBD surgery [21–23].

After the implantation of the biomaterial, the implant site was covered with a collagen membrane (mucoderm®, Botiss Biomaterials, Zossen, Germany). In the control group, only a collagen membrane was used to cover a defect, but no bone biomaterial.

2.3. Surgical Protocol and Bone Sample Harvesting

Surgical instruments sterilized in an autoclave and cooled to room temperature were used to perform the CCBD and implant the healing-promoting material. The work surface for performing the surgical procedures on the animal was disinfected once with 70% ethanol.

The animals were anesthetized using ketamine (80 mg/kg) and xylazine (5 mg/kg body weight). For pain relief, an intraperitoneal injection of tramadol (10 mg/kg, Henry Schein, Melville, NY, USA) was administered. A local anesthetic of 0.3–0.4 mL of 1% lidocaine was applied subcutaneously at the incision site. During surgery, each animal received a subcutaneous injection of sterile saline (0.9% NaCl, Henry Schein, NY, USA) at a rate of 10 mL/kg/h to compensate for visible and invisible fluid loss during and after the procedure. Blood oxygen levels were continuously monitored using a pulse oximeter (MouseSTAT, Pulse Oximeter & Heart Rate Monitor Module, Kent Scientific Corporation, Torrington, CT, USA), while the depth of anesthesia and analgesia was assessed by observing the animal's response to external stimuli. The surgical area on the animal's head was prepared by shaving the fur with an electric trimmer designed for small animals (MOSER 1556 AKKU, professional cordless hair trimmer, BIOSEB in vivo Research Instruments, Schönwalde-Glien, Germany), covering the region from the muzzle between the eyes to the back of the skull. The area was then sterilized using iodine sticks (Impregnated Swabstick Dynarex 10% Strength Povidone-Iodine Individual Packet, Dynarex, Orangeburg, New York, NY, USA). An incision was made in the skin at the prepared site, followed by the application of a sterile drape. A 1.5 cm periosteal incision was performed across the skull, from the nasal bone to the bregma, exposing the calvarial bone after retracting the periosteum. A trephine with an outer diameter of 8 mm (Helmut Zepf, Seitingen-Oberflacht, Germany) was used to drill the fronto-parietal complex at 1500 rpm, creating an intracranial defect. Throughout the drilling process, sterile saline was applied dropwise, approximately one drop every two seconds, to both the eyelid and calvaria. The low drilling speed and continuous moisturizing were essential to prevent thermal injury that could damage surrounding tissue. Gentle pressure with an elevator was used to lift a part of the bone at the injury edge, detaching it from the underlying dura. The defect site was thoroughly rinsed with sterile physiological solution to clear away bone fragments and dust generated during drilling. To standardize the defect site, a rat head holder (Model 920-E Rat Head Holder, David Kopf Instruments, Los Angeles, CA, USA) and a tissue marking instrument (Biopsy Punch, Kai Medical, Tokyo, Japan) were utilized. Since a trephine with a diameter of 8 mm was used for surgical bone removal and defect formation, the defects themselves were standardized. When drilling into the calvaria, care was taken to ensure that the instrument did not penetrate too deeply. As the thickness of the calvaria of the experimental animals was approximately 1 mm, the markings on the eyelid were used as a guide to depth and the entire drilling process was continuously monitored for depth. To avoid injury to the skull or brain, the force applied was less than the weight of the drill. As the calvaria becomes transparent on the defect during drilling, the dura and the surface of the brain can be more easily recognized. As the defect approached the appropriate thickness, the drilled part of the bone would move slightly downward, indicating that the drilled calvaria had almost reached full thickness. A specific type of biomaterial was then implanted into the surgically created calvarial bone defect (see Table 1). The biomaterial was weighed using a precision balance (ME-T Precision Balance, Mettler Toledo, Columbus, OH, USA) to ensure a consistent amount of 20 mg for each animal. The granulated biomaterial had a uniform particle size of 0.5 to

1 mm across all groups. To complete the procedure, the implantation site was covered with a collagen membrane.

The skin was sutured over the biomaterial and collagen membrane using either single or continuous sutures (3-0 USP, Hu-Friedy Perma Sharp Sutures, Irvine, CA, USA). After completing the procedure, the surgical site was thoroughly cleaned with sterile saline or diluted hydrogen peroxide (3%) to remove any remaining blood. The animals were then placed in cages equipped with heating pads (heating pads for rats—20.5 × 12 cm, DC temperature controller, FHC, South Gate, CA, USA) to warm them up quickly and safely, helping to minimize postoperative trauma. In this way, the animals' body temperature was permanently maintained in the range of 36.5 to 37.5 °C. The animals in each of the 4 groups were randomly divided into a further 4 groups of 5 animals, depending on the time of sacrifice. The animals were sacrificed at intervals of 7, 15, 21 and 30 days after the operation.

Sacrifice was carried out in the usual humane manner with three times the anesthetic dose, i.e., with ketamine (240 mg/kg) and xylazine (15 mg/kg of body weight). After sacrificing the experimental animals, tissue samples were taken from the entire fronto-parietal–occipital complex at the site of the pre-existing CCBD in which the biomaterial was embedded. Depending on the research method, the tissue was subjected to different preparation methods.

2.4. Micro-CT Analysis of 3D Parameters

The tissue samples were preserved in 4% paraformaldehyde at 4 °C prior to transport. Before being transported, the samples were placed in a 70% alcohol solution until they could be scanned using a micro-CT device.

The samples were then scanned with a micro-CT scanner (Skyscan 1076, Bruker, Kontich, Belgium) at a resolution of 18 μm, with a 0.40-degree rotation and a 0.025 mm titanium filter. The average image section was set to 2. The resulting images were reconstructed using NRecon software v. 2.0 (Bruker, Kontich, Belgium) and analyzed with CTAn software v.1.8. (Bruker, Kontich, Belgium). For the analysis, a circular area with a diameter of 8 mm was drawn along the edges of the original defect edge. To distinguish the new bone formation from the used biomaterial, specific thresholds were used for each biomaterial, while they remained the same for the new bone formation. The threshold for BX and BX + Mg was 200–255, and for AD, 110–255, while for new bone formation, it was 50–255. The value range of the applied threshold differentiated the individual biomaterials according to different densities. The value of the biomaterial in relation to new bone formation was then determined by subtraction. Based on that, the following parameters were calculated: the ratio of bone volume to trabecular volume (bone volume/total volume, BV/TV, %) and the percentage of residual biomaterial (RB, %).

2.5. Real-Time Polymerase Chain Reaction (RT-PCR) Analysis

The relative expression levels of the following proteins were analyzed using the RT-PCR method on bone samples: vascular endothelial growth factor (VEGF), osteocalcin (OC), receptor activator of nuclear factor κB (RANK) and sclerostin (SOST). The protocols recommended by the manufacturer were followed when performing the RT-PCR method.

RNA isolation was performed as follows. Tissue samples from rat calvarial halves with defect and biomaterial were stored at −80 °C after sacrifice. On the day of isolation, the tissue samples were taken at −80 °C and placed in liquid nitrogen. During the mechanical homogenization of the bone in the crucible, the bone was cooled with liquid nitrogen, after which 30 mg of bone tissue from each sample was separated into each individual 1.5 mL test tube and then refrozen in liquid nitrogen. The "mini" NucleoSpin® RNA protocol (Macherey-Nagel GmbH & Co. KG, Düren, Germany) was used for RNA isolation. Complementary DNA (cDNA) was obtained by the reverse transcription of total RNA using the High Capacity cDNA Reverse Transcription Kit (Applied Biosystems, Foster City, CA, USA). The amount of messenger RNA for VEGF, RANK, SOST and OC primers was determined with specific Taqman probes using the 7300 Fast Real-Time PCR System (Applied Biosystems, Foster City, CA, USA).

A reaction in which water was added instead of reverse transcriptase ("no RT control", NRTC) was used as a negative control that excluded the multiplication of genomic DNA. Contamination with exogenous nucleic acids was excluded using a control in which the cDNA was replaced with water. Amplification was performed with the 7300 Real Time PCR System under standardized conditions: incubation, 50 °C/2 min/1 cycle; initial denaturation, 95 °C/10 min/1 cycle; and 40 amplification cycles at 95 °C for 15 s (denaturation) and at 60 °C for 1 min (maturation/elongation). The amplification results were analyzed using the 7300 System SDS Software v1.3 computer application. The reactions of all samples were performed in duplicate, and the result was considered reliable only when there were no differences between the amplification curves.

The comparative cycle threshold (CT) method (ΔΔCT) was used for the relative representation of gene expression according to the following formula: fold-change (FC) = $2^{-\Delta\Delta CT}$ [24]. The results are presented as a relative change in gene expression compared to the control sample. The abbreviation CT stands for the cycle in which the signal of the target genes rises above the detection limit for the first time. The first delta in the formula indicates the difference between the values for the housekeeping gene and the genes of interest, i.e., the level of gene transcriptional activity of each sample normalized with the values obtained for the housekeeping gene GAPDH used as endogenous control for the sample of interest, while the second delta indicates the difference between the values of the treated group and the untreated control groups.

2.6. Immunohistochemical Analysis

The expression of protetctive cytokines (VEGF) and osteoinductive proteins and their intracellular messenger molecules (BMP-2/4 and SMAD-1/5/8) was determined by immunohistochemical analysis.

The primary antibodies listed below were used for the immunohistochemical analyzes:

- BMP-2/4 (sc-137087 SCBT, Santa Cruz Biotechnology, Dallas, TX, USA, SAD), 1:200;
- VEGF (ab 231260 abcam, Abcam, Cambridge, UK), 1:200;
- SMAD-1/5/8 (#95115 CST, Cell Signaling Technology, Danvers, MA, USA, SAD), 1:100.

The secondary antibodies were obtained from the manufacturer Dako (2° Real En Vision Detection System rabbit/mouse, Glostrup, Denmark). The tissue sections were first deparaffinized using a xylene solution and then dehydrated through a series of ethyl alcohol solutions with decreasing concentrations (100%, 96% and 75%). Next, the sections were washed three times with PBS solution and heated in a 10 mM citrate buffer (pH 6.0) at 65 °C for 20 min. Afterward, the sections were washed with a 0.3% Triton X-100 solution in PBS at room temperature. Endogenous peroxidase was then inactivated by treating the sections with 0.3% hydrogen peroxide in methanol for 30 min. Following a PBS wash, the tissue sections were treated with 10% normal serum for 60 min, chosen according to the species of the secondary antibody carrier. The primary antibodies dissolved in PBS solution were incubated in a humid chamber at a temperature of 4 °C for 18 h. After washing the tissue sections in PBS solution (pH 7.4), a secondary antibody was added depending on the primary antibody. After 60 min, the tissue section was washed in PBS solution and treated with streptavidin. 3,3-diaminobenzidine in hydrogen peroxide (DAB) was used for visualization. The tissue section was then washed with water and counterstained with hematoxylin. The tissue was rehydrated in ethyl alcohols of increasing concentrations (75%, 96% and 100%). After clarification with xylene, the specimen was placed in entalan. The intensity of the immunohistochemical staining was evaluated with the computer program ImageJ v 1.54 (available from: https://imagej.net/ij/).

2.7. Statistical Analysis

Statistical analysis was carried out using the Statistica 11.1 software (StatSoft, Inc., Tulsa, OK, USA). After confirming the normal distribution of the data, differences between the biomaterial groups were assessed using repeated measures analysis of variance (ANOVA). Tukey's Highly Significant Difference Test was employed for post hoc analysis to identify specific differences between the groups. Additionally, multiple regression analysis was conducted to evaluate the impact of predictor variables on bone volume. The results were considered statistically significant at $p < 0.01$ and $p < 0.05$.

3. Results

3.1. Mirco-CT Analysis

A series of images of the calvaria of animals from different groups and on different days of healing clearly shows the extent of CCBD closure (Figure 1).

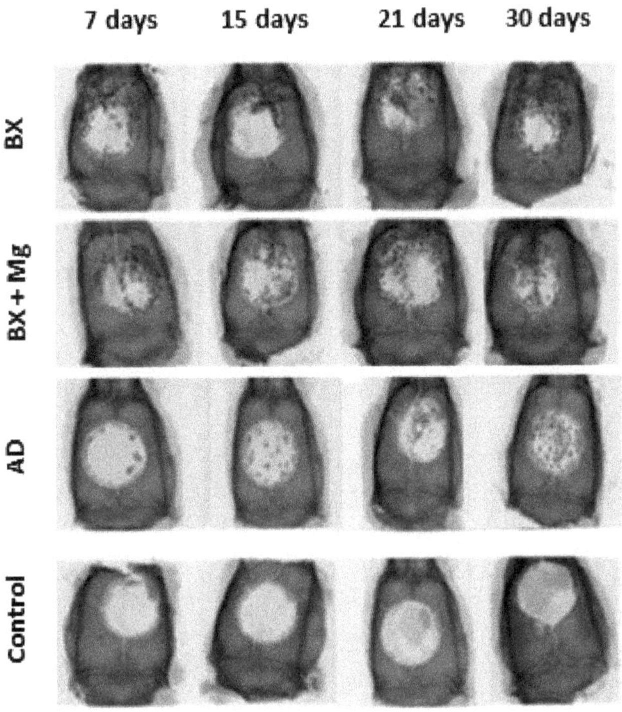

Figure 1. Visualization of the fronto-parieto-occipital complex of the calvaria of rats, imaged with a micro-CT device. The images show the progression of CCBD closure over the days of healing.

The quantitative results of the micro-CT analysis are summarized in Table 2. They show statistically significant differences in the micro-CT bone morphometry parameters depending on the biomaterial and the day. In general, the values of bone volume measured by micro-CT increased with increasing time intervals, while the values of residual biomaterial decreased during the healing period. On the 30th day of healing, the AD group had the highest bone volume and the lowest residual biomaterial values.

Table 2. The quantitative results of the micro-CT analysis.

		Biomaterials			
		BX [3] (N = 5)	BX + Mg [4] (N = 5)	AD [5] (N = 5)	Control [6] (N = 5)
Days	**μCT Parametrs**	**Mean ± SD**	**Mean ± SD**	**Mean ± SD**	**Mean ± SD**
7	BV/TV [1] (%) RB [2] (%)	19,795 ± 1943 [a,g] 27,847 ± 1468 [f]	19,813 ± 1643 [a,e,g] 24,626 ± 1520 [f]	18,935 ± 1601 [a,g] 29,459 ± 1164 [c,g]	6952 ± 1002 [f] /
15	BV/TV (%) RB (%)	24,238 ± 1943 [a,f] 30,989 ± 1468 [b,c,g]	23,848 ± 1643 [a,f] 20,904 ± 1520 [f]	24,967 ± 1601 [a,g] 22,886 ± 1164 [g]	9906 ± 1002 /
21	BV/TV (%) RB (%)	31,033 ± 1943 [a] 22,912 ± 1468 [b]	31,885 ± 1643 [a] 21,986 ± 1520 [f,b]	37,909 ± 1601 [a,c,f] 12,916 ± 1164	11,921 ± 1002 /
30	BV/TV (%) RB (%)	33,015 ± 1943 [a] 19,905 ± 1468 [b,c]	39,501 ± 1643 [a,d] 12,726 ± 1520	48,994 ± 1601 [a,c] 11,996 ± 1164	16,938 ± 1002 /

Legend: [1] bone volume/total volume; [2] residual biomaterial; [3] bovine xenograft; [4] bovine xenograft + magnesium; [5] autologous dentin; [6] control. Statistical significance ($p < 0.05$) compared to [a] control; [b] AD; [c] BX + Mg; [d] BX; [e] 15th day; [f] 30th day; [g] 21st and 30th day.

3.2. The Results of Real-Time Polymerase Chain Reaction (RT-PCR) Analysis

Figure 2 summarizes the results of the gene expression of the bone remodeling factors osteocalcin (OC), receptor activator of nuclear factor κB (RANK) and sclerostin (SOST), as well as the gene expression of the vascular endothelial growth factor (VEGF).

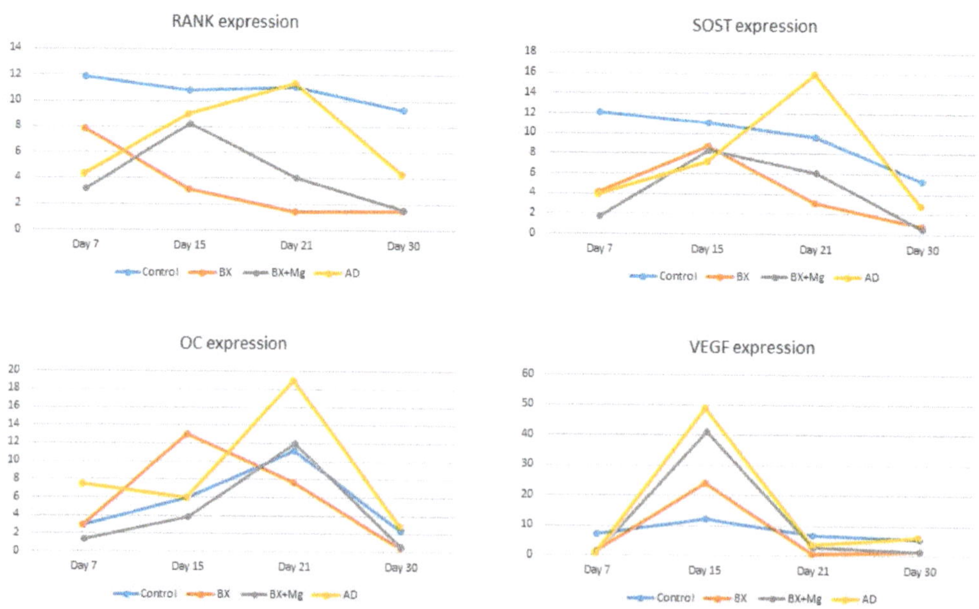

Figure 2. The expressions of RANK, SOST and OC were highest in the AD group on day 21 of healing. The VEGF expression was also highest in the AD group, but on the 15th day of healing.

3.3. Results of Immunohistochemical Analysis

3.3.1. Immunohistochemical Analysis of BMP-2/4 Expression

More intense BMP-2/4 immunopositivity was observed in mesenchymal cells in transition during the earlier stages of healing. From mid-healing to later stages, there are numerous immunopositive osteoblasts, their precursors and osteocytes. Multinucleated cells are also visible, appearing on day 7 in the BX group and on days 7, 15 and 21 in the BX + Mg group (Figure 3). The relative values of immunohistochemical BMP-2/4 staining intensity for each biomaterial examined are shown in Figure 3. The general trend shows the highest values in the BX group, with the highest single value measured on day 21 of healing (r = 173.419).

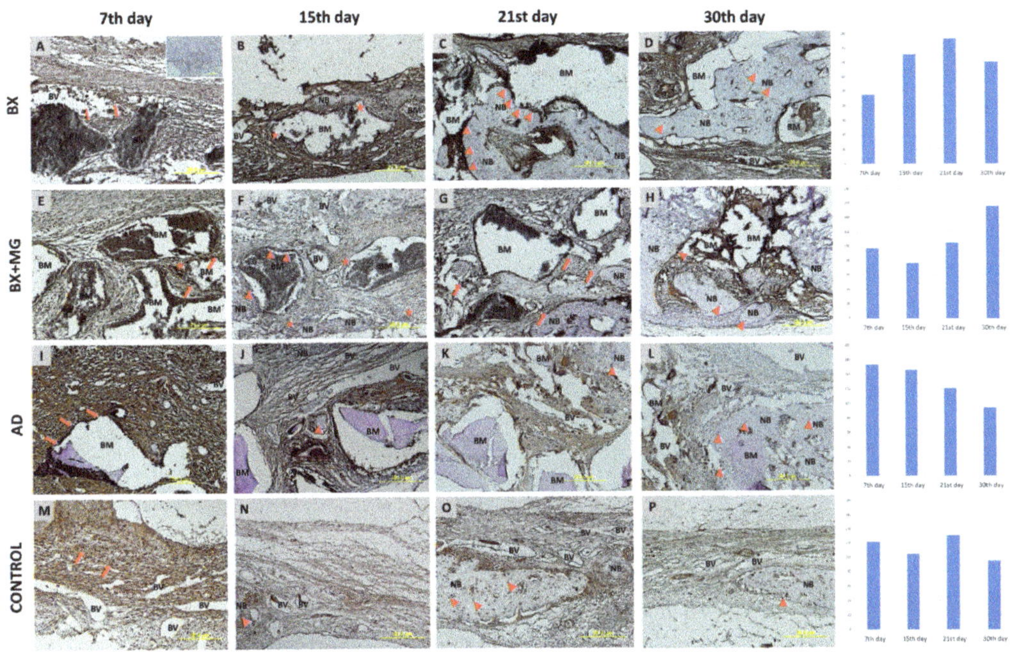

Figure 3. BMP-2/4 immunohistochemical staining of coronal sections of calvarial bone defects in the following groups: BX—bovine xenograft (**A–D**); BX + MG—bovine xenograft and magnesium (**E–H**); AD—autologous dentin (**I–L**); and control group (**M–P**) on the 7th, 15th, 21st and 30th day. Legend of abbreviations. The BMP-2/4 immunohistochemical staining of the coronal sections of the calvarial bone defects are shown for the following groups: BX—bovine xenograft (**A–D**); BX + MG—bovine xenograft; BM—biomaterial; NB—new bone formation; BV—blood vessel. Triangles (▲) indicate osteoblasts and osteocytes in lacunae, asterisks (★) indicate immunopositive multinucleated cells, and arrows (↑) indicate sites of bone bridging and apposition. On the right side of the immunohistochemistry slides, there are diagrams showing the immunohistochemical staining for each biomaterial broken down by day, on days 7, 15, 21 and 30 of healing.

3.3.2. Immunohistochemical Analysis of SMAD-1/5/8 Expression

SMAD-1/5/8 intracellular signaling molecules track the expression of BMP-2/4. Similar to the immunolocalization observed for BMP-2/4, SMAD-1/5/8 is present in mesenchymal cells, osteoblasts and osteocytes and is also found in multinucleated cells, sporadically in connective tissue and around degrading biomaterial particles (Figure 4). The relative values of the immunohistochemical staining intensity of SMAD-1/5/8 for each biomaterial examined are shown in Figure 4. The general trend indicates the highest immunohistochemical staining values in the AD group, with the highest single value recorded on the 15th day of healing (r = 155.404).

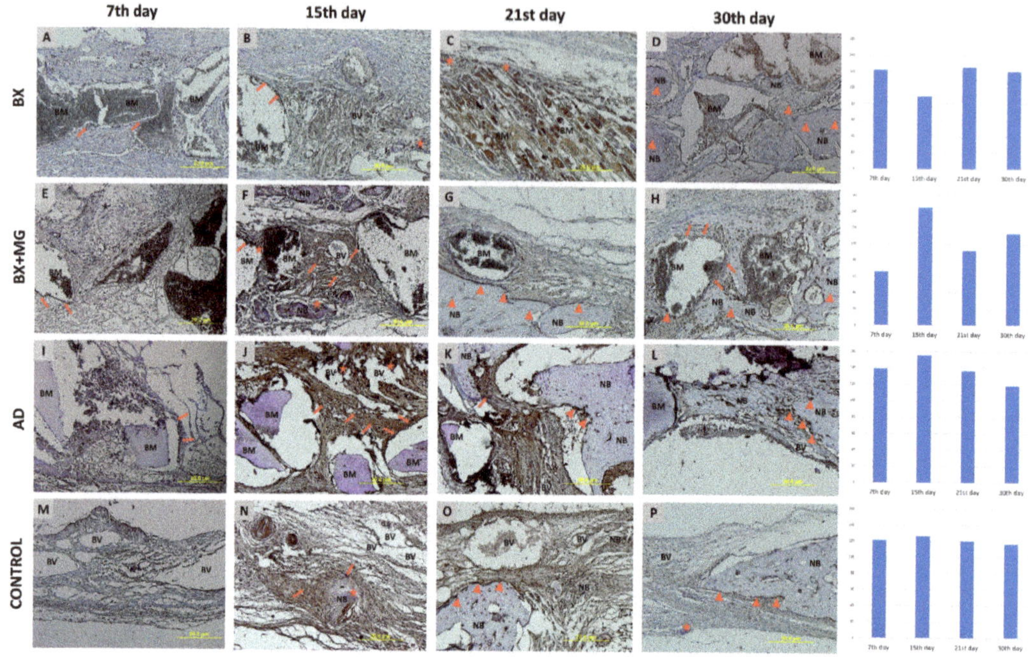

Figure 4. SMAD-1/5/8 immunohistochemical staining of coronal sections of calvarial bone defects for the following groups: BX—bovine xenograft (**A–D**); BX + MG—bovine xenograft and magnesium (**E–H**); AD—autologous dentin (**I–L**); and control group (**M–P**) on the 7th, 15th, 21st and 30th day. Legend of abbreviations: BM—biomaterial; NB—new bone formation; BV—blood vessel. Triangles (▲) indicate osteoblasts and osteocytes in lacunae, asterisks (★) indicate immunopositive multinucleated cells, and arrows (↑) indicate sites of bone bridging and apposition. On the right side of the immunohistochemistry slides, there are diagrams showing the immunohistochemical staining for each biomaterial broken down by day, on days 7, 15, 21 and 30 of healing.

3.3.3. Immunohistochemical Analysis of VEGF Expression

A visible expression of the VEGF is observed in pluripotent mesenchymal cells at sites of neovascularization and at sites where blood vessels infiltrate biomaterial particles that are degraded (Figure 5). The relative values of the immunohistochemical VEGF staining intensity for each biomaterial examined are shown in Figure 5. The general trend indicates the highest immunohistochemical staining values in the BX group, with the highest single value recorded on day 15 of healing (r = 184.955).

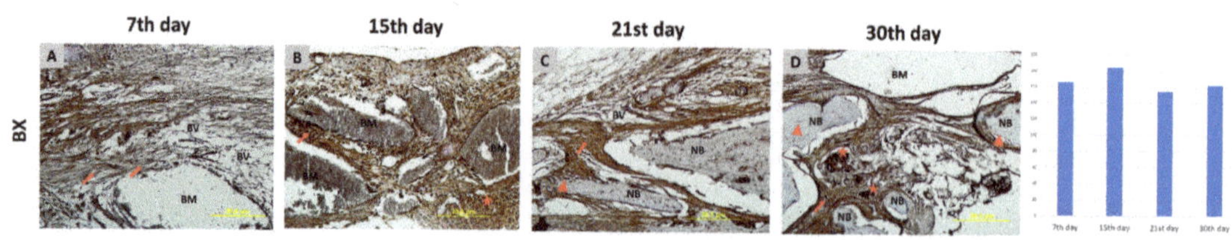

Figure 5. *Cont.*

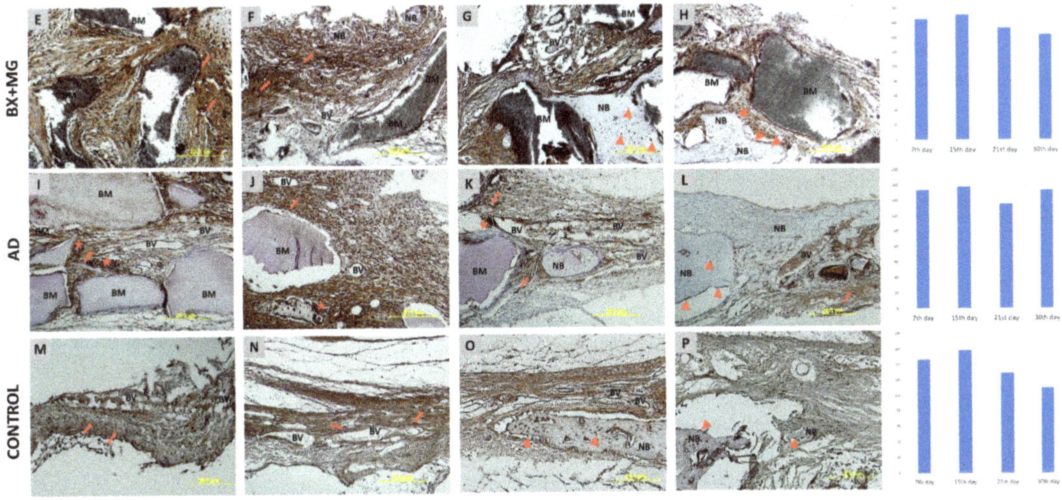

Figure 5. VEGF immunohistochemical staining of coronal sections of calvarial bone defects for the following groups: BX—bovine xenograft (**A–D**); BX + MG—bovine xenograft and magnesium (**E–H**); AD—autologous dentin (**I–L**); and control group (**M–P**) on the 7th, 15th, 21st and 30th day. Legend of abbreviations: BM—biomaterial; NB—new bone formation; BV—blood vessel. Triangles (▲) indicate osteoblasts and osteocytes in lacunae, asterisks (★) indicate immunopositive multinucleated cells, and arrows (↑) indicate sites of bone bridging and apposition. On the right side of the immunohistochemistry slides; there are diagrams showing the immunohistochemical staining for each biomaterial broken down by day, on days 7, 15, 21 and 30 of healing.

4. Discussion

This study investigated the osteoconductive and osteoinductive properties of different biomaterials in models of critical cranial bone defects (CCBDs). The focus was on the comparison of bovine xenograft (BX), bovine xenograft with magnesium (BX + Mg) and autologous dentin (AD) for bone regeneration using micro-CT and immunohistochemical analysis, as well as on the expression of key genes involved in bone remodeling.

The micro-CT analysis generally demonstrated that all biomaterials promoted bone formation over time, with AD showing the highest bone volume (BV/TV) and the lowest residual biomaterial (RB) on the 30th day of healing (Table 2). The AD group showed statistically significantly higher BV/TV values on day 21 than the BX, BX + Mg and control groups. When analyzing the healing dynamics, we observed the highest bone growth in the AD group between day 15 and 21, which means that the bone is intensively formed and remodeled during this period. The basis for this statement lies in the fact that AD has both osteoconductive and osteoinductive properties [25]. In this context, AD as an implantation material in bone defects is characterized by the presence of morphogenetic bone proteins that promote bone formation. Therefore, it has very similar properties to native bone in terms of both organic and inorganic substances [26,27].

The addition of magnesium to bovine xenograft (BX + Mg) resulted in significantly higher bone volume compared to BX alone, especially at day 30, suggesting that magnesium enhances the regenerative properties of BX by creating a favorable environment for bone tissue formation and possibly influencing the activity of osteoblasts and osteoclasts, thus promoting balanced bone remodeling (Table 2). The use of magnesium implants found its basis in clinical research as early as the 1930s [28]. The complete resorption of magnesium and the short postoperative recovery without pain encouraged Lambotte to pioneer the use of magnesium implants for supracondylar fractures in children, as they healed quickly. Although magnesium implants are degraded in vivo by corrosion, they have no harmful effects on the tissue. On the contrary, they stimulate osteoblast activity around the implant, leading to the complete replacement of the implant by bone tissue [29,30].

The study by Perić Kačarević et al. [31] investigated magnesium implants and their biological interactions with bone tissue. Their research revealed that the calcium phosphate corrosive layer that forms on magnesium implants after implantation helps to slow down the corrosion process, allowing direct contact with bone tissue. During corrosion, metallic magnesium undergoes oxidation, releasing magnesium ions as corrosion products. This process causes the protective layer of magnesium hydroxide to dissolve, at least in localized areas, which permits the corrosion to continue until the implant is fully degraded. The study also indicated that an alkaline pH is unlikely to negatively impact bone regeneration, as bone graft materials made of pure hydroxyapatite decompose at slightly alkaline pH levels and have been effectively used in bone regeneration for many years. Similar findings were found by Jung et al. [32]. The biomaterial was almost completely resorbed within only 3 months. In a recent experimental study, a magnesium alloy-enriched biomaterial was implanted in the distal condyle of an animal. After three months, the magnesium-enriched material had decomposed, with much of the original magnesium alloy having disappeared. Simultaneously, a fibrous capsule had formed around the surgical site. Histological analysis showed that the magnesium scaffolds had not caused any significant damage to the surrounding tissue. This study shows that even fast-degrading magnesium scaffolds maintain good biocompatibility and trigger an appropriate inflammatory response in vivo. Consequently, magnesium alloy-based implants hold promise for applications in oral implantology [33].

Based on the available literature investigating the osteoconductive properties of magnesium-enriched xenogeneic biomaterials, our study on a 5 mm calvarial defect stands out with the highest bone volume values for the BX + Mg group in all observed time intervals compared to the other two BX groups [20]. This finding can be fully equated with the results for a 8 mm defect. The observed residual biomaterial is lower in the BX + Mg group than in the BX group, suggesting also that magnesium contributes to a better degradation of the biomaterial.

There is also a publication in which magnesium was used in combination with porcine bone [17]. In the calvaria of 14 adult male New Zealand rabbits, defects with a diameter of 7 mm were created and filled with the following biomaterials: untreated porcine bone, BioOss® (Geistlich, Wolhusen, Switzerland) and porcine bone containing Mg. The percentage of new bone formation was analyzed histomorphometrically 2 and 4 weeks after implantation. The results showed that in the magnesium-filled group, the percentage of new bone formation was 11.8% at 2 weeks and 22.3% at 4 weeks, with statistically significantly higher results in this group compared to the other two groups. However, we cannot objectively compare our study with the aforementioned study because we did not analyze our samples histomorphometrically.

The RT-PCR results offer insights into the molecular mechanisms involved in bone regeneration. The analysis focused on the relative expression levels of vascular endothelial growth factor (VEGF), osteocalcin (OC), receptor activator of nuclear factor κB (RANK) and sclerostin (SOST). VEGF, which plays a key role in angiogenesis, was found to be upregulated in the AD and BX + Mg groups, indicating enhanced vascularization—a crucial factor for bone healing and regeneration [34]. In the study conducted by Hassumi and colleagues, the relative level of gene expression for several proteins in the alveolar bone was analyzed by RT-PCR [35]. The levels of these genes were compared on days 7, 14 and 28. RANK was found to have the highest value on day 28, which is comparable to our finding that RANK in the AD group reaches the highest value on day 21 (Figure 2), while in the other groups, the highest values are recorded in the middle of the healing process (day 15). In addition, in the abovementioned study, the highest levels of gene expression for OC were found on day 28. Our study shows that the highest gene expression levels of OC are found in the AD group on day 21 (Figure 2). The BX + Mg and the control groups also showed the highest gene expression levels of OC on day 21, and the BX group, on day

15. Osteocalcin is a protein secreted by mature osteoblasts, and since it binds to calcium, it is the best indicator of bone formation [36]. This result indicates that most bone tissue is formed between days 15 and 21. The lower expression of SOST, an inhibitor of bone formation, in the AD group is consistent with the higher bone volume observed (Table 2) in this group, as this indicates less inhibition of osteogenesis.

Immunohistochemical analysis for VEGF, BMP-2/4 and SMAD-1/5/8 proteins provided further confirmation of the molecular findings. Following the increased stimulation of osteoblasts, SMAD-1/5/8 are activated in the BMP signaling pathway and promote osteogenesis by regulating osteogenic genes [5]. The expression of SMAD-1/5/8 follows the expression of BMP-7, which was confirmed by our results where the tested biomaterials showed an increase in levels in the middle of the healing cycle, with the highest levels occurring in the AD group on day 15. VEGF promotes neovascularization and is expressed at low levels at the beginning of osteoblast differentiation, then strongly during terminal differentiation, and reaches a maximum during the mineralization process [37]. In all biomaterials tested, an increase in the measured values was observed in the middle of the healing cycle, with the highest measured value in the BX group on day 15. In the study conducted by Pires et al. [38], the effectiveness of synthetic hydroxyapatite and xenografts, both in their pure form and enriched with the mononuclear fraction of bone marrow, was compared for the regeneration of critical-sized bone defects in the calvaria of rats. This comparison was made using histomorphometric and immunohistochemical analyses (anti-VEGF, anti-osteopontin). Forty rats were divided into five groups based on the biomaterials used: synthetic hydroxyapatite, xenograft, synthetic hydroxyapatite enriched with the mononuclear fraction of bone marrow, xenograft enriched with the mononuclear fraction and a control group with no intervention. The animals had 8 mm critical-sized bone defects created, and after eight weeks, they were euthanized. The data analysis revealed a significant increase in new bone matrix formation in all experimental groups compared to the control group [38].

In this study, we chose micro-CT analysis to evaluate the quantitative bone parameters because of its high sensitivity and accuracy in quantifying bone volume and residual biomaterial [39]. The parameters evaluated, including BV/TV and RB, are widely recognized indicators of bone regeneration and provide a reliable measure of the efficacy of the materials tested. While histochemical analysis could have provided complementary information, the quantitative nature of micro-CT provides a comprehensive assessment that meets the aims of this research. Future studies could incorporate histologic analysis alongside the quantitative data from micro-CT to further improve the understanding of bone tissue properties.

It is important to consider the possible impact of tooth extraction on the immune response in the AD group. It is known that any wound that occurs, such as after a tooth extraction, triggers an acute immune response that could activate the wound healing pathways and promote tissue regeneration [40]. This immune activation, which occurred immediately prior to CCBD formation, may have influenced the regenerative environment and contributed to the enhanced healing observed in this group. In particular, the release of cytokines and the recruitment of immune cells after extraction may have accelerated the early stages of bone repair [40,41]. Although this potential influence is consistent with our findings, further studies would be required to specifically investigate the role of such immune modulation in combination with AD grafts. Another focus of future research could be the evaluation of the gene expression of genes of the early healing phase, such as runt-related transcription factor 2 (RUNX2), Osterix, etc. These genes play a crucial role in the early stages of bone formation and could provide a more precise understanding of the timing of bone healing [42]. Expanding gene analysis in future studies could lead to a more comprehensive characterization of the molecular events underlying the healing process.

In conclusion, this study provides comprehensive evidence of the osteogenic properties of AD and BX + Mg in a CCBD model. The increased bone volume, reduced amount of residual biomaterial and favorable gene expression profiles highlight the potential of BX + Mg for clinical application.

Author Contributions: Conceptualization, M.Č., A.T.J.R., S.Z.C. and Ž.P.K.; methodology, O.C.P. and A.T.J.R.; software A.T.J.R.; formal analysis, O.C.P. and A.T.J.R.; investigation, O.C.P., A.T.J.R., H.O., B.F., L.B., T.G., T.Č.-M. and L.M.; writing—original draft preparation, M.Č., A.T.J.R., O.C.P. and Ž.P.K.; writing—review and editing H.O., B.F., L.B., T.G., T.Č.-M., S.Z.C. and L.M.; supervision, Ž.P.K.; project administration, O.C.P. and A.T.J.R. All authors have read and agreed to the published version of the manuscript.

Funding: This research was funded by the Croatian Science Foundation under number IP-2020-02-7875 and by the University of Rijeka under number UNIRI-biomed-18-280.

Institutional Review Board Statement: The study was conducted according to the guidelines of the Declaration of Helsinki and approved by the Ethics Committee of the University of Rijeka and the Ministry of Agriculture (EP 302/2021).

Informed Consent Statement: Not applicable.

Data Availability Statement: The data presented in this article are available upon request from the corresponding author.

Conflicts of Interest: Ž.P.K. is a part-time employee of botiss GmbH. The rest of the authors declare no conflicts of interest.

References

1. Rowe, P.; Koller, A.; Sharma, S. Physiology, Bone Remodeling. In *StatPearls*; StatPearls Publishing: Treasure Island, FL, USA, 2024.
2. Albrektsson, T.; Johansson, C. Osteoinduction, Osteoconduction and Osseointegration. *Eur. Spine J.* **2001**, *10* (Suppl. S2), S96–S101. [CrossRef]
3. Kopf, J.; Petersen, A.; Duda, G.N.; Knaus, P. BMP2 and Mechanical Loading Cooperatively Regulate Immediate Early Signalling Events in the BMP Pathway. *BMC Biol.* **2012**, *10*, 37. [CrossRef]
4. Rath, B.; Nam, J.; Deschner, J.; Schaumburger, J.; Tingart, M.; Grässel, S.; Grifka, J.; Agarwal, S. Biomechanical Forces Exert Anabolic Effects on Osteoblasts by Activation of SMAD 1/5/8 through Type 1 BMP Receptor. *Biorheology* **2011**, *48*, 37–48. [CrossRef]
5. Zou, M.-L.; Chen, Z.-H.; Teng, Y.-Y.; Liu, S.-Y.; Jia, Y.; Zhang, K.-W.; Sun, Z.-L.; Wu, J.-J.; Yuan, Z.-D.; Feng, Y.; et al. The Smad Dependent TGF-β and BMP Signaling Pathway in Bone Remodeling and Therapies. *Front. Mol. Biosci.* **2021**, *8*. [CrossRef]
6. Chen, Y.; Yin, Y.; Luo, M.; Wu, J.; Chen, A.; Deng, L.; Xie, L.; Han, X. Occlusal Force Maintains Alveolar Bone Homeostasis via Type H Angiogenesis. *J. Dent. Res.* **2023**, *102*, 1356–1365. [CrossRef]
7. Rodan, G.A. Bone Homeostasis. *Proc. Natl. Acad. Sci. USA* **1998**, *95*, 13361–13362. [CrossRef]
8. Hansson, S.; Halldin, A. Alveolar Ridge Resorption after Tooth Extraction: A Consequence of a Fundamental Principle of Bone Physiology. *J. Dent. Biomech.* **2012**, *3*, 1758736012456543. [CrossRef]
9. Sodek, J.; Mckee, M.D. Molecular and Cellular Biology of Alveolar Bone. *Periodontology* **2000**, *24*, 99–126. [CrossRef]
10. Pesic, Z.; Pejcic, A. Biomaterials in Dentistry—Implantology and Guided Bone Regeneration. In *Biomaterials in Clinical Practice: Advances in Clinical Research and Medical Devices*; Zivic, F., Affatato, S., Trajanovic, M., Schnabelrauch, M., Grujovic, N., Choy, K.L., Eds.; Springer International Publishing: Cham, Switzerland, 2018; pp. 697–739. ISBN 978-3-319-68025-5.
11. Inchingolo, A.M.; Patano, A.; Di Pede, C.; Inchingolo, A.D.; Palmieri, G.; de Ruvo, E.; Campanelli, M.; Buongiorno, S.; Carpentiere, V.; Piras, F.; et al. Autologous Tooth Graft: Innovative Biomaterial for Bone Regeneration. Tooth Transformer® and the Role of Microbiota in Regenerative Dentistry. A Systematic Review. *J. Funct. Biomater.* **2023**, *14*, 132. [CrossRef]
12. Minetti, E.; Palermo, A.; Ferrante, F.; Schmitz, J.H.; Lung Ho, H.K.; Dih Hann, S.N.; Giacometti, E.; Gambardella, U.; Contessi, M.; Celko, M.; et al. Autologous Tooth Graft after Endodontical Treated Used for Socket Preservation: A Multicenter Clinical Study. *Appl. Sci.* **2019**, *9*, 5396. [CrossRef]
13. Mazor, Z.; Horowitz, R.A.; Prasad, H.; Kotsakis, G.A. Healing Dynamics Following Alveolar Ridge Preservation with Autologous Tooth Structure. *Int. J. Periodontics Restor. Dent.* **2019**, *39*, 697–702. [CrossRef]
14. Pang, K.-M.; Um, I.-W.; Kim, Y.-K.; Woo, J.-M.; Kim, S.-M.; Lee, J.-H. Autogenous Demineralized Dentin Matrix from Extracted Tooth for the Augmentation of Alveolar Bone Defect: A Prospective Randomized Clinical Trial in Comparison with Anorganic Bovine Bone. *Clin. Oral Implant. Res.* **2017**, *28*, 809–815. [CrossRef]
15. Mahesh, L.; Mascarenhas, G.; Bhasin, M.T.; Guirado, C.; Juneja, S. Histological Evaluation of Two Different Anorganic Bovine Bone Matrixes in Lateral Wall Sinus Elevation Procedure: A Retrospective Study. *Natl. J. Maxillofac. Surg.* **2020**, *11*, 258–262. [CrossRef]

16. Kloss, F.R.; Kämmerer, P.W.; Kloss-Brandstätter, A. First Clinical Case Report of a Xenograft–Allograft Combination for Alveolar Ridge Augmentation Using a Bovine Bone Substitute Material with Hyaluronate (Cerabone® Plus) Combined with Allogeneic Bone Granules (Maxgraft®). *J. Clin. Med.* **2023**, *12*, 6214. [CrossRef]
17. Park, J.-W.; Ko, H.-J.; Jang, J.-H.; Kang, H.; Suh, J.-Y. Increased New Bone Formation with a Surface Magnesium-Incorporated Deproteinized Porcine Bone Substitute in Rabbit Calvarial Defects. *J. Biomed. Mater. Res. A* **2012**, *100*, 834–840. [CrossRef]
18. de Freitas Silva, L.; de Carvalho Reis, E.N.R.; Barbara, T.A.; Bonardi, J.P.; Garcia, I.R.; de Carvalho, P.S.P.; Ponzoni, D. Assessment of Bone Repair in Critical-Size Defect in the Calvarium of Rats after the Implantation of Tricalcium Phosphate Beta (β-TCP). *Acta Histochem.* **2017**, *119*, 624–631. [CrossRef]
19. Kasuya, S.; Inui, S.; Kato-Kogoe, N.; Omori, M.; Yamamoto, K.; Inoue, K.; Ito, Y.; Nakajima, Y.; Hirata, A.; Ueno, T. Evaluation of Guided Bone Regeneration Using the Bone Substitute Bio-Oss® and a Collagen Membrane in a Rat Cranial Bone Defect Model. *J. Hard Tissue Biol.* **2018**, *27*, 79–84. [CrossRef]
20. Jerbić Radetić, A.T.; Zoričić Cvek, S.; Tomas, M.; Erjavec, I.; Oguić, M.; Perić Kačarević, Ž.; Cvijanović Peloza, O. CSBD Healing in Rats after Application of Bovine Xenogeneic Biomaterial Enriched with Magnesium Alloy. *Int. J. Mol. Sci.* **2021**, *22*, 9089. [CrossRef]
21. Cvijanović Peloza, O.; Jerbić Radetić, A.T.; Baričić, M.; Bukovac, L.; Zoričić Cvek, S. Dynamics of CSBD Healing after Implementation of Dentin and Xenogeneic Bone Biomaterial. *Materials* **2023**, *16*, 1600. [CrossRef]
22. Bormann, K.-H.; Suarez-Cunqueiro, M.M.; Sinikovic, B.; Kampmann, A.; von See, C.; Tavassol, F.; Binger, T.; Winkler, M.; Gellrich, N.-C.; Rücker, M. Dentin as a Suitable Bone Substitute Comparable to SS-TCP--an Experimental Study in Mice. *Microvasc. Res.* **2012**, *84*, 116–122. [CrossRef]
23. Rücker, M.; Laschke, M.W.; Junker, D.; Carvalho, C.; Schramm, A.; Mülhaupt, R.; Gellrich, N.-C.; Menger, M.D. Angiogenic and Inflammatory Response to Biodegradable Scaffolds in Dorsal Skinfold Chambers of Mice. *Biomaterials* **2006**, *27*, 5027–5038. [CrossRef]
24. Rao, X.; Huang, X.; Zhou, Z.; Lin, X. An Improvement of the 2ˆ(–Delta Delta CT) Method for Quantitative Real-Time Polymerase Chain Reaction Data Analysis. *Biostat. Bioinforma Biomath.* **2013**, *3*, 71–85.
25. Shive, M.S.; Anderson, J.M. Biodegradation and Biocompatibility of PLA and PLGA Microspheres. *Adv. Drug Deliv. Rev.* **1997**, *28*, 5–24. [CrossRef]
26. Pinholt, E.M.; Bang, G.; Haanaes, H.R. Alveolar Ridge Augmentation by Osteoinduction in Rats. *Scand. J. Dent. Res.* **1990**, *98*, 434–441. [CrossRef]
27. Ike, M.; Urist, M.R. Recycled Dentin Root Matrix for a Carrier of Recombinant Human Bone Morphogenetic Protein. *J. Oral Implantol.* **1998**, *24*, 124–132. [CrossRef]
28. Witte, F. The History of Biodegradable Magnesium Implants: A Review. *Acta Biomater.* **2010**, *6*, 1680–1692. [CrossRef]
29. Witte, F.; Kaese, V.; Haferkamp, H.; Switzer, E.; Meyer-Lindenberg, A.; Wirth, C.J.; Windhagen, H. In Vivo Corrosion of Four Magnesium Alloys and the Associated Bone Response. *Biomaterials* **2005**, *26*, 3557–3563. [CrossRef]
30. Witte, F.; Fischer, J.; Nellesen, J.; Crostack, H.-A.; Kaese, V.; Pisch, A.; Beckmann, F.; Windhagen, H. In Vitro and in Vivo Corrosion Measurements of Magnesium Alloys. *Biomaterials* **2006**, *27*, 1013–1018. [CrossRef]
31. Rider, P.; Kačarević, Ž.P.; Elad, A.; Rothamel, D.; Sauer, G.; Bornert, F.; Windisch, P.; Hangyási, D.; Molnar, B.; Hesse, B.; et al. Analysis of a Pure Magnesium Membrane Degradation Process and Its Functionality When Used in a Guided Bone Regeneration Model in Beagle Dogs. *Materials* **2022**, *15*, 3106. [CrossRef]
32. Jung, O.; Hesse, B.; Stojanovic, S.; Seim, C.; Weitkamp, T.; Batinic, M.; Goerke, O.; Kačarević, Ž.P.; Rider, P.; Najman, S.; et al. Biocompatibility Analyses of HF-Passivated Magnesium Screws for Guided Bone Regeneration (GBR). *Int. J. Mol. Sci.* **2021**, *22*, 12567. [CrossRef]
33. Witte, F.; Ulrich, H.; Rudert, M.; Willbold, E. Biodegradable Magnesium Scaffolds: Part 1: Appropriate Inflammatory Response. *J. Biomed. Mater. Res. A* **2007**, *81*, 748–756. [CrossRef] [PubMed]
34. Hu, K.; Olsen, B.R. The Roles of Vascular Endothelial Growth Factor in Bone Repair and Regeneration. *Bone* **2016**, *91*, 30–38. [CrossRef] [PubMed]
35. Hassumi, J.S.; Mulinari-Santos, G.; da S. Fabris, A.L.; Jacob, R.G.M.; Gonçalves, A.; Rossi, A.C.; Freire, A.R.; Faverani, L.P.; Okamoto, R. Alveolar Bone Healing in Rats: Micro-CT, Immunohistochemical and Molecular Analysis. *J. Appl. Oral. Sci.* **2018**, *26*, e20170326. [CrossRef] [PubMed]
36. Kanbur, N.O.; Derman, O.; Sen, T.A.; Kinik, E. Osteocalcin. A Biochemical Marker of Bone Turnover during Puberty. *Int. J. Adolesc. Med. Health* **2002**, *14*, 235–244. [CrossRef] [PubMed]
37. Deckers, M.M.; Karperien, M.; van der Bent, C.; Yamashita, T.; Papapoulos, S.E.; Löwik, C.W. Expression of Vascular Endothelial Growth Factors and Their Receptors during Osteoblast Differentiation. *Endocrinology* **2000**, *141*, 1667–1674. [CrossRef]
38. da S. Pires, J.L.; de Carvalho, J.J.; Pereira, M.J.D.S.; da S. Brum, I.; Nascimento, A.L.R.; Dos Santos, P.G.P.; Frigo, L.; Fischer, R.G. Repair of Critical Size Bone Defects Using Synthetic Hydroxyapatite or Xenograft with or without the Bone Marrow Mononuclear Fraction: A Histomorphometric and Immunohistochemical Study in Rat Calvaria. *Materials* **2021**, *14*, 2854. [CrossRef]
39. Verna, C.; Dalstra, M.; Wikesjö, U.M.E.; Trombelli, L. Carles Bosch Healing Patterns in Calvarial Bone Defects Following Guided Bone Regeneration in Rats. A Micro-CT Scan Analysis. *J. Clin. Periodontol.* **2002**, *29*, 865–870. [CrossRef]
40. Gurtner, G.C.; Werner, S.; Barrandon, Y.; Longaker, M.T. Wound Repair and Regeneration. *Nature* **2008**, *453*, 314–321. [CrossRef]

41. Devlin, H. Early Bone Healing Events Following Rat Molar Tooth Extraction. *Cells Tissues Organs* **2000**, *167*, 33–37. [CrossRef]
42. Komori, T. Regulation of Bone Development and Extracellular Matrix Protein Genes by RUNX2. *Cell Tissue Res.* **2010**, *339*, 189–195. [CrossRef]

Journal of
Functional Biomaterials

MDPI

Systematic Review

Evaluating Bioassays for the Determination of Simvastatin's Osteogenic Activity: A Systematic Review

Lara Steiner Back [1], Isabella Schönhofen Manso [1], Mariane Beatriz Sordi [1], Gabriel Leonardo Magrin [1], Águedo Aragonês [1], Ricardo de Souza Magini [1], Reinhard Gruber [2,*] and Ariadne Cristiane Cabral Cruz [1,3,*]

1 Post-Graduation Program of Dentistry, Center for Education and Research on Dental Implants, Federal University of Santa Catarina, Florianópolis 88053-701, Brazil; lara_back@msn.com (L.S.B.); isabellamanso@yahoo.com.br (I.S.M.); marianesordi@hotmail.com (M.B.S.); gabriel.magrin@ufsc.br (G.L.M.); aguedo@terra.com.br (Á.A.); ricardo.magini@gmail.com (R.d.S.M.)

2 Department of Oral Biology, University Clinic of Dentistry, Medical University of Vienna, 1090 Vienna, Austria

3 Applied Virology Laboratory, Federal University of Santa Catarina, Florianópolis 88053-701, Brazil

* Correspondence: reinhard.gruber@meduniwien.ac.at (R.G.); ariadne.cruz@ufsc.br (A.C.C.C.); Tel.: +43-699-10718472 (R.G.); +55-48-3721-3407 (A.C.C.C.)

Abstract: Objective: Osteogenic differentiation is a complex process, and its analysis requires several biomarkers. Allied with this, there are no standardized bioassays to monitor the activity of simvastatin in osteogenesis in vitro. Therefore, identifying the most efficient and sensitive bioassays may enhance the quality of in vitro studies, bridging the gap with in vivo findings, saving time and resources, and benefiting the community. This systematic review aimed to determine the most efficient bioassay for simvastatin's osteogenic activity in vitro, in terms of sensitivity. Materials and Methods: In vitro studies evaluating undifferentiated mesenchymal cells treated with simvastatin were considered eligible. References were selected in a two-phase process. Electronic databases and the grey literature were screened up to September 2023. The Office of Health Assessment and Translation (OHAT) tool was used to assess the risk of bias. Certainty in cumulative evidence was evaluated using the Grading of Recommendations, Assessment, Development, and Evaluation (GRADE) criteria. Data were analyzed considering extracellular matrix mineralization, alkaline phosphatase, and the expression of potential target genes, such as bone morphogenetic protein-2 (BMP-2), collagen type I, Runt-related transcription factor 2, osterix, osteocalcin, and osteopontin. Results: Fourteen studies were included. A "probably low" or a "definitely low" risk of bias was assigned to the included studies. The simvastatin concentration ranged from 0.1 nM to 10 μM. Considering a minimum 4-fold increase, simvastatin caused robust mineralization of the extracellular matrix in four studies (4.0-, 4.4-, 5.0-, and 39.5-fold). Moreover, simvastatin substantially increased BMP-2 expression in mesenchymal cells in three studies (4-, 11-, and 19-fold). Conclusion: Therefore, mineralization of the extracellular matrix and BMP-2 expression in mesenchymal cells are the most efficient bioassays for determining the osteogenic activity of simvastatin in vitro (high certainty level). These findings provide a standardized approach that can enhance the reliability and comparability of in vitro studies, bridging the gap with in vivo research and optimizing resources in the field of bone regeneration.

Keywords: simvastatin; mesenchymal stromal cell; osteogenesis; osteoblast; mineralization

1. Introduction

Bone defects caused by trauma, tumors, tooth loss, or periodontal disease are challenging for clinicians. The regeneration of these defects frequently requires bone grafting before dental implant placement and implant-supported rehabilitation [1,2]. Autologous bone grafts have limitations, including the morbidity of the donor area, limited bone availability for harvesting, and difficulty in predicting resorption over time [3]. In turn, immunogenic issues and the risk of disease transmission are limitations concerning allogeneic bone grafts [4]. Regarding xenogeneic grafts, it is imperative to thoroughly remove proteins during the biomaterial processing to mitigate the risk of immunologic complications and potential contamination [5]. As an alternative approach, tissue engineering takes advantage of bone substitutes to interact with anabolic signaling molecules and mesenchymal cells [6–8]. Strategies in bone tissue engineering necessitate a profound understanding of how mesenchymal cells respond to anabolic signals, with simvastatin emerging as one of the most promising pharmacological compounds.

Simvastatin, a member of the statin family [9], is a 3-hydroxy-3-methylglutaryl coenzyme A (HMG-CoA) reductase inhibitor prescribed to prevent cardiovascular diseases [10]. Besides lowering lipids, Mundy et al. [9] demonstrated that simvastatin exerts a bone anabolic effect using calvaria defect models. This effect is mediated through reductions in isoprenoid intermediates, such as farnesyl pyrophosphate and geranylgeranyl pyrophosphate, which are essential for the prenylation of small GTP-binding proteins like Ras, Rho, and Rac. The inhibition of RhoA, in particular, has been associated with the stimulation of bone morphogenetic protein type 2 (BMP-2) expression. Furthermore, simvastatin influences the Wnt/β-catenin signaling pathway, which interacts synergistically with BMP signaling to promote osteoblast differentiation. These mechanisms collectively enhance the activity of Runx2, a master transcription factor for osteoblast differentiation, further supporting its role in bone regeneration [11]. Additionally, drug delivery systems embedding simvastatin into bone substitutes potentially affect simvastatin release kinetics and activity, and have garnered growing attention from the scientific community in recent years [12–16]. Therefore, biomaterials with different compositions have been proposed as drug delivery systems for simvastatin [12,13,17]. Consequently, several in vitro studies have been performed as preliminary evaluations.

It is important to remember that osteogenic differentiation is a complex process that requires the analyses of several early and late biomarkers, including alkaline phosphatase, bone morphogenetic protein-2 (BMP-2), collagen type I, Runt-related transcription factor 2, osterix, osteonectin, osteocalcin, and osteopontin, as well as extracellular matrix mineralization. The variability and inconsistency of results across bioassays make it challenging to determine which assays are the most reliable and efficient for drawing accurate conclusions regarding osteogenic differentiation. Furthermore, the lack of standardized bioassays to specifically monitor the activity of simvastatin in osteogenesis in vitro adds to this challenge. The purpose of this study is to identify the most efficient and sensitive bioassays to enhance the quality of in vitro studies, bridging the gap with in vivo findings, saving time and resources, and ultimately benefiting the broader scientifical community.

Considering the effect of simvastatin in osteogenesis in vitro, various types of mesenchymal cells have been evaluated, such as osteogenic cell lines [18–20], cells from murine (bone marrow-derived mesenchymal cells [15,16,21–24]), from human (adipose-, bone marrow-, and human amniotic membrane epithelial-derived mesenchymal cells [14,25–27]), and embryonic stem cells [28,29]. Based on these cell types, the following outcome variables have previously been evaluated: matrix mineralization [14–16,22,23,25–30], alkaline phosphatase (ALP) [14–16,21,23–30], bone morphogenetic protein type-2 (BMP-2) [14,15,23–25], collagen type I (COL1) [16,22,28,29,31], osteocalcin (BGLAP) [14–16,23,24,28], Runt-related transcription factor 2 (RUNX2) [14–16,21,28,29,31], osterix (OSX) [21,28,29], and osteopontin [24]. As can be seen, there is a large spectrum of mesenchymal cells and outcome parameters to measure simvastatin activity in vitro. Therefore, through a systematic review, we aimed to identify the most sensitive outcome variables to be used as simvastatin bioassays for osteogenesis in vitro. To estimate the x-fold increase in each bioassay induced by simvastatin, data from the simvastatin-treated group were normalized against the control group. The control group could consist of either an osteogenic medium or a non-osteogenic medium, both without simvastatin. A minimum 4-fold increase was considered robust.

2. Materials and Methods

2.1. Study Design

The systematic review protocol was based on the Cochrane Handbook for Systematic Reviews of Interventions [32] and Preferred Reporting Items for Systematic Review and Meta-Analysis (PRISMA 2020) checklist guidelines [33]. The protocol was registered on the Open Science Framework registration platform (10.17605/OSF.IO/GBK5S) on 27 August 2020.

2.2. Inclusion Criteria

The PICOS factors (population, intervention, comparison, outcome, and type of studies) used in this systematic review were as follows: Population (P)—undifferentiated mesenchymal cells. Intervention (I)—simvastatin treatment. Comparison (C)—no simvastatin treatment. Outcome (O)—bioassay efficiency. Type of studies (S)—in vitro studies. The focus question was as follows: "What is the most efficient bioassay to evaluate in vitro simvastatin activity during osteogenesis?" Studies were considered eligible when they evaluated undifferentiated mesenchymal cells treated with simvastatin. No publication time restriction was applied.

2.3. Exclusion Criteria

The following exclusion criteria were considered: (1) Studies investigating other cell types, such as differentiated cells. (2) Studies assessing drug or biomaterial associations that do not present separate data for simvastatin alone. (3) Studies that do not present an untreated control group with the same cell type. (4) Studies that do not present a control group without simvastatin treatment. (5) Studies that do not evaluate osteogenic differentiation. (6) Studies that are not in vitro (clinical studies, animal studies, conference abstracts, letters, pilot studies, review articles, case reports, protocols, short communications, personal opinions, posters, and book chapters). (7) Studies that do not specify the culture medium. (8) Studies that do not quantify alkaline phosphatase (ALP) or extracellular matrix mineralization for test and control groups. (9) Full text not available (book chapters, conference abstracts, expert opinions, letters, and literature reviews). (10) Duplicated data (e.g., dissertations/theses in which corresponding published articles were available). (11) Studies not published in the Latin alphabet.

2.4. Search Strategy

A search strategy based on the PICOS structure using MeSH terms and keywords was developed. Also, Embase, Literature of Latin American and Caribbean Health Sciences (LILACS), PubMed, SCOPUS, and Web of Science were used as electronic databases. Additionally, the grey literature was consulted with a search strategy for Google Scholar web search (first 100 references) and ProQuest. The electronic search was performed until September 2023 (Appendix A). A hand search on the reference list of identified records was also performed. All records were exported to reference manager software (Mendeley Desktop, Elsevier, London, UK version v1.19.8), and duplicates were removed.

2.5. Study Selection

A two-phase selection process was performed. Two independent reviewers (L.S.B. and I.S.M.) selected the references using online software (Rayyan, Qatar Computing Research Institute, Qatar). In phase one, both reviewers read the titles and abstracts independently while applying the eligibility criteria. The reviewers performed a full-text reading using the eligibility criteria in phase two. A third reviewer (M.B.S.) cross-checked the retrieved information in both phases. The final selection was always based on the full-text publication. Articles that met the eligibility criteria proceeded for data extraction.

2.6. Data Collection Process and Data Items

Two independent reviewers (L.S.B. and I.S.M.) collected data from the selected articles. Once selected, they cross-checked the retrieved information with a third reviewer (M.B.S.). Any disagreement was discussed among them. The following data were extracted for each included study: author, year of publication, country, cell type, the origin of cell lines, cell treatment, simvastatin concentration, evaluation methods, experimental time, and main results. If data were missing or unclear, we contacted the corresponding authors of the included studies to resolve or clarify the issue.

2.7. Risk-of-Bias Assessment in Individual Studies

The Office of Health Assessment and Translation (OHAT) tool was employed to evaluate the risk of bias (RoB) among individual studies [34], with adaptations made to in vitro studies [35]. Briefly, regarding selection bias under the domains "Randomization of the exposure levels" and "Allocation concealment", all studies using homogeneous cell suspensions were considered as having a definitely low risk of bias. Additionally, confounding bias is not a relevant key item for experimental animal or in vitro studies. Consequently, questions 3 and 4 were not applied. In sequence, two reviewers (L.S.B. and I.S.M.) independently assessed the risk of bias in the included studies. Discrepancies between the reviewers were resolved by discussion until agreement. The possible answers were "definitely low", "probably low", "probably high", or "definitely high" risk of bias, following specific criteria detailed in the protocol. After addressing all the questions, these scores were subsequently categorized into three tiers: Tier 1 (high quality), Tier 2 (moderate quality), or Tier 3 (low quality). Tier 1 is defined as having a "++" (definitely low) or "+" (probably low) risk of bias in all key domains, as well as a "++" (definitely low) or "+" (probably low) risk of bias for ≥50% of the other domains. Tier 2 comprises studies that do not meet the criteria for placement in either Tier 1 or Tier 3. Tier 3 represents a "- -" (definitely high) or "-" (probably high) risk of bias in all key domains, as well as "- -" (definitely high) or "-" (probably high) risk of bias for ≥50% of the other domains.

2.8. Summary Measures and Synthesis of Results

Information concerning gene expression analysis (levels of osteogenic markers), bone protein expression, the quantitative staining of extracellular matrix mineralization or calcium, and ALP activity or staining were considered during the evaluation of outcomes. A meta-analysis was planned if the data from the included studies were deemed homogeneous. To estimate the x-fold increase in each bioassay, data from the SIM-treated group were normalized against the SIM control group, which could consist of either an osteogenic medium (OM) or a non-osteogenic medium (NOM) without SIM.

2.9. Certainty in Cumulative Evidence and Risk of Bias Across Studies

Methodological heterogeneity was assessed by comparing the variability in study design and the individual RoB of the included studies. The certainty in cumulative evidence was assessed using the Grading of Recommendations, Assessment, Development, and Evaluation (GRADE) criteria [36], considering in vitro studies instead of randomized clinical trials [36]. Two authors (L.S.B. and A.C.C.C) methodically appraised the selected studies, evaluating risk of bias, inconsistency, indirectness, imprecision, and publication bias, and categorized the outcomes across the included articles as having a "high", "moderate", "low", or "very low" quality of evidence, according to the analysis of each study. The specific reasons for each judgment were considered: High certainty—we are very confident that the true effect lies close to that of the estimated effect. Moderate certainty—we are moderately confident in the effect estimate. Therefore, the true effect is likely to be close to the estimate of the effect, but there is a possibility that it is substantially different. Low certainty—our confidence in the estimate is limited. Therefore, the true effect may be substantially different from the estimate of the effect. Very low certainty—we have very little confidence in the effect estimate. The true effect is likely to be substantially different from the estimate of the effect. The following thresholds were considered for overall GRADE assessments: (1) "not serious" if concerns were present in less than 25% of the included studies; (2) "serious" if concerns were present in between 25% and 50% of the included studies; and (3) "very serious" if concerns were present in more than 50% of the included studies. A high level of certainty means the authors have a lot of confidence that the true effect is similar to the estimated effect. Moderate certainty indicates the authors believe the true effect is probably close to the estimated effect. Low certainty suggests the true effect might be markedly different from the estimated effect, while very low certainty indicates the true effect is probably markedly different from the estimated effect.

3. Results

3.1. Study Selection

In phase one, 2088 records were retrieved from an electronic search of databases and the grey literature after duplicate removal. A comprehensive evaluation of titles and abstracts was performed, resulting in 77 potentially suitable references. After full-text analyses, 14 articles were included in this systematic review according to the inclusion and exclusion criteria. Step-by-step process details are shown in Figure 1. The articles excluded and the reasons for exclusion are detailed in Appendix B.

3.2. Study Characteristics

The included studies, published between 2009 and 2023, were conducted in seven countries: China [14,16,21,24,26], Germany [30], South Korea [28,31], Switzerland [25], the United Kingdom [29], Taiwan [15,22,23], and Iran [27].

Undifferentiated mesenchymal cell lines from different origins and species were used, including human adipose tissue [14,26], human bone marrow [25], murine bone marrow [15,16,21,23,24], murine embryonic cells [28,29], human gingiva-derived cells [31], and human amniotic membrane epithelial-derived cells [27]. Most studies employed two-dimensional cultures, with one study using three-dimensional spheroids [31]. Cells were treated with non-osteogenic medium (NOM) [14,25,26,29], osteogenic medium (OM) [15,16,21,24–29,31], or NOM switching to OM [22,23]. Osteogenic differentiation was evaluated by ALP (activity or histochemical staining), gene expression, extracellular matrix mineralization (Von Kossa staining, Alizarin Red staining, calcium quantification, or phosphorous quantification), and protein expression. Simvastatin molarity was determined. Figure 2 contains the data of each study, showing the x-fold increase determined by each bioassay. The main characteristics of the included studies are provided in Table 1.

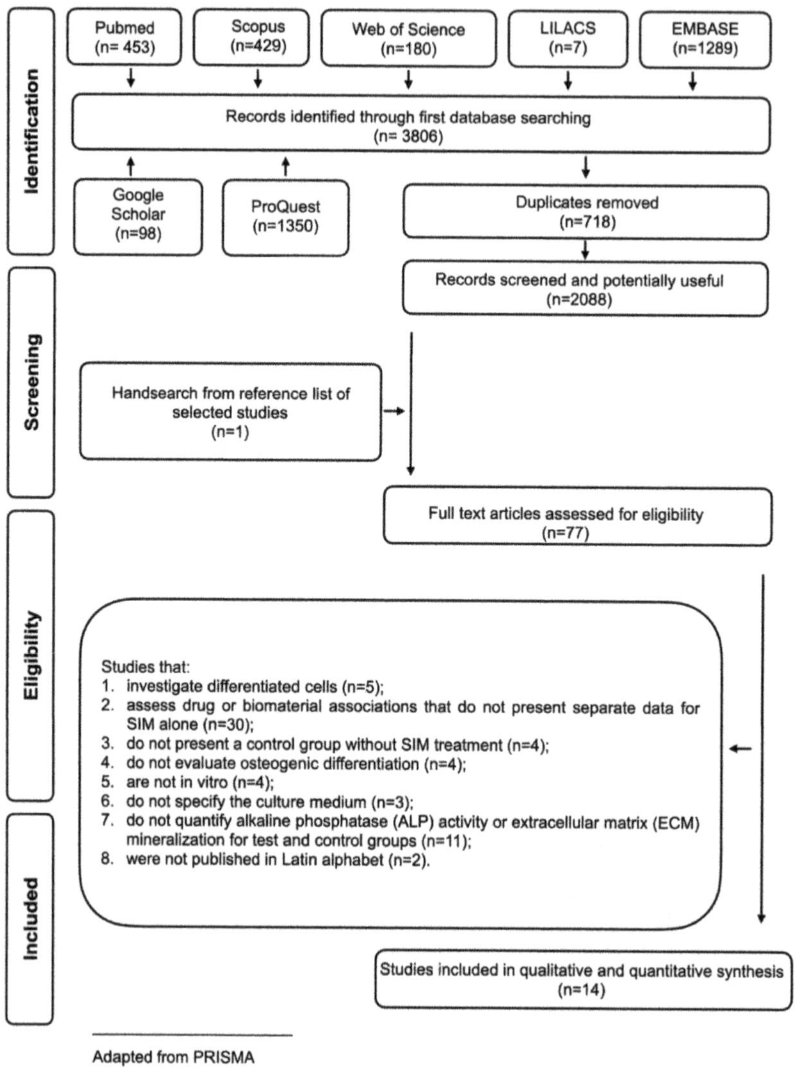

Figure 1. Flow diagram of literature search and selection criteria adapted from PRISMA. References were selected in a two-phase process. Electronic databases (Embase, LILACS, LIVIVO, PubMed, SCOPUS, and Web of Science) and grey literature databases (Google Scholar, Open Grey, and ProQuest) were searched up to September 2023.

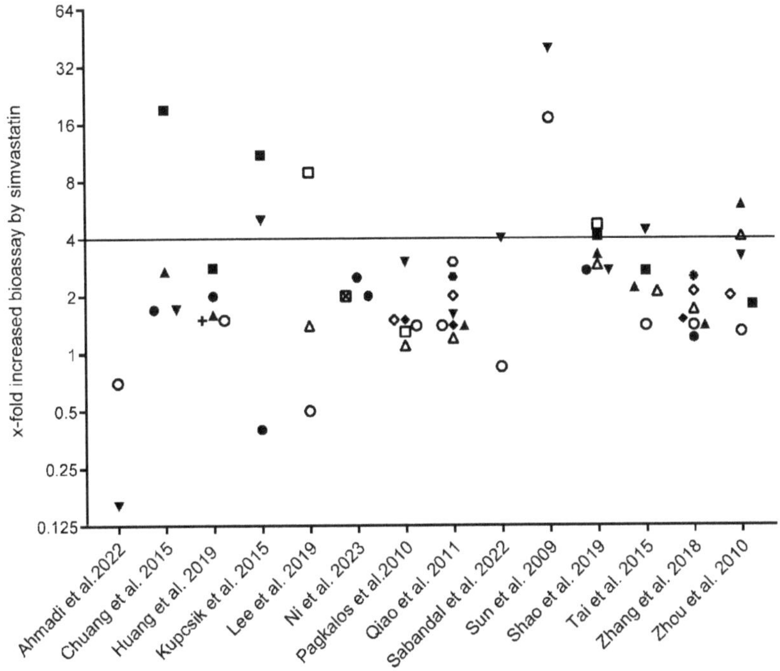

Figure 2. Summary of data showing the x-fold increase determined by each bioassay. The data of the simvastatin-treated group are normalized using the same treatment with no simvastatin (osteogenic medium or non-osteogenic medium). Studies are in alphabetic order. Ahmadi et al. (2022) [27], Chuang et al. (2015) [23], Huang et al. (2019) [24], Kupcsik et al. (2009) [25], Lee et al. (2019) [31], Ni et al. (2023) [16], Pagkalos et al. (2010) [29], Qiao et al. (2011) [28], Sabandal et al. (2022) [30], Shao et al. (2019) [22], Sun et al. (2009) [26], Tai et al. (2015) [15], Zhang et al. (2018) [21], and Zhou et al. (2010) [14]. Each geometric shape represents one bioassay. ■: BMP-2 transcript; □: COL1 transcript; •: ALP transcript; ○: ALP; ▲: BGLAP transcript; △: RUNX2 transcript; ▼: Mineralization; +: BMP-9 transcript; ♦: OSX transcript; ◇: BGLAP; ⊠: IBSP transcript; ✱: RUNX2.

Table 1. Summary of descriptive characteristics of included studies (n = 14).

Author (Year) Country	Cell Type (Origin)	Treatment Type	SIM Concentration	Evaluation Methods (Period)	mRNA Expression	Western Blotting/ Immunohisto-chemistry/ Biochemistry	Mineralization
Ahmadi et al. (2022) [27] Iran	Human amniotic epithelial stem cells	T3(o) T4(o) SIM-OM compared to OM	100 nM	ALP activity (Day 14) Calcium quantification (Day 14) Phosphorous quantification (Day 14)	Not evaluated	ALP Activity ↓	Calcium quantification ↓ Phosphorous quantification ↓
Chuang et al., 2015 [23] Taiwan	mBMSCs (Murine bone marrow—D1 Cells)	T5(q) T6(q) SIM-OM compared to OM	0.5 μM	RT-PCR (Days 1, 3, and 5) ALP Staining (Day 5) Alizarin Red staining (5 days after OM induction)	ALP transcript ↑ BGLAP transcript ↑ BMP-2 transcript ↑	ALP ↑ BMP-2 ↑	Alizarin Red staining ↑

Table 1. *Cont.*

Author (Year) Country	Cell Type (Origin)	Treatment Type	SIM Concentration	Evaluation Methods (Period)	mRNA Expression	Western Blotting/ Immunohisto-chemistry/ Biochemistry	Mineralization
Huang et al. (2019) [24] China	mBMSCs (Murine bone marrow)	T3(e) T4(e) SIM-OM compared to OM	0.025, 0.10, 0.25, and 1.0 μM	RT-PCR (Days 3 and 7) ALP activity (Days 3 and 7) Alizarin Red staining (Days 3, 7, and 14)	ALP transcript ↑ BGLAP transcript ↑ BMP-2 transcript ↑ BMP-9 transcript ↑	ALP activity ↑	Alizarin Red staining not quantified
Kupcsik et al. (2009) [25] Switzerland	hBMSCs (Human bone marrow)	T1(a) T2(a) T3(f) SIM-NOM compared to NOM	1 and 5 μM	RT-PCR (Days 4, 11, and 18) ALP activity (Day 11) Von Kossa staining (Day 25) Calcium deposition (Days 18 and 25)	ALP transcript ↓	ALP activity ↓	Calcium quantification ↑ Von Kossa staining not quantified
Lee et al. (2019) [31] South Korea	GMSCs spheroids (Human gingiva)	T3(j) T4(j) SIM-OM compared to OM	1 and 10 μM	RT-PCR (Day 7) ALP activity (Day 14) Alizarin Red staining (Days 7 and 14)	COL1 transcript ↑ RUNX2 transcript ↑	ALP activity ↓	Alizarin Red staining not quantified
Ni et al. (2023) [16] China	mBMSCs (Murine bone marrow)	T3(p) T4(p) SIM-OM compared to OM	100 nM	RT-PCR (Days 7 and 14) ALP staining (Days 7 and 14) Alizarin Red staining(Days 7 and 14) Von Kossa staining (Days 7 and 14)	COL1 transcript ↑ RUNX2 transcript ↑ BGLAP transcript ↑ IBSP transcript ↑	ALP ↑ COL1 ↑ RUNX2 ↑ BGLAP ↑	Alizarin Red staining ↑ Von Kossa staining ↑
Pagkalos et al. (2010) [29] United Kingdom	mESCs (Murine embryonic cells, E14Tg2α)	T1(b) T2(b) T3(g) SIM-NOM compared to NOM	0.1, 1, 10, and 100 nM	RT-PCR (Day 12) ALP activity (Days 4, 8, 12, 16, and 21) Alizarin Red staining (Days 6, 11, 12, 16, and 21)	BGLAP transcript ↑ COL1 transcript ↑ RUNX2 transcript ↑ OSX transcript ↑	ALP ↑	Alizarin Red staining ↑
Qiao et al. (2011) [28] Korea	mESCs (Murine embryonic cells, D3 line ATCC)	T3(h) T4(h) SIM-OM compared to OM	1, 10, 100, 200 nM	RT-PCR (Day 4) ALP activity (Days 4 and 7) Alizarin Red staining (Days 7 and 14) Western blot (Day 7)	BGLAP transcript ↑ COL1 transcript ↑ RUNX2 transcript ↑ OSX transcript ↑	ALP ↑	Alizarin Red staining ↑
Sabandal et al. (2022) [30] Germany	hADSC	T3 (i) T4 (i) SIM-OM compared to OM	0.01, 0.1, 1 and 2 μM	ALP activity (days 18, 21, 28) Alizarin Red staining (days 18, 21, 28)	Not evaluated	ALP ↓	Alizarin Red staining ↑
Shao et al. (2019) [22] Taiwan	mBMSCs (Murine bone marrow, D1 Cells)	T5(r) T6(r) SIM-OM compared to OM	0.1, 0.2 and 0.5 μM	RT-PCR (Day 3) ALP activity (1 day after OM induction—day 4) Alizarin Red staining (5 days after OM induction—day 7) Calcium deposition (5 days after OM induction—day 7)	ALP transcript ↑ BMP2 transcript ↑ COL1 transcript ↑	ALP ↑	Alizarin Red staining ↑

Table 1. *Cont.*

Author (Year) Country	Cell Type (Origin)	Treatment Type	SIM Concentration	Evaluation Methods (Period)	mRNA Expression	Western Blotting/ Immunohisto-chemistry/ Biochemistry	Mineralization
Sun et al. (2009) [26] China	ADSCs (Human adipose tissue)	T1(c) T2(c) T3(k) SIM-NOM compared to NOM	1 μM	RT-PCR (Days 1, 3, 6, 7, 9, and 12) ALP staining (Days 7, 14, 21, and 28) Von kossa staining (Days 7, 14, 21, and 28)	RT-PCR data were not quantified	ALP ↑	Von Kossa staining ↑
Tai et al. (2015) [15] Taiwan	mBMSCs (Murine bone marrow, D1 Cells)	T3(n) T4(n) SIM-OM compared to OM	1 μM	RT-PCR (12, 24, and 48 h) ALP staining (Day 3) Alizarin Red staining (Day 5)	BGLAP transcript ↑ BMP-2 transcript ↑ RUNX2 transcript ↑	ALP ↑	Alizarin Red staining ↑
Zhang et al. (2018) [21] China	mBMSCs (Murine bone marrow)	T3(l) T4(l) SIM-OM compared to OM	0.3 nM	RT-PCR (Day 7) ALP activity (Day 7) ALP staining (Day 7) Alizarin Red staining (Day 7) Western blot (Day 7)	ALP transcript ↑ BGLAP transcript A ↑ RUNX2 transcript ↑ OSX transcript ↑	ALP ↑ BGLAP ↑ OPN ↑ RUNX2 ↑	Alizarin Red staining ↑
Zhou et al. (2010) [14] China	ADSCs (Human adipose tissue)	T1(d) T2(d) T3(m) SIM-NOM compared to NOM	0.01, 0.1, and 1 μM	RT-PCR (Day 3) ALP activity (Days 6 and 14) Alizarin Red staining (Day 14) Osteocalcin radioimmunoassay	BGLAP transcript ↑ BMP2 transcript ↑ RUNX2 transcript A ↑	ALP ↑ BGLAP ↑	Alizarin Red staining ↑

Treatment type composition: T1—non-osteogenic medium (NOM). T2—simvastatin–NOM (SIM-NOM). T3—osteogenic medium (OM). T4—SIM-OM. T5—NOM changing to OM. T6—SIM-NOM changing to OM. NOM: (a) minimum essential medium alfa (α-MEM) + 10% FBS. (b) α-MEM + 15% FBS + 1% penicillin/streptomycin + 10 mM β-glycerophosphate. (c) α-MEM + 10% FBS + 1% penicillin/streptomycin. (d) DMEM + 10% FBS + 100 U/mL penicillin + 100 mg/mL streptomycin. OM: (e) Roswell Park Memorial Institute 1640 Medium (RPMI1640) + 10% FBS + 0.2 mmol/L ascorbic acid + 10 nmol/L dexamethasone + 10 mM β-glycerophosphate. (f) α-MEM + 10% FBS + non-essential amino acids + 10 mmol/L β-glycerophosphate + 0.1 mmol/L ascorbic acid + 10 nmol/L dexamethasone. (g) α-MEM + 15% FBS + 1% penicillin/streptomycin + 10 mM β-glycerophosphate + 50 μg/mL ascorbic acid + 1 μM dexamethasone. (h) α-MEM + 5% FBS + 50 μg/mL ascorbic acid + 1 μM dexamethasone + 3 mmol/L β-glycerophosphate. (i) α-MEM + 10% bovine calf serum + 16 ng/mL^{-1} dexamethasone + 10 nM β-glycerophosphate + 1.4 mM ascorbic acid. (j) α-MEM + 15% FBS + 100 U/mL penicillin + 100 μg/mL streptomycin + 200 mmol/L L-glutamine + 10 mmol/L ascorbic acid + 38 μg/mL dexamethasone. (k) α-MEM + 10% FBS + 1% penicillin/streptomycin + 0.1 μM dexamethasone + 10 μmol/L β-glycerophosphate + 50 μM ascorbic acid. (l) DMEM + 10% FBS + 1% penicillin/streptomycin + 10 mmol/L β-glycerophosphate + 50 μg/mL ascorbic acid. (m) DMEM + 10% FBS + 100 U/mL penicillin + 100 mg/mL streptomycin + 100 nM dexamethasone + 0.2 mM ascorbic acid + 10 nM β-glycerophosphate. (n) DMEM + 12% FBS + 100 μg/mL ascorbic acid + 100 μg/mL non-essential amino acids + 100 μg/mL penicillin/streptomycin + 100 nM dexamethasone + 0.2 mM 1-ascorbic acid-2-phophate. (o) DMEM + 10% FBS + 1% L- glutamine + 1% penicillin/streptomycin + 50 μM ascorbic acid + 10 nM 1.25(OH)D3 + 0.1 μM dexamethasone + 10 mM of β-glycerophosphate. (p) α-MEM + 10% FBS + 1% penicillin + 1% streptomycin + 0.2 mM ascorbic acid + 10mM β-glycerophosphate + 10 μM dexamethasone. NOM changing to OM: (q) MEM + 10% FBS + 100 U/mL of penicillin + 100 mg/mL non-essential amino acids + 50 μg/mL ascorbic acid/DMEM + 10% FBS + 100 U/mL of penicillin + 100 mg/mL non-essential amino acids + 50 μg/mL ascorbic acid + 10 mmol/L β-glycerophosphate + 0.1 μM dexamethasone. (r) DMEM + 10% FBS + 100 U/mL of penicillin/streptomycin + 100 μg/mL sodium ascorbate + 100 mg/mL non-essential amino acids/DMEM + 10% FBS + 100 U/mL of penicillin/streptomycin + 100 μg/mL sodium ascorbate + 100 mg/mL non-essential amino acids + 50 μM ascorbic acid + 10 mM β-glycerophosphate + 0.1 μM dexamethasone (NOM changes to OM on day 3. ALP assay was performed one day after medium changes. Alizarin Red staining and calcium deposition analysis were performed five days after medium changes). Legend: ADSCs—adipose-derived stem cells, ALP—alkaline phosphatase, BGLAP—osteocalcin, BMP-2—Bone Morphogenetic Protein-2, GMSCs—gingiva-derived stem cells, hBMSCs—human bone marrow stem cells, mBMSCs—murine bone marrow stem cells, mESCs—murine embryonic stem cells, OPN—osteopontin, OSX—osterix, RUNX2—Runt-related transcription factor 2, SIM—simvastatin, SMSCs—sinus maxillary stem cells, ↑—increase, ↓—decrease.

3.3. *Risk of Bias in Individual Studies*

Most of the questions were assessed as having a "probably low" or a "definitely low" risk of bias for all included studies. Therefore, as shown in Table 2, all included articles, except one [26], were of high quality (Tier 1) [14–16,21–31]. This one article was categorized as Tier 3 [26]. Despite this, all included studies indicated a "probably high" risk of bias for the question regarding the blinding of researchers during the study [14–16,21–31]. Blinding minimizes bias by ensuring that researchers are unaware of treatment allocation, which helps prevent conscious or unconscious influences on outcomes. Also, one study [26] did not report the complete outcome data without attrition or exclusion from analysis, and confidence in the outcome assessment was considered to have a "probably high" risk of bias. Additionally, the same study potentially threatened internal validity since the statistical analysis was not described.

Table 2. The risk of bias in individual studies was assessed by the adapted Office of Health Assessment and Translation (OHAT) risk-of-bias tool. Ahmadi et al. (2022) [27], Chuang et al. (2015) [23], Huang et al. (2019) [24], Kupcsik et al. (2009) [25], Lee et al. (2019) [31], Ni et al. (2023) [16], Pagkalos et al. (2010) [29], Qiao et al. (2011) [28], Sabandal et al. (2022) [30], Shao et al. (2019) [22], Sun et al. (2009) [26], Tai et al. (2015) [15], Zhang et al. (2018) [21], and Zhou et al. (2010) [14]. After addressing all the questions, these scores were subsequently categorized into three tiers: Tier 1 (high quality), Tier 2 (moderate quality), or Tier 3 (low quality). Tier 1 is defined as having a "++" (definitely low) or "+" (probably low) risk of bias in all key domains, as well as a "++" (definitely low) or "+" (probably low) risk of bias for ≥50% of the other domains. Tier 2 comprises studies that do not meet the criteria for placement in either Tier 1 or Tier 3. Tier 3 represents a "- -" (definitely high) or "-" (probably high) risk of bias in all key domains AND "- -" (definitely high) or "-" (probably high) risk of bias for ≥50% of the other domains. Adapted with permission from Back, 2021 [37].

Bias Domains and Questions	Ahmadi et al. (2022) [27]	Chuang et al. (2015) [23]	Huang et al. (2019) [24]	Kupcsik et al. (2009) [25]	Lee et al. (2019) [31]	Ni et al. (2023) [16]	Pagkalos et al. (2010) [29]	Qiao et al. (2011) [28]	Sabandal et al. (2022) [30]	Shao et al. (2019) [22]	Sun et al. (2009) [26]	Tai et al., 2015 [15]	Zhang et al. (2018) [21]	Zhou et al. (2010) [14]
Selection Bias														
1. Was administered dose or exposure level adequately randomized?	++	++	++	++	++	++	++	++	++	++	++	++	++	++
2. Was allocation to study groups adequately concealed?	++	++	++	++	++	++	++	++	++	++	++	++	++	++
Performance Bias														
5. Were experimental conditions identical across study groups?	++	++	++	++	++	++	++	++	++	++	++	++	++	++
6. Were the research personnel and human subjects blinded to the study group during the study?	-	-	-	-	-	-	-	-	-	-	-	-	-	-
Attrition/Exclusion Bias														
7. Were outcome data complete without attrition or exclusion from the analysis?	++	++	++	++	++	++	++	++	++	++	-	++	++	+
8. Can we be confident in the exposure characterization?	++	++	++	++	++	++	++	+	++	++	++	++	++	++
9. Can we be confident in the outcome assessment?	++	++	++	++	++	++	++	++	++	++	-	++	++	+
Selective Reporting Bias														
10. Were all measured outcomes reported?	++	++	++	++	++	++	++	++	++	++	++	++	++	++

Table 2. *Cont.*

Bias Domains and Questions	Ahmadi et al. (2022) [27]	Chuang et al. (2015) [23]	Huang et al. (2019) [24]	Kupcsik et al. (2009) [25]	Lee et al. (2019) [31]	Ni et al. (2023) [16]	Pagkalos et al. (2010) [29]	Qiao et al. (2011) [28]	Sabandal et al. (2022) [30]	Shao et al. (2019) [22]	Sun et al. (2009) [26]	Tai et al., 2015 [15]	Zhang et al. (2018) [21]	Zhou et al. (2010) [14]
	Other Sources of Bias													
11. Were there no other potential threats to internal validity (e.g., statistical methods were appropriate, and researchers adhered to the study protocol)?	++	++	++	++	++	++	++	++	++	++	-	++	++	+
Overall	tier 1	tier 1	tier 1	tier 1	tier 1	tier 1	tier 1	tier 1	tier 1	tier 1	tier 3	tier 1	tier 1	tier 1
	Answer Format													
++	Definitely low risk of bias: There is direct evidence of low-risk-of-bias practices. (May include a specific example of low-risk-of-bias practices).													
+	Probably low risk of bias: There is indirect evidence of low-risk-of-bias practices, OR it is deemed that deviations from low-risk-of-bias practices for these criteria during the study would not appreciably bias results, including consideration of direction and magnitude of bias.													
-	Probably high risk of bias: There is indirect evidence of high-risk-of-bias practices OR insufficient information (e.g., not reported or "NR") provided about the relevant risk-of-bias practices.													
- -	Definitely high risk of bias: There is direct evidence of high-risk-of-bias practices. (May include a specific example of high-risk-of-bias practices).													

3.4. Certainty in Cumulative Evidence and Risk of Bias Across Studies

Sources of variability across studies were mostly related to used simvastatin dose and applied methodologies. The certainty in cumulative evidence assessed by modified GRADE criteria was considered low for OPN [28], IBSP [16], and BMP-9 [24]. For ALP [14,15,21–26,28–31], RUNX2 [15,21,22,28,29,31], and OSX [21,28,29], the certainty in cumulative evidence was judged moderate. High certainty in cumulative evidence was addressed to extracellular matrix mineralization [14,15,21–26,28–31], BMP-2 [14,15,23–25], COL1 [22,28,29,31], and BGLAP [14,15,21,23,24,28,29]. According to the GRADE tool, the risk of bias across studies was considered as "not serious" since the concerns were present in less than 25% of the included studies. The level of certainty was considered high, meaning that the authors have a lot of confidence that the true effect is similar to the estimated effect. A summary of the findings is presented in Table 3.

Table 3. Modified Grading of Recommendations, Assessment, Development, and Evaluation (GRADE) analysis.

Outcome **Osteogenic Differentiation of Mesenchymal Cells**	**Object of Analysis**	**Relative Importance**	**Number of Studies**	**Certainty of the Evidence (GRADE)**
Extracellular matrix mineralization	Von Kossa staining Alizarin Red staining Calcium content Phosphate content	Critical	14	⊕⊕⊕⊕ HIGH [a,b]
ALP	Transcript ALP activity ALP staining	Critical	14	⊕⊕⊕ MODERATE [a,b,c]

Table 3. *Cont.*

Outcome Osteogenic Differentiation of Mesenchymal Cells	Object of Analysis	Relative Importance	Number of Studies	Certainty of the Evidence (GRADE)
BMP-2	Transcript Protein expression	Critical	5	⊕⊕⊕⊕ HIGH [a,b]
BMP-9	Transcript	Critical	1	⊕⊕ LOW [d,e,f]
IBSP	Transcript	Critical	1	⊕⊕ LOW
RUNX2	Transcript Protein expression	Critical	7	⊕⊕⊕ MODERATE [a,b,e]
COL1	Transcript Protein expression	Critical	5	⊕⊕⊕⊕ HIGH [a,b]
BGLAP	Transcript Protein expression	Critical	8	⊕⊕⊕⊕ HIGH [a,b]
OPN	Protein expression	Critical	1	⊕⊕ LOW [d,e,f]
OSX	Transcript	Critical	3	⊕⊕⊕ MODERATE [b,c,e]

GRADE Working Group grades of evidence: High certainty = we are very confident that the true effect lies close to that of the estimated effect. Moderate certainty = we are moderately confident in the effect estimate. The true effect is likely to be close to the estimate of the effect, but there is a possibility that it is substantially different. Low certainty = our confidence in the estimate is limited. The true effect may be substantially different from the estimate of the effect. Very low certainty = we have very little confidence in the effect estimate. The true effect is likely to be substantially different from the estimate of effect. Explanations: ([a]) Included studies presented methodological and statistical heterogeneity. ([b]) There is no standardization of methods, doses, or treatment regimens. ([c]) Studies from the same authors or conducted in the same institution presenting similar results. ([d]) Collectively, studies presented a high risk of bias. ([e]) Estimates were not sufficiently supported by the presented experiments. ([f]) The study presented a moderate risk of bias. Legend: ALP—alkaline phosphatase; BGLAP—osteocalcin; BMP-2—Bone Morphogenetic Protein-2; BMP-9—Bone Morphogenetic Protein-9; IBSP—Integrin Binding Sialoprotein; OPN—osteopontin; OSX—osterix; RUNX2—Runt-related transcription factor 2.

3.5. Synthesis of Results

Considering a minimum 4-fold increase, as indicated in Figure 2, simvastatin caused a robust extracellular mineralization in four studies (4-, 4.4-, 5-, and 39.5-fold) [15,25,26,30]; two of them compared the effects of simvastatin to the osteogenic medium [15,30], while two compared the simvastatin effect to the non-osteogenic medium [25,26]. BMP-2 expression was at least 4-fold increased by simvastatin in three studies (4.1-, 11-, and 19-fold) [22,24,25]. Occasionally, simvastatin reached a 4-fold increase in COL1 [22,31], BGLAP [14], RUNX2 [14], and ALP activity [26]. Therefore, it is concluded that the extracellular matrix mineralization of mesenchymal cells and BMP-2 expression may be used as the most efficient bioassays for simvastatin in vitro. A high heterogeneity among studies concerning the intervention protocol, applied assays, culture medium, and simvastatin concentration was observed. For instance, simvastatin was tested in the presence [14,15,21–26,28,29,31] and absence [14,25,26,29] of osteogenic differentiation medium. The simvastatin concentrations used in the included studies are listed in Table 4. The bioassays evaluated were ALP [15,16,21–28,31], BGLAP [14–16,21,22,24], BMP-2 [14,22–24], BMP-9 [24], COL1 [16,22,28,29,31], RUNX2 [14,21,28,29,31], OSX [21,28,29], OPN [21], IBSP [16], and ECM mineralization by Alizarin Red [14–16,21–24,28,29], Von Kossa [16,26], and calcium quantification [25,27]. The overall impact of simvastatin on the expression of osteogenic differentiation markers was moderate. A meta-analysis was not conducted due to the heterogeneous data across the included studies. Heterogeneity among studies can significantly impact the data reliability and interpretation and make it challenging to directly compare findings. This variability may lead to inconsistent or conflicting conclusions, reducing the overall confidence.

Table 4. Simvastatin concentrations used in the included studies. Ahmadi et al. (2022) [27], Chuang et al. (2015) [23], Huang et al. (2019) [24], Kupcsik et al. (2009) [25], Lee et al. (2019) [31], Ni et al. (2023) [16], Pagkalos et al. (2010) [29], Qiao et al. (2011) [28], Sabandal et al. (2022) [30], Shao et al. (2019) [22], Sun et al. (2009) [26], Tai et al. (2015) [15], Zhang et al. (2018) [21], and Zhou et al. (2010) [14].

Simvastatin Concentration	Reference
0.1 nM	[29]
0.3 nM	[21]
1 nM	[28,29]
10 nM	[14,22,28,29]
20 nM	[22]
100 nM	[14,16,24,27–29]
0.3 µM	[22]
0.5 µM	[23]
1 µM	[14,15,24–26,31]
5 µM	[25]
10 µM	[31]

4. Discussion

Simvastatin, embedded within a bone substitute, can support bone regeneration, which requires mesenchymal cell differentiation towards bone-forming osteoblasts. With advances in material engineering, different bone graft compositions have been proposed as simvastatin drug delivery systems [12,13]. Consequently, several preliminary studies evaluating simvastatin for promoting osteogenesis have been performed in vitro [14,22–24,31]. However, there is no consensus on which bioassay is more efficient to measure simvastatin activity in vitro. Based on this systematic review, there is accumulating evidence that mineralization of the extracellular matrix and BMP-2 expression are the most efficient bioassays, in terms of sensitivity, to determine the osteogenic activity of simvastatin in vitro. It is relevant to mention that BMP-2 expression, in addition to its sensitivity, is a fast assay due to BMP-2 being an early osteogenic marker. Therefore, the findings of this systematic review may serve as a reference for future studies assessing in vitro osteogenesis promoted by simvastatin.

It has been demonstrated that phosphoinositide 3-kinase (PI3K) contributes to the simvastatin-induced activation of mitogen-activated protein kinases (MAPK), which, together with Akt kinase, regulate BMP-2 expression [38]. The PI3K pathway is activated upon simvastatin treatment, leading to the phosphorylation and activation of Akt (protein kinase B). This activation promotes downstream signaling cascades that enhance the transcriptional activity of BMP-2. The PI3K/Akt pathway also contributes to cell survival and proliferation, which may further support osteogenic differentiation by creating an environment conducive to bone morphogenetic signaling. Similarly, simvastatin influences the MAPK pathway, which involves three main subfamilies: extracellular signal-regulated kinases (ERKs), c-Jun N-terminal kinases (JNKs), and p38 MAPK. Simvastatin has been found to activate the p38 MAPK pathway, which is particularly important for osteogenic differentiation. The activation of p38 MAPK facilitates the phosphorylation of transcription factors, such as Runx2, that interact with the BMP-2 promoter, thereby enhancing BMP-2 gene expression. In some cases, simvastatin also activates ERK1/2, which may act synergistically with p38 to amplify BMP-2 transcription. This information supports BMP-2 as a solid simvastatin target gene. Therefore, these data suggest that BMP-2 is a robust target gene that can complement in vitro mineralization assays to determine simvastatin activity in mesenchymal cells.

We further observed seven studies reporting matrix mineralization. Four studies demonstrated that the mineralization increased approximately 4-fold after 2-3 weeks [14,24,28,30]. In one study, simvastatin increased mineralization after only one week [26]. In support of these observations, simvastatin increased cytosolic and mitochondrial calcium levels in breast cancer cells [39]. Calcium is further linked to BMP-2 expression since extracellular calcium promotes the BMP-2-related osteogenic differentiation of mesenchymal cells [40]. Additionally, simvastatin upregulates Runx2, ALP, osteocalcin, and osteopontin, all of which are critical for matrix maturation and mineralization. Also, simvastatin promotes the activation of the Wnt/β-catenin signaling pathway, which plays a central role in osteoblast differentiation and mineralization [14,21,26,28]. This information could explain why the in vitro mineralization of mesenchymal cells and BMP-2 expression were the most efficient bioassays for simvastatin-induced osteogenic differentiation in vitro. Conversely, less sensitive to simvastatin were the classical markers of osteogenic differentiation, e.g., COL1, BGLAP, RUNX2, and ALP activity performed on day 7 [21,24,28] and day 3 [14,22], as well as after 12 h [15].

Although systematic reviews are designed to answer clinical questions, this review was applied to in vitro studies [41,42]. It is worth mentioning that systematic reviews attempt to identify, appraise, and synthesize all the evidence that meets eligibility criteria to answer a specific research question. Additionally, the systematic review purposes to reveal limitations in the conduct of previous studies that might be addressed in future ones, as well as to propose standardized methods to optimize future studies and allow a comparison among studies [33]. Therefore, this study demonstrated that the mineralization of the extracellular matrix and BMP-2 expression are the most efficient bioassays to determine the osteogenic activity of simvastatin in vitro. Additionally, it was demonstrated that not all the studies mentioned performed this analysis.

Regarding the risk of bias, most questions of the OHAT tool were considered as having a "probably low" or "definitely low" risk of bias. No or minor deviations from the true effect estimation were estimated, providing confidence for interpreting the results [43]. Conversely, all included studies [14–16,21–31] failed to report the blinding of the treatment group, which is considered to indicate a "probably high" risk of bias; therefore, conscious or unconscious influences on the outcomes could have occurred. Furthermore, one study [26] presented a "definitely high" risk of bias. Moreover, a high heterogeneity between studies was identified due to the use of different cell lines from murine and human species, as well as different culture conditions and observation periods. All of these may lead to inconsistent or conflicting conclusions and imply that the certainty level of the conclusion is not definitive.

Given the complexity of the osteogenesis process, this research aimed to identify the most efficient bioassay for evaluating in vitro simvastatin activity during osteogenesis. This identification not only aids in data interpretation, but could also suggest additional experiments to enhance comprehensiveness within this topic. It is imperative to underscore that understanding the findings in studies related to in vitro osteogenesis extends beyond the selection of an optimal marker of osteogenic differentiation. Evaluating the specific focus of each study involves not only assessing the outcome, but also understanding the effect of that outcome on osteogenesis as a whole.

Concerning the limitations of this systematic review, the primary aim of this study was to identify the most effective bioassays for assessing the impact of simvastatin on undifferentiated mesenchymal cells during osteogenesis, rather than systematically addressing the variances between different treatments and cell types. However, it was observed that mineralization seems to be an efficient biomarker of osteogenesis for both osteogenic and non-osteogenic media. This was suggested when comparing four studies [15,25,26,30], where, considering a minimum 4-fold increase, simvastatin induced a robust in vitro mineralization of mesenchymal cells. Among these, two studies compared the effects of simvastatin to an osteogenic medium [15,30], while the other two compared simvastatin's effect to a non-osteogenic medium [25,26]. Regarding BMP-2 expression, two studies used an osteogenic medium [22,24], while one evaluated a non-osteogenic medium [25]. As mentioned, the investigations used undifferentiated mesenchymal cell lines from different origins and species. According to these findings, BMP-2 expression was increased at least 4-fold by simvastatin in three studies, all evaluating bone-marrow-derived cells, irrespective of species (rat [24], mouse [22], or human [25]). Regarding mineralization, two studies focused on bone-marrow-derived mesenchymal cells [15,25], while the other two investigated adipose-derived cells [26,30].

The variability in methodologies among the included studies—such as differences in simvastatin concentrations, cell types, and experimental protocols—represents a limitation for data interpretation in this study. A key purpose of systematic reviews is to identify limitations in previous studies, which can guide future research with the proposal of standardized methods to optimize study designs and enable meaningful comparisons across studies. In this context, it is essential to emphasize that non-immortalized human cell lineages offer a more suitable model for human studies. These cells are human-derived and have not undergone any immortalization processes that might alter their cellular mechanisms. Additionally, if the purpose is to evaluate the osteogenic potential of a substance, it would be more appropriate to use an osteogenic cell culture medium as the control of the differentiation process.

Further investigations are warranted to confirm whether these cells exhibit heightened sensitivity to simvastatin or if the enhancement in BMP-2 expression is specific to these lineages. Hence, additional studies concerning bioassays for determining the simvastatin effect could be conducted, evaluating the influence of different sources of mesenchymal cells and types of cell culture medium. Additionally, relevant information pertaining to high doses of simvastatin and the effect of this on the healing of bone, soft tissue, and cartilage has been published [44,45]. It was not possible to correlate these high doses of simvastatin and their effects with the findings of this review due to the heterogeneity of the cell lineages and methodological protocols previously mentioned. Consequently, future studies are warranted to evaluate the effect of the dose of simvastatin on various human tissues.

5. Conclusions

In conclusion, based on a high level of certainty, extracellular matrix mineralization and BMP-2 expression are the most effective bioassays to determine simvastatin's impact on osteogenesis in vitro. These findings provide a standardized approach that can enhance the reliability and comparability of in vitro studies, bridging the gap with in vivo research and optimizing resources in the field of bone regeneration and tissue engineering applications. However, it is important to consider the high heterogeneity among studies, which may influence the broader applicability of these results. Future studies concerning bioassays determining the effects of simvastatin are warranted in order to evaluate the influence of different sources of mesenchymal cells and types of cell culture medium.

Author Contributions: Conceptualization, L.S.B., I.S.M., M.B.S., G.L.M., Á.A., R.d.S.M., R.G. and A.C.C.C.; methodology, L.S.B., I.S.M., R.G., A.C.C.C.; formal analysis, L.S.B., I.S.M., R.G., A.C.C.C.; investigation, L.S.B., I.S.M., R.G., A.C.C.C.; writing—original draft preparation, L.S.B., I.S.M., R.G., A.C.C.C.; writing—review and editing, L.S.B., I.S.M., R.G. and A.C.C.C.; supervision, R.G. and A.C.C.C.; project administration, R.G. and A.C.C.C. All authors have read and agreed to the published version of the manuscript.

Funding: This research received no external funding. Isabella Schönhofen Manso was supported by a scholarship from Fundação de Amparo à Pesquisa e Inovação do Estado de Santa Catarina (FAPESC—Research and Innovation Support Foundation of Santa Catarina State), Brazil. Gabriel Leonardo Magrin was supported with scholarships from Coordenação de Aperfeiçoamento de Pessoal de Nível Superior, Brazil (CAPES—Coordination for the Improvement of Higher Education Personnel), Brazil, and Fundação de Amparo à Pesquisa e Inovação do Estado de Santa Catarina (FAPESC—Research and Innovation Support Foundation of Santa Catarina State), Brazil. The University Clinic of Dentistry of the Medical University of Vienna supported the open access publishing.

Institutional Review Board Statement: Not applicable.

Informed Consent Statement: Not applicable.

Acknowledgments: The authors would like to thank Carlos Flores-Mir for his support during the elaboration of this systematic review and Karyn Munik Lehmkuhl for the instructions concerning the search strategy elaboration.

Conflicts of Interest: The authors declare no conflict of interest.

Abbreviations

The following abbreviations are used in this manuscript:

MDPI	Multidisciplinary Digital Publishing Institute
DOAJ	Directory of open access journals
TLA	Three-letter acronym
LD	Linear dichroism

Appendix A

Data search strategy for the electronic databases (Embase, LILACS, LIVIVO, PubMed, SCOPUS, and Web of Science) and grey literature databases (Google Scholar and ProQuest). Adapted with permission from Back, 2021 [37].

Database	Search Query
PubMed	"Stem Cells" OR "Stem Cell" OR "Progenitor Cells" OR "Progenitor Cell" OR "Mother Cells" OR "Mother Cell" OR "Colony Forming Unit" OR "Colony Forming Units" OR "undifferentiated cells" OR Pericytes OR Pericyte OR "Rouget Cells" AND "Simvastatin" OR Zocor OR "MK 733" OR "MK733" OR Synvinolin OR "Hydroxymethylglutaryl CoA Reductase Inhibitors" OR "HMG CoA Reductase Inhibitors" OR "HMG CoA Statins" OR "Hydroxymethylglutaryl CoA Inhibitors" OR Statins OR "Hydroxymethylglutaryl Coenzyme A Inhibitors" AND "Osteogenesis" OR "bone formation" OR Ossification OR Osteoclastogenesis OR "osteoinduction" OR osteoinduct OR "Cell Differentiation" OR "Cell Differentiations" OR "Cell Growth Processes" OR "Bone Morphogenetic Proteins" OR "Bone Morphogenetic Protein" OR "Alkaline Phosphatase" OR Osteocalcin OR "Core Binding Factor Alpha 1 Subunit" OR "Runx2 Protein" AND "Osteogenesis" OR "bone formation" OR Ossification OR Ossifications OR Osteoclastogenesis OR osteoinduction OR osteoinducting OR osteoinductive OR osteoinductivity OR "Cell Differentiation" OR "Cell Differentiations" OR "Cell Growth Processes" OR "Bone Morphogenetic Proteins" OR "Bone Morphogenetic Protein" OR "Alkaline Phosphatase" OR Osteocalcin OR "Core Binding Factor Alpha 1 Subunit" OR "Runx2 Protein"
LILACS (English, Portuguese, and Spanish)	("Stem Cells" OR "Stem Cell" OR "Progenitor Cells" OR "Progenitor Cell" OR "Mother Cells" OR "Mother Cell" OR "Colony Forming Unit" OR "Colony Forming Units" OR "Undifferentiated Cells" OR Pericytes OR Pericyte OR "Rouget Cells" OR "Células-Tronco" OR "Células Básicas" OR "Células Mãe" OR "Células Precursoras" OR "Células Primitivas" OR "Células Primordiais" OR "Células Progenitoras" OR "Células Tronco" OR "Unidades Formadoras de Colônias" OR "Pericitos" OR "Células de Rouget" OR "Células Madre" OR "Células Básicas" OR "Células Primitivas" OR "Células Primordiales" OR "Células Progenitoras" OR "Células Troncales" OR "Unidades que Forman Colonias") AND ("Simvastatin" OR Zocor OR "MK 733" OR "MK733" OR Synvinolin OR "Hydroxymethylglutaryl CoA Reductase Inhibitors" OR "HMG CoA Reductase Inhibitors" OR "HMG CoA Statins" OR "Hydroxymethylglutaryl CoA Inhibitors" OR Statins OR "Hydroxymethylglutaryl Coenzyme A Inhibitors" OR "Sinvastatina" OR "Inibidores da Hidroximetilglutaril-CoA Redutases" OR "Estatinas" OR "Estatinas de HMG-CoA" OR "Inibidor de Hidroximetilglutaril-CoA Redutases" OR "Inibidores de HMG-CoA Redutases" OR "Simvastatina" OR "Inhibidores de Hidroximetilglutaril-CoA Reductasas" OR "Estatinas HMG-CoA" OR "Inhibidor de Hidroximetilglutaril-CoA Reductasas" OR "Inhibidor de Hidroximetilglutaril-CoA-Reductasa" OR "Inhibidor de las Hidroximetilglutaril-CoA Reductasas" OR "Inhibidores de HMG CoA Reductasa") AND (Osteogenesis OR "bone formation" OR Ossification OR Osteoclastogenesis OR "osteoinduction" OR osteoinduct OR "Cell Differentiation" OR "Cell Differentiations" OR "Cell Growth Processes" OR "Bone Morphogenetic Proteins" OR "Bone Morphogenetic Protein" OR "Alkaline Phosphatase" OR Osteocalcin OR "Core Binding Factor Alpha 1 Subunit" OR "Runx2 Protein" OR Osteogênese OR "Formação dos Ossos" OR "Formação Óssea" OR Ossificação OR "Ossificação Endocondral" OR "Ossificação Fisiológica" OR Osteoclastogênese OR Osteoindução OR "Diferenciação Celular" OR "Processos de Crescimento Celular" OR "Processos de Crescimento da Célula" OR "Proteínas Morfogenéticas Ósseas" OR "Proteínas Morfogenéticas do Osso" OR "Proteínas Morfogênicas do Osso" OR "Proteínas Morfogênicas Ósseas" OR "Proteínas Ósseas Morfogenéticas" OR "Fosfatase Alcalina" OR Osteocalcina OR "Gla-Proteína Óssea" OR "Proteína de Ligação a Cálcio Dependente de Vitamina K" OR "Proteína de Ligação a Cálcio Vitamina K-Dependente" OR "Proteína Óssea Dependente de Vitamina K" OR "Proteína Óssea Vitamina K-Dependente" OR "Subunidade alfa 1 de Fator de Ligação ao Core" OR Osteogénesis OR "Formación del Hueso" OR Osificación OR "Osificación Endocondral" OR "Osificación Fisiológica" OR Osteoclastogénesis OR Osteoinducción OR "Diferenciación Celular" OR "Procesos de Crecimiento Celular" OR "Proteínas Morfogenéticas Óseas" OR "Proteínas Morfogenéticas de Hueso" OR "Proteínas Óseas Morfogenéticas" OR "Fosfatasa Alcalina" OR "Gla-Proteína Ósea" OR "Proteína Gla del Hueso" OR "Proteína de Unión a Calcio Dependiente de Vitamina K" OR "Proteína de Unión a Calcio Vitamina K-Dependiente" OR "Proteína Ósea Dependiente de Vitamina K" OR "Proteína Ósea Vitamina K-Dependiente" OR "Subunidad alfa 1 del Factor de Unión al Sitio Principal" OR "Subunidad alfa del CBF")

Database	Search Query
SCOPUS	(TITLE-ABS-KEY("Stem Cells" OR "Stem Cell" OR "Progenitor Cells" OR "Progenitor Cell" OR "Mother Cells" OR "Mother Cell" OR "Colony Forming Unit" OR "Colony Forming Units" OR "undifferentiated cells" OR Pericytes OR Pericyte OR "Rouget Cells") AND TITLE-ABS-KEY(Simvastatin OR Zocor OR "MK 733" OR "MK733" OR Synvinolin OR "Hydroxymethylglutaryl CoA Reductase Inhibitors" OR "HMG CoA Reductase Inhibitors" OR "HMG CoA Statins" OR "Hydroxymethylglutaryl CoA Inhibitors" OR Statins OR "Hydroxymethylglutaryl Coenzyme A Inhibitors") AND TITLE-ABS-KEY(Osteogenesis OR "bone formation" OR Ossification OR Osteoclastogenesis OR "osteoinduction" OR osteoinduct OR "Cell Differentiation" OR "Cell Differentiations" OR "Cell Growth Processes" OR "Bone Morphogenetic Proteins" OR "Bone Morphogenetic Protein" OR "Alkaline Phosphatase" OR Osteocalcin OR "Core Binding Factor Alpha 1 Subunit" OR "Runx2 Protein")
EMBASE	('stem cells'/exp OR 'stem cells' OR 'stem cell'/exp OR 'stem cell' OR 'progenitor cells' OR 'progenitor cell'/exp OR 'progenitor cell' OR 'mother cells' OR 'mother cell'/exp OR 'mother cell' OR 'colony forming unit'/exp OR 'colony forming unit' OR 'colony forming units' OR 'undifferentiated cells' OR 'pericytes'/exp OR pericytes OR 'pericyte'/exp OR pericyte OR 'rouget cells') AND ('simvastatin'/exp OR simvastatin OR 'zocor'/exp OR zocor OR 'mk 733'/exp OR 'mk 733' OR 'mk733'/exp OR 'mk733' OR 'synvinolin'/exp OR synvinolin OR 'hydroxymethylglutaryl coa reductase inhibitors'/exp OR 'hydroxymethylglutaryl coa reductase inhibitors' OR 'hmg coa reductase inhibitors'/exp OR 'hmg coa reductase inhibitors' OR 'hmg coa statins' OR 'hydroxymethylglutaryl coa inhibitors' OR 'statins'/exp OR statins OR 'hydroxymethylglutaryl coenzyme a inhibitors') AND ('osteogenesis'/exp OR osteogenesis OR 'bone formation'/exp OR 'bone formation' OR 'ossification'/exp OR ossification OR ossifications OR 'osteoclastogenesis'/exp OR osteoclastogenesis OR 'osteoinduction'/exp OR osteoinduction OR osteoinducting OR osteoinductive OR 'osteoinductivity'/exp OR osteoinductivity OR 'cell differentiation'/exp OR 'cell differentiation' OR 'cell differentiations' OR 'cell growth processes'/exp OR 'cell growth processes' OR 'bone morphogenetic proteins'/exp OR 'bone morphogenetic proteins' OR 'bone morphogenetic protein'/exp OR 'bone morphogenetic protein' OR 'alkaline phosphatase'/exp OR 'alkaline phosphatase' OR 'osteocalcin'/exp OR osteocalcin OR 'core binding factor alpha 1 subunit'/exp OR 'core binding factor alpha 1 subunit' OR 'runx2 protein'/exp OR 'runx2 protein')
Web of Science	TS=("Stem Cells" OR "Stem Cell" OR "Progenitor Cells" OR "Progenitor Cell" OR "Mother Cells" OR "Mother Cell" OR "Colony Forming Unit" OR "Colony Forming Units" OR "undifferentiated cells" OR Pericytes OR Pericyte OR "Rouget Cells") AND TS=(Simvastatin OR Zocor OR "MK 733" OR "MK733" OR Synvinolin OR "Hydroxymethylglutaryl CoA Reductase Inhibitors" OR "HMG CoA Reductase Inhibitors" OR "HMG CoA Statins" OR "Hydroxymethylglutaryl CoA Inhibitors" OR Statins OR "Hydroxymethylglutaryl Coenzyme A Inhibitors") AND TS=(Osteogenesis OR "bone formation" OR Ossification OR Osteoclastogenesis OR "osteoinduction" OR osteoinduct OR "Cell Differentiation" OR "Cell Differentiations" OR "Cell Growth Processes" OR "Bone Morphogenetic Proteins" OR "Bone Morphogenetic Protein" OR "Alkaline Phosphatase" OR Osteocalcin OR "Core Binding Factor Alpha 1 Subunit" OR "Runx2 Protein")
ProQuest	noft("Stem Cells" OR "Stem Cell" OR "Progenitor Cells" OR "Progenitor Cell" OR "Mother Cells" OR "Mother Cell" OR "Colony Forming Unit" OR "Colony Forming Units" OR "undifferentiated cells" OR Pericytes OR Pericyte OR "Rouget Cells") AND ("Simvastatin" OR Zocor OR "MK 733" OR "MK733" OR Synvinolin OR "Hydroxymethylglutaryl CoA Reductase Inhibitors" OR "HMG CoA Reductase Inhibitors" OR "HMG CoA Statins" OR "Hydroxymethylglutaryl CoA Inhibitors" OR Statins OR "Hydroxymethylglutaryl Coenzyme A Inhibitors") AND (Osteogenesis OR "bone formation" OR Ossification OR Osteoclastogenesis OR "osteoinduction" OR osteoinduct OR "Cell Differentiation" OR "Cell Differentiations" OR "Cell Growth Processes" OR "Bone Morphogenetic Proteins" OR "Bone Morphogenetic Protein" OR "Alkaline Phosphatase" OR Osteocalcin OR "Core Binding Factor Alpha 1 Subunit" OR "Runx2 Protein")
Google Scholar	"Undifferentiated Cells" AND "Simvastatin" AND "Cell Differentiation" AND "Bone Formation"

Appendix B. Articles Excluded and the Reasons for Exclusion (n = 63)

Author (Year)	Reason for Exclusion
Aly et al. (2018)	8
Arpornmaeklong et al. (2017)	2
Baadani et al. (2023)	2
Baek et al. (2005)	8
Biouki et al. (2019)	2
Chi et al. (2019)	7
Cui et al. (2013)	5
Feng et al. (2020)	7
Huang et al. (2014)	2
Huang et al. (2019)	5
Hungaro et al. (2017)	2
Janz et al. (2014)	8

Author (Year)	Reason for Exclusion
Jin et al. (2011)	11
Kamada et al. (2017)	8
Kim et al. (2011)	1
Lai et al. (2018)	2
Lai et al. (2019)	1
Lee et al. (2016)	2
Lee et al. (2018)	2
Li et al. (2018)	2
Liu et al. (2009)	8
Liu et al. (2014)	2
Liu et al. (2014)	8
Liu et al. (2016)	8
Matos et al. (2015)	2
Niu et al. (2015)	8
Oh et al. (2021)	2
Park et al. (2012)	1
Pullisaar et al. (2014)	2
Qi et al. (2013)	2
Qing et al. (2019)	4
Rostami et al. (2020)	2
Samiei et al. (2016)	2
Simões et el. (2013)	2
Song et al. (2003)	8
Song et al. (2008)	8
Sukul et al. (2015)	2
Tai et al. (2010)	6
Tai et al. (2011)	6
Tai et al. (2013)	6
Wadagaki et al. (2011)	2
Wang et al. (2014)	2
Wang et al. (2016)	6
Wang et al. (2018)	2
Wang et al. (2020)	1
Wu et al. (2014)	4
Wu et al. (2020)	2
Wu et al. (2021)	2

Author (Year)	Reason for Exclusion
Xiaolin et al. (2023)	11
Xu et al. (2015)	4
Xu et al. (2018)	5
Xue et al. (2019)	2
Yu et al. (2017)	2
Yuan et al. (2019)	2
Yun et al. (2013)	8
Zanette et al. (2015)	5
Zhang et al. (2015)	2
Zhang et al. (2018)	2
Zhang et al. (2018)	2
Zhang et al. (2019)	4
Zhao & Liu (2014)	7
Zheng et al. (2010)	1
Zijah et al. (2016)	2
Xiaolin et al. (2023)	11
Baadani et al. (2023)	2

Legend: (1) Studies investigating other cell types, such as differentiated cells. (2) Studies assessing drug or biomaterial associations that do not present separate data for simvastatin alone. (3) Studies that do not present an untreated control group with the same cell type. (4) Studies that do not present a control group without simvastatin treatment. (5) Studies that do not evaluate osteogenic differentiation. (6) Studies that are not in vitro (clinical studies, animal studies, conference abstracts, letters, pilot studies, review articles, case reports, protocols, short communications, personal opinions, posters, and book chapters). (7) Studies that do not specify the culture medium. (8) Studies that do not quantify alkaline phosphatase (ALP) or extracellular matrix mineralization for test and control groups. (9) Full text not available (book chapters, conference abstracts, expert opinions, letters, and literature reviews). (10) Duplicated data (e.g., dissertations/theses in which corresponding published articles were available). (11) Studies not published in the Latin alphabet. Adapted with permission from Back, 2021 [37].

References

1. Amaral Valladão, C.A.; Freitas Monteiro, M.; Joly, J.C. Guided Bone Regeneration in Staged Vertical and Horizontal Bone Augmentation Using Platelet-Rich Fibrin Associated with Bone Grafts: A Retrospective Clinical Study. *Int. J. Implant. Dent.* **2020**, *6*, 72. [CrossRef]
2. Wessels, R.; Vervaeke, S.; Seyssens, L.; Eghbali, A.; Cosyn, J. A 5-Year Cohort Study on Early Implant Placement with Guided Bone Regeneration or Alveolar Ridge Preservation with Connective Tissue Graft. *Clin. Implant. Dent. Relat. Res.* **2020**, *22*, 697–705. [CrossRef]
3. Dimitriou, R.; Mataliotakis, G.I.; Angoules, A.G.; Kanakaris, N.K.; Giannoudis, P.V. Complications Following Autologous Bone Graft Harvesting from the Iliac Crest and Using the RIA: A Systematic Review. *Injury* **2011**, *42* (Suppl. 2), S3–S15. [CrossRef] [PubMed]
4. Betz, R.R. Limitations of Autograft and Allograft: New Synthetic Solutions. *Orthopedics* **2002**, *25*, S561–S570. [CrossRef] [PubMed]

5. Sun, X.; Liu, C.; Shi, Y.; Li, C.; Sun, L.; Hou, L.; Wang, X. The Assessment of Xenogeneic Bone Immunotoxicity and Risk Management Study. *Biomed. Eng. Online* **2019**, *18*, 108. [CrossRef]
6. Retzepi, M.; Donos, N. Guided Bone Regeneration: Biological Principle and Therapeutic Applications. *Clin. Oral Implants Res.* **2010**, *21*, 567–576. [CrossRef] [PubMed]
7. Nyman, S. Bone Regeneration Using the Principle of Guided Tissue Regeneration. *J. Clin. Periodontol.* **1991**, *18*, 494–498. [CrossRef] [PubMed]
8. Han, J.; Menicanin, D.; Gronthos, S.; Bartold, P.M. Stem Cells, Tissue Engineering and Periodontal Regeneration. *Aust. Dent. J.* **2014**, *59*, 117–130. [CrossRef] [PubMed]
9. Mundy, G.; Garrett, R.; Harris, S.; Chan, J.; Chen, D.; Rossini, G.; Boyce, B.; Zhao, M.; Gutierrez, G. Stimulation of Bone Formation in Vitro and in Rodents by Statins. *Science (1979)* **1999**, *286*, 1946–1949. [CrossRef] [PubMed]
10. Goldstein, J.L.; Brown, M.S. Regulation of the Mevalonate Pathway. *Nature* **1990**, *343*, 425–430. [CrossRef] [PubMed]
11. Jadhav, S.B.; Jain, G.K. Statins and Osteoporosis: New Role for Old Drugs. *J. Pharm. Pharmacol.* **2006**, *58*, 3–18. [CrossRef] [PubMed]
12. Oh, J.-S.; Park, J.-S.; Lee, E.-J. Enhanced Effect of Polyethyleneimine-Modified Graphene Oxide and Simvastatin on Osteogenic Differentiation of Murine Bone Marrow-Derived Mesenchymal Stem Cells. *Biomedicines* **2021**, *9*, 501. [CrossRef] [PubMed]
13. Wu, S.-C.; Chang, C.-H.; Chang, L.-H.; Wu, C.-W.; Chen, J.-W.; Chen, C.-H.; Lin, Y.-S.; Chang, J.-K.; Ho, M.-L. Simvastatin Enhances the Chondrogenesis But Not the Osteogenesis of Adipose-Derived Stem Cells in a Hyaluronan Microenvironment. *Biomedicines* **2021**, *9*, 559. [CrossRef] [PubMed]
14. Zhou, Y.; Ni, Y.; Liu, Y.; Zeng, B.; Xu, Y.; Ge, W. The Role of Simvastatin in the Osteogenesis of Injectable Tissue-Engineered Bone Based on Human Adipose-Derived Stromal Cells and Platelet-Rich Plasma. *Biomaterials* **2010**, *31*, 5325–5335. [CrossRef]
15. Tai, I.C.; Wang, Y.H.; Chen, C.H.; Chuang, S.C.; Chang, J.K.; Ho, M.L. Simvastatin Enhances Rho/Actin/Cell Rigidity Pathway Contributing to Mesenchymal Stem Osteogenic Differentiation. *Int. J. Nanomed.* **2015**, *10*, 5881–5894. [CrossRef]
16. Ni, Q.; Zhu, J.; Li, Z.; Li, B.; Wang, H.; Chen, L. Simvastatin Promotes Rat Achilles Tendon-Bone Interface Healing by Promoting Osteogenesis and Chondrogenic Differentiation of Stem Cells. *Cell Tissue Res.* **2023**, *391*, 339–355. [CrossRef]
17. Sun, T.; Huang, J.; Zhang, W.; Zheng, X.; Wang, H.; Liu, J.; Leng, H.; Yuan, W.; Song, C. Simvastatin-Hydroxyapatite Coatings Prevent Biofilm Formation and Improve Bone Formation in Implant-Associated Infections. *Bioact. Mater.* **2023**, *21*, 44–56. [CrossRef]
18. Sugiyama, M.; Kodama, T.; Konishi, K.; Abe, K.; Asami, S.; Oikawa, S. Compactin and Simvastatin, but Not Pravastatin, Induce Bone Morphogenetic Protein-2 in Human Osteosarcoma Cells. *Biochem. Biophys. Res. Commun.* **2000**, *271*, 688–692. [CrossRef]
19. Thunyakitpisal, P.D.; Chaisuparat, R. Simvastatin, an HMG-CoA Reductase Inhibitor, Reduced the Expression of Matrix Metalloproteinase-9 (Gelatinase B) in Osteoblastic Cells and HT1080 Fibrosarcoma Cells. *J. Pharmacol. Sci.* **2004**, *94*, 403–409. [CrossRef] [PubMed]
20. Ruiz-Gaspa, S.; Nogues, X.; Enjuanes, A.; Monllau, J.C.; Blanch, J.; Carreras, R.; Mellibovsky, L.; Grinberg, D.; Balcells, S.; Díez-Perez, A.; et al. Simvastatin and Atorvastatin Enhance Gene Expression of Collagen Type 1 and Osteocalcin in Primary Human Osteoblasts and MG-63 Cultures. *J. Cell Biochem.* **2007**, *101*, 1430–1438. [CrossRef]
21. Zhang, M.; Bian, Y.-Q.; Tao, H.-M.; Yang, X.-F.; Mu, W.-D. Simvastatin Induces Osteogenic Differentiation of MSCs via Wnt/β-Catenin Pathway to Promote Fracture Healing. *Eur. Rev. Med. Pharmacol. Sci.* **2018**, *22*, 2696–2905.
22. Shao, P.-L.; Wu, S.-C.; Lin, Z.-Y.; Ho, M.-L.; Chen, C.-H.; Wang, C.-Z. Alpha-5 Integrin Mediates Simvastatin-Induced Osteogenesis of Bone Marrow Mesenchymal Stem Cells. *Int. J. Mol. Sci.* **2019**, *20*, 506. [CrossRef] [PubMed]
23. Chuang, S.-C.; Chen, C.-H.; Fu, Y.-C.; Tai, I.-C.; Li, C.-J.; Chang, L.-F.; Ho, M.-L.; Chang, J.-K. Estrogen Receptor Mediates Simvastatin-Stimulated Osteogenic Effects in Bone Marrow Mesenchymal Stem Cells. *Biochem. Pharmacol.* **2015**, *98*, 453–464. [CrossRef] [PubMed]
24. Huang, Y.; Lin, Y.; Rong, M.; Liu, W.; He, J.; Zhou, L. 20(S)-Hydroxycholesterol and Simvastatin Synergistically Enhance Osteogenic Differentiation of Marrow Stromal Cells and Bone Regeneration by Initiation of Raf/MEK/ERK Signaling. *J. Mater. Sci. Mater. Med.* **2019**, *30*, 87. [CrossRef]
25. Kupcsik, L.; Meurya, T.; Flury, M.; Stoddart, M.; Alini, M. Statin-Induced Calcification in Human Mesenchymal Stem Cells Is Cell Death Related. *J. Cell Mol. Med.* **2009**, *13*, 4465–4473. [CrossRef]
26. Sun, S.-H.; Lee, I.-K.; Lee, J.-W.; Shim, I.-S.; Kim, S.-H.; Kim, K.-S. Simvastatin Induces Osteogenic Differentiation and Suppresses Adipogenic Differentiation in Primarily Cultured Human Adipose-Derived Stem Cells. *Biomol. Ther.* **2009**, *17*, 353–361. [CrossRef]
27. Ahmadi, A.; Ebadi, S.S.; Tayebi, T.; Ebadi, S.A.; Sarzaeem, M.M.; Niknejad, H. Osteogenic Differentiation Effect of BMP-9 with Phenamil and Simvastatin on Intact Human Amniotic Epithelial Stem Cells. *Iran. Biomed. J.* **2022**, *26*, 463–474. [CrossRef]
28. Qiao, L.J.; Kang, K.L.; Heo, J.S. Simvastatin Promotes Osteogenic Differentiation of Mouse Embryonic Stem Cells via Canonical Wnt/β-Catenin Signaling. *Mol. Cells* **2011**, *32*, 437–444. [CrossRef] [PubMed]
29. Pagkalos, J.; Cha, J.M.; Kang, Y.; Heliotis, M.; Tsiridis, E.; Mantalaris, A. Simvastatin Induces Osteogenic Differentiation of Murine Embryonic Stem Cells. *J. Bone Miner. Res.* **2010**, *25*, 2470–2478. [CrossRef] [PubMed]

30. Sabandal, M.M.I.; Schäfer, E.; Petsching, S.; Jung, S.; Kleinheinz, J.; Sielker, S. Pleiotropic Effects on Proliferation and Mineralization of Primary Human Adipose Tissue-Derived Stromal Cells Induced by Simvastatin. *Open Biol.* **2022**, *12*, 210337. [CrossRef] [PubMed]
31. Lee, H.; Lee, H.; Na, C.-B.; Park, J.-B. The Effects of Simvastatin on Cellular Viability, Stemness and Osteogenic Differentiation Using 3-Dimensional Cultures of Stem Cells and Osteoblast-like Cells. *Adv. Clin. Exp. Med.* **2019**, *28*, 699–706. [CrossRef] [PubMed]
32. Needleman, I.G. A Guide to Systematic Reviews. *J. Clin. Periodontol.* **2002**, *29*, 6–9. [CrossRef] [PubMed]
33. Page, M.J.; McKenzie, J.E.; Bossuyt, P.M.; Boutron, I.; Hoffmann, T.C.; Mulrow, C.D.; Shamseer, L.; Tetzlaff, J.M.; Akl, E.A.; Brennan, S.E.; et al. The PRISMA 2020 Statement: An Updated Guideline for Reporting Systematic Reviews. *BMJ* **2021**, *372*, n71. [CrossRef] [PubMed]
34. National Toxicology Program—Office of Health Assessment and Translation NTP-OHAT. *Handbook for Conducting a Literature-Based Health Assessment Using OHAT Approach for Systematic Review and Evidence Integration*; National Toxicology Program—US Department of Health and Human Services: Research Triangle Park, NC, USA, 2019.
35. Romeo, S.; Zeni, O.; Sannino, A.; Lagorio, S.; Biffoni, M.; Scarfi, M.R. Genotoxicity of Radiofrequency Electromagnetic Fields: Protocol for a Systematic Review of in Vitro Studies. *Environ. Int.* **2021**, *148*, 106386. [CrossRef] [PubMed]
36. Guyatt, G.; Oxman, A.D.; Akl, E.A.; Kunz, R.; Vist, G.; Brozek, J.; Norris, S.; Falck-Ytter, Y.; Glasziou, P.; DeBeer, H.; et al. GRADE Guidelines: 1. Introduction-GRADE Evidence Profiles and Summary of Findings Tables. *J. Clin. Epidemiol.* **2011**, *64*, 383–394. [CrossRef] [PubMed]
37. Back, L.S. Efeitos In Vitro da Sinvastatina na Osteoindução de Células Indiferenciadas: Uma Revisão Sistemática. Master's Thesis, Universidade Federal de Santa Catarina, Florianópolis, Brazil, 2021.
38. Ghosh-Choudhury, N.; Mandal, C.C.; Choudhury, G.G. Statin-Induced Ras Activation Integrates the Phosphatidylinositol 3-Kinase Signal to Akt and MAPK for Bone Morphogenetic Protein-2 Expression in Osteoblast Differentiation. *J. Biol. Chem.* **2007**, *282*, 4983–4993. [CrossRef]
39. Abdoul-Azize, S.; Buquet, C.; Li, H.; Picquenot, J.-M.; Vannier, J.-P. Integration of Ca2+ Signaling Regulates the Breast Tumor Cell Response to Simvastatin and Doxorubicin. *Oncogene* **2018**, *37*, 4979–4993. [CrossRef] [PubMed]
40. Aquino-Martínez, R.; Artigas, N.; Gámez, B.; Rosa, J.L.; Ventura, F. Extracellular Calcium Promotes Bone Formation from Bone Marrow Mesenchymal Stem Cells by Amplifying the Effects of BMP-2 on SMAD Signalling. *PLoS ONE* **2017**, *12*, e0178158. [CrossRef] [PubMed]
41. Hammel, C.; Pandis, N.; Pieper, D.; Faggion, C.M. Methodological Assessment of Systematic Reviews of In-Vitro Dental Studies. *BMC Med. Res. Methodol.* **2022**, *22*, 110. [CrossRef]
42. Elshafay, A.; Omran, E.S.; Abdelkhalek, M.; El-Badry, M.O.; Eisa, H.G.; Fala, S.Y.; Dang, T.; Ghanem, M.A.T.; Elbadawy, M.; Elhady, M.T.; et al. Reporting Quality in Systematic Reviews of in Vitro Studies: A Systematic Review. *Curr. Med. Res. Opin.* **2019**, *35*, 1631–1641. [CrossRef]
43. Guyatt, G.H.; Oxman, A.D.; Vist, G.E.; Kunz, R.; Falck-Ytter, Y.; Alonso-Coello, P.; Schünemann, H.J. GRADE: An Emerging Consensus on Rating Quality of Evidence and Strength of Recommendations. *BMJ* **2008**, *336*, 924–926. [CrossRef] [PubMed]
44. Gupta, S.; Del Fabbro, M.; Chang, J. The Impact of Simvastatin Intervention on the Healing of Bone, Soft Tissue, and TMJ Cartilage in Dentistry: A Systematic Review and Meta-Analysis. *Int. J. Implant. Dent.* **2019**, *5*, 17. [CrossRef] [PubMed]
45. Leutner, M.; Butylina, M.; Matzhold, C.; Klimek, P.; Cuhaj, C.; Bellach, L.; Baumgartner-Parzer, S.; Reiter, B.; Preindl, K.; Kautzky, A.; et al. Simvastatin Therapy in Higher Dosages Deteriorates Bone Quality: Consistent Evidence from Population-Wide Patient Data and Interventional Mouse Studies. *Biomed. Pharmacother.* **2023**, *158*, 114089. [CrossRef] [PubMed]

Journal of
Functional Biomaterials

MDPI

Article

Cu-MOF-Decorated 3D-Printed Scaffolds for Infection Control and Bone Regeneration

Ting Zhu [1], Qi Ni [1], Wenjie Wang [2], Dongdong Guo [2], Yixiao Li [1], Tianyu Chen [2], Dongyang Zhao [1], Xingyu Ma [1] and Xiaojun Zhang [1,*]

[1] School of Medicine, Northwest University, Xi'an 710069, China
[2] Key Laboratory of Resource Biology and Biotechnology Western China, Ministry of Education, Provincial Key Laboratory of Biotechnology, College of Life Sciences, Northwest University, Xi'an 710069, China
* Correspondence: zhangxj@nwu.edu.cn

Abstract: Infection control and bone regeneration remain critical challenges in bone defect treatment. We developed a 3D-printed scaffold incorporating copper-based metal–organic framework-74 (Cu-MOF-74) within a polycaprolactone/hydroxyapatite composite. The synthesized Cu-MOF-74 exhibited a well-defined crystalline structure and rod-like morphology, as confirmed by TEM, EDS, FTIR, and XRD analyses. The scaffolds exhibited hierarchical pores (100–200 μm) and demonstrated tunable hydrophilicity, as evidenced by the water contact angles decreasing from $103.3 \pm 2.02°$ (0% Cu-MOF-74) to $63.60 \pm 1.93°$ (1% Cu-MOF-74). A biphasic Cu^{2+} release profile was observed from the scaffolds, reaching cumulative concentrations of 98.97 ± 3.10 ppm by day 28. Antimicrobial assays showed concentration-dependent efficacy, with 1% Cu-MOF-74 scaffolds achieving $90.07 \pm 1.94\%$ and $80.03 \pm 2.17\%$ inhibition against *Staphylococcus aureus* and *Escherichia coli*, respectively. Biocompatibility assessments using bone marrow-derived mesenchymal stem cells revealed enhanced cell proliferation at Cu-MOF-74 concentrations $\leq 0.2\%$, while concentrations $\geq 0.5\%$ induced cytotoxicity. Osteogenic differentiation studies highlighted elevated alkaline phosphatase activity and mineralization in scaffolds with 0.05–0.2% Cu-MOF-74 scaffolds, particularly at 0.05% Cu-MOF-74 scaffolds, which exhibited the highest calcium deposition and upregulation of bone sialoprotein and osteopontin expression. These findings demonstrate the dual functional efficacy of Cu-MOF-74/PCL/HAp scaffolds in promoting both infection control and bone regeneration. These optimized Cu-MOF-74 concentrations (0.05–0.2%) effectively balance antimicrobial and osteogenic properties, presenting a promising strategy for bone defect repair in clinical applications.

Keywords: copper; metal–organic frameworks; 3D printing scaffold; antibacterial properties; osteoinductive potential; bone regeneration

1. Introduction

Bone defects caused by trauma, infection, or tumor resection present a significant clinical challenge, often requiring complex interventions to achieve effective bone regeneration [1]. Annually, millions of patients worldwide undergo surgical procedures to address bone defects, leading to substantial social and economic burdens [2]. Autologous bone grafts, considered to be the gold standard for bone regeneration, are limited by donor site morbidity and insufficient material availability [3–5]. Allogeneic bone grafts, while a viable alternative, pose risks of immune rejection, disease transmission, and suboptimal integration with host tissues [4,6]. Synthetic bone substitutes, such as bone cements, are typically restricted to small, non-load-bearing defects due to their limited porosity and degradability, which are crucial for vascularization and long-term bone regeneration [7,8]. Therefore, the development of innovative therapeutic strategies that combine structural integrity, bioactivity, and antibacterial properties is urgently needed to overcome these limitations.

Tissue engineering has emerged as a promising strategy for bone defect repair, with scaffolds designed to mimic the natural bone microenvironment and promote regenerative processes. Scaffolds provide a three-dimensional (3D) architecture that supports cell adhesion, proliferation, and differentiation, while simultaneously delivering bioactive cues to enhance tissue regeneration [9–12]. Among the various scaffold fabrication techniques, 3D printing has gained significant attention due to its capacity to create complex, patient-specific structures with precise control over porosity, geometry, and material composition [12,13]. Polycaprolactone (PCL), a biodegradable polyester approved by the FDA for medical applications, is frequently employed in 3D printing owing to its favorable mechanical properties and processability [14]. However, PCL lacks intrinsic bioactivity, necessitating the incorporation of bioactive components such as hydroxyapatite (HAp) and metal ions to enhance its osteogenic potential [15].

To enhance the bioactivity of PCL scaffolds, HAp, a primary mineral component of natural bone, has been extensively utilized in bone tissue engineering due to its excellent biocompatibility and osteoconductive properties [16]. The incorporation of HAp into PCL scaffolds has been shown to enhance osteogenic activity and mechanical strength, rendering it a suitable composite for bone regeneration [15]. However, a critical limitation of HAp/PCL scaffolds is the absence of intrinsic antibacterial properties, which is a significant concern given the high risk of infection associated with bone defect repair, particularly in cases of trauma or complex fractures. Postoperative infections can severely compromise bone healing, leading to complications and implant failure [17]. To address this limitation, researchers have explored incorporating metal ions, such as copper (Cu^{2+}), into scaffolds to impart both antibacterial and osteogenic functionalities [18–20].

Copper, an essential trace element, plays a crucial role in numerous physiological processes, including angiogenesis, enzymatic activity, and bone metabolism [21]. Recent studies have demonstrated that copper-doped scaffolds can enhance osteogenic differentiation and exhibit antibacterial activity, positioning them as promising candidates for bone defect repair [19,20]. However, uncontrolled copper release can induce cytotoxicity, highlighting the need for controlled release mechanisms to ensure therapeutic efficacy and safety [22]. Metal–organic frameworks (MOFs), particularly copper-based MOFs (Cu-MOFs), offer a compelling strategy by enabling the sustained release of Cu^{2+} while maintaining structural integrity and biocompatibility [23,24]. Cu-MOFs are porous crystalline materials composed of metal ions coordinated with organic ligands, providing a high surface area and tunable physicochemical properties [24]. Among Cu-MOFs, Cu-MOF-74 has attracted considerable interest due to its high density of open metal sites, which are known to enhance antibacterial and osteogenic properties [25,26]. The porous structure of Cu-MOF-74 facilitates controlled copper ion release, mitigating the risk of cytotoxicity associated with a burst release. Furthermore, the inherent properties of Cu-MOF-74 itself may contribute to enhanced biocompatibility and osteogenic potential within the scaffold. Given the critical challenge of bacterial infection in bone reconstruction—particularly in trauma settings, where compromised soft tissue and vascularization increase susceptibility to colonization [27]—the incorporation of Cu-MOF-74 into bone scaffolds represents a rational approach to address both regenerative and antimicrobial requirements.

In this study, we developed Cu-MOF-74/HAp/PCL composite scaffolds using 3D printing. We hypothesized that incorporating Cu-MOF-74 into HAp/PCL scaffolds would enhance their antibacterial and osteogenic properties, rendering them suitable for bone defect repair and infection control. The scaffolds were characterized for their structural, mechanical, and biological properties, and their antibacterial and osteogenic potential was evaluated in vitro. Our findings demonstrate that Cu-MOF-74/HAp/PCL scaffolds exhibit excellent biocompatibility, antibacterial activity, and osteogenic potential, offering a promising solution for bone defect repair and regeneration.

2. Materials and Methods

2.1. Materials

PCL was purchased from Engineering for Life Technology (Suzhou, China). HAp, monohydrated copper acetate ($Cu(OAc)_2{\cdot}H_2O$), 2,5-dihydroxyterephthalic acid (DHTP), tetrahydrofuran (THF), and methanol were purchased from Macklin Biochemical Technology (Shanghai, China;. α-MEM medium was purchased from Hyclone Company, Logan, UT, USA. FITC-labeled phalloidin, Live/Dead Cell Staining Kits, Cell Counting Kit-8 (CCK-8), Alizarin Red S staining solution, BCA protein concentration test kits, and alkaline phosphatase (ALP) assay kits were provided by Beyotime Biotechnology (Shanghai, China). Fetal bovine serum (FBS) was provided by Cell-Box Biotechnology (Changsha, China). Primary antibodies, bone sialoprotein (BSP), osteopontin (OPN), and osteocalcin (OCN), were purchased from Proteintech Group (Wuhan, China).

2.2. Synthesis and Characterization of Cu-MOF-74

2.2.1. Synthesis of Cu-MOF-74

Cu-MOF-74 was synthesized using a modified method reported by Zheng et al. [24]. Briefly, 0.4 g of $Cu(OAc)_2{\cdot}H_2O$ was dissolved in 10 mL of methanol to obtain a 40 g/L solution. Then, 0.2 g of DHTP was added to 5 mL of methanol, followed by the dropwise addition of the $Cu(OAc)_2{\cdot}H_2O$ solution. The mixture was magnetically stirred for 24 h, and the resulting reddish-brown crystalline solid was collected by filtration. The resulting Cu-MOF-74 was washed with methanol four times to remove impurities and dried in a vacuum at 85 °C for 24 h.

2.2.2. Morphological Characterization of Cu-MOF-74

The microstructure of Cu-MOF-74 was observed using scanning electron microscopy (SEM) (Hitachi, S-4800, Tokyo, Japan) and transmission electron microscopy (TEM) (JEOL, JEM-2100Plus, Tokyo, Japan).

2.2.3. Physicochemical Characterization of Cu-MOF-74

The crystal structures of the synthesized nanoparticles were analyzed using X-ray diffraction (XRD) (Bruker-D8, Bremen, Germany). Fourier-transform infrared (FTIR) spectra (AVATAR-FTIR-360, Thermo Fisher Scientific, Waltham, MA, USA) were recorded in the range of 4000–400 cm^{-1}. Elemental mapping of Cu-MOF-74 was conducted using an energy-dispersive spectrometer (EDS) (Bruker QUANTAX 400, Bremen, Germany).

2.3. Fabrication and Characterization of Cu-MOF-74/HAp/PCL Scaffolds

2.3.1. Fabrication of the Scaffolds

To prepare the printing bio-ink, 20 g of PCL and 4 g of HAp were dispersed in 200 mL of THF. Subsequently, Cu-MOF-74 was incorporated into the mixture at concentrations of 0%, 0.05%, 0.1%, 0.2%, 0.5%, and 1% by weight relative to PCL (Table 1). After thorough mixing to ensure homogeneity, the resulting blends were dried in a fume hood to remove the solvent, cut into small pieces, loaded into a printing cartridge, and melted at 65 °C within the printer. The scaffolds were printed using an extrusion-based pneumatic 3D printer (Engineering For Life Technology, Suzhou, China) with the following parameters: 6–10 MPa air pressure, 0.35 mm nozzle diameter, 30 mm/s printing speed, 65–70 °C extrusion temperature, 20 °C platform temperature, and 0.2 mm layer height. The scaffolds were designed using Bioplotter CAD/CAM software (EFL_PotatoE 6601 V1.2.6) and printed layer by layer with a pore size of 100 μm, 5 layers, and dimensions of 10 × 10 × 1 mm^3.

Table 1. List of compositions of 3D-printed scaffolds.

Name	Weight			Cu-MOF-74/PCL (*w*/*w*%)
	PCL (g)	HAp (g)	Cu-MOF-74 (g)	
0 Cu	20	4	0	0
0.05 Cu	20	4	0.01	0.05
0.1 Cu	20	4	0.02	0.1
0.2 Cu	20	4	0.04	0.2
0.5 Cu	20	4	0.1	0.5
1 Cu	20	4	0.2	1

2.3.2. Structural Properties of the Scaffolds

The scaffolds were coated with gold and imaged using a field-emission scanning electron microscope (JEOL, JSM-7610Fplus, Tokyo, Japan).

2.3.3. Hydrophilicity of the Scaffold Surfaces

The hydrophilicity of the scaffolds was evaluated by measuring the static water contact angle. Deionized water (20 μL) was dropped onto the scaffold surface, and the contact angle was calculated using imaging software (Java 1.8.0_322).

2.3.4. Cu^{2+} Release Performance

The release dynamics of Cu^{2+} from the Cu-MOF-74/HAp/PCL composite scaffolds were quantified using inductively coupled plasma mass spectrometry (ICP-MS). The scaffolds were immersed in phosphate-buffered saline (PBS) at a ratio of 0.5 cm^2/mL and incubated for 28 days. At predetermined timepoints (1, 4, 7, 14, 21, and 28 days), 3 mL of PBS supernatant was collected, and 3 mL of fresh PBS was added to each sample. The Cu^{2+} concentrations were determined via ICP-MS to construct ion release profiles.

2.4. Antibacterial Performance Testing

The antibacterial efficacy of the Cu-MOF-74/HAp/PCL composite scaffolds was assessed against a Gram-positive bacterium, *Staphylococcus aureus* (*S. aureus*, ATCC 25923), and a Gram-negative bacterium, *Escherichia coli* (*E. coli*, ATCC 25922). Specifically, 500 μL of the bacterial suspension, diluted to a concentration of 1 × 10^5 CFU/mL, was added onto the sterilized Cu-MOF-74/HAp/PCL composite scaffolds, placed in a 24-well plate, and incubated at 37 °C for 12 h. Non-adherent bacteria were rinsed using PBS. Ultrasonic treatment (40 Hz, 10 min) was applied to dislodge bacteria adhering to the composite scaffolds. The bacterial solutions were diluted 1000 times, and 60 μL of the diluted solution was evenly spread on LB solid medium and incubated for another 12 h. The colony counts on the LB agar plates indirectly quantified the bacterial CFUs on the scaffold surfaces. The bacterial survival rate was calculated as follows:

$$R = \frac{B - A}{B} \times 100\%,$$

where

- A represents the average number of surviving bacteria on the Cu-MOF-74/HAp/PCL scaffolds;
- B represents the average number on 0 Cu scaffolds.

2.5. Cell Biocompatibility Analysis

2.5.1. Cell Culture

Primary bone marrow mesenchymal stem cells (BMSCs) were obtained from one-week-old Sprague-Dawley rats. All animal experiments were performed in accordance with the relevant regulations of the Ministry of Health of the People's Republic of China and the Guidelines on the Good Treatment of Laboratory Animals. The experiments were reviewed and approved by the Laboratory Animal Ethics Review Committee of Northwest University, China. Femurs and tibias were isolated using ophthalmic scissors and tweezers, with muscle and connective tissue removed. After rinsing with PBS, the bone marrow was flushed with α-MEM medium supplemented with 10% FBS and 1% penicillin–streptomycin to create a cell suspension. The suspension was then cultured in flasks at 37 °C under 5% CO_2.

2.5.2. Cell Attachment Assay

The growth and spreading morphology of BMSCs were characterized using F-actin staining. Briefly, the scaffolds were seeded with 5×10^4 BMSCs and incubated at 37 °C for 1 day. After washing with PBS, the scaffolds were fixed in 4% paraformaldehyde overnight. The samples were permeabilized with 0.1% Triton X-100 for 15 min, stained with FITC-labeled phalloidin solution for 30 min, and observed using a Zeiss LSM 710 confocal microscope (Carl Zeiss AG, Oberkochen, Germany).

2.5.3. Cell Viability Analysis

BMSCs were seeded onto the Cu-MOF-74/HAp/PCL composite scaffolds at a concentration of 1×10^5 cells per scaffold. After 3 d of incubation, the cells were treated with live/dead cell staining solution for 30 min, washed three times with PBS, and subsequently observed using laser confocal microscopy. Live cells fluoresced green, while dead cells fluoresced red.

2.5.4. Cell Proliferation Assay

BMSCs were cultured on Cu-MOF-74/HAp/PCL composite scaffolds at a density of 1×10^5 cells per scaffold for 7 d. At predetermined time intervals (days 1, 4, and 7), the scaffolds were aseptically transferred to new 24-well plates, and 500 μL of working solution containing 10% CCK-8 reagent and 90% cell culture medium was added. Following 2 h of incubation, 100 μL of supernatant from each sample was transferred to a 96-well plate. Absorbance measurements were performed at a wavelength of 450 nm using a microplate reader (BioTek Synergy H1, Agilent, Santa Clara, CA, USA).

2.6. Osteogenic Differentiation Analysis

2.6.1. Cellular ALP Activity Assay

BMSCs (1×10^5 cells per scaffold) were cultured on Cu-MOF-74/HAp/PCL composite scaffolds with α-MEM medium supplemented with 10% FBS and 1% penicillin–streptomycin. At days 7 and 14 post-induction, the cells were washed twice with PBS, lysed with RIPA buffer, and quantified using an ALP assay kit according to the manufacturer's protocol. Briefly, 50 μL of cell lysate was mixed with 100 μL of p-nitrophenyl phosphate (pNPP) substrate and incubated at 37 °C for 30 min. The reaction was terminated with 50 μL of 0.1 N NaOH, and absorbance was measured at 405 nm using a microplate reader (BioTek Synergy H1).

2.6.2. Mineralization Assessment by Tetracycline Labeling

To evaluate extracellular matrix mineralization, BMSC-seeded scaffolds (1×10^5 cells per scaffold) were cultured for 21 d, with the medium replaced every 3 d. The samples were stained with 5 mg/mL tetracycline hydrochloride for 1 h at 37 °C, followed by three washes with PBS. Fresh α-MEM medium was added, and the samples were further incubated for 1 h at 37 °C under 5% CO_2. After fixation with 4% paraformaldehyde for 30 min, the mineralization patterns were visualized using a confocal laser scanning microscope (Leica, TCS SP8 MP, Heidelberg, Germany).

2.6.3. Calcium Deposition Quantification via Alizarin Red S Staining

After 21 d of incubation, scaffolds were fixed with 4% paraformaldehyde overnight at 4 °C and stained with 2% Alizarin Red S for 30 min. Excess dye was removed by five sequential PBS washes under gentle agitation. For quantification, mineralized nodules were dissolved in 10% cetylpyridinium chloride in 10 mM sodium phosphate buffer for 30 min. The solution was centrifuged at 11,000 rpm for 5 min, and the absorbance of the supernatant was measured at 562 nm.

2.6.4. Osteogenic Protein Expression Profiling

BMSCs were cultured onto Cu-MOF-74/HAp/PCL scaffolds for 21 d. The scaffolds were rinsed three times with PBS, sectioned, and placed in 1.5 μL EP tubes. Total cellular proteins were extracted using RIPA buffer containing 1% protease inhibitor on ice for 5 min, and then centrifuged at 12,000 rpm (4 °C) for 60 s. The protein concentration was determined using a BCA protein assay kit. Proteins were separated by 10% SDS-PAGE and transferred to PVDF membranes. Membranes were blocked with 5% non-fat dry milk in Tris-buffered saline with 0.1% Tween-20 for 2 h, and then incubated with primary antibodies against BSP, OPN, OCN, and β-tubulin overnight at 4 °C. After incubation with HRP-conjugated secondary antibodies for 1 h, the protein bands were detected and quantified with Image Lab Software (Java 1.8.0_322).

2.7. Statistical Analysis

The experiments were independently replicated a minimum of three times, and the data are presented as the mean ± standard deviation. Data were analyzed using one-way analysis of variance (ANOVA) utilizing GraphPad Prism 8.2.1. A p-value of less than 0.05 was considered statistically significant (* $p < 0.05$; ** $p < 0.01$).

3. Results

3.1. Structure and Physical Characteristics of Cu-MOF-74

The morphological and structural properties of the synthesized Cu-MOF-74 were systematically characterized using advanced analytical techniques. TEM analysis revealed well-defined crystalline structures with uniform rod-like morphologies, measuring 300–500 nm in length and 50–100 nm in width (Figure 1A). EDS elemental mapping (Figure 1B–F) demonstrated a homogeneous distribution of constituent elements, with mass percentages of 64.54% C, 0.94% N, 18.20% O, and 16.33% Cu, and corresponding atomic percentages of 78.62% C, 0.98% N, 16.64% O, and 3.76% Cu. These values align closely with the theoretical stoichiometry of Cu-MOF-74, confirming the successful synthesis of the nanoscale metal–organic framework.

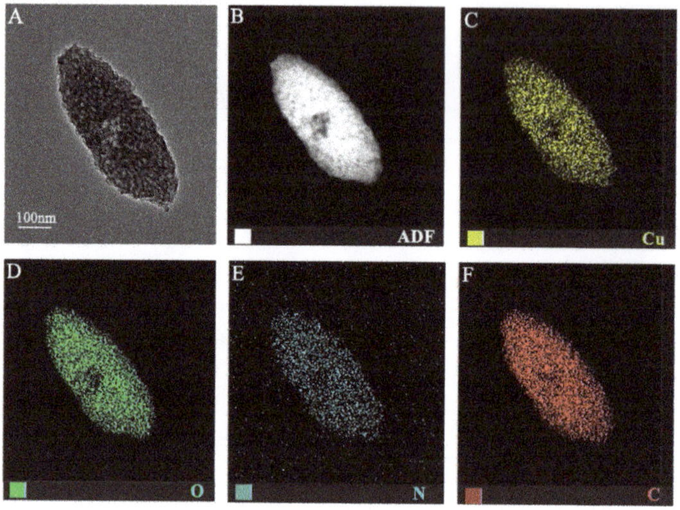

Figure 1. TEM image of Cu-MOF-74 (**A**) and the corresponding elemental mapping (**B–F**).

The FTIR spectrum, as depicted in Figure 2A, exhibited a broad absorption band in the range of 3000–3650 cm^{-1}, attributed to O-H stretching vibrations from coordinated water molecules and free hydroxyl groups. Characteristic peaks were observed at 1292 cm^{-1} (C-O stretching of carboxyl groups), 1496 cm^{-1} (C=O stretching), and 1774 cm^{-1} (O-H bending vibrations), indicating the successful coordination between the -COOH groups of DHTP and Cu^{2+} ions. A prominent peak at 1652 cm^{-1} was assigned to the asymmetric stretching of $-COO^{-}$ groups in the organic ligand, while the peak at 849 cm^{-1} corresponded to C-H bending vibrations in the aromatic ring of DHTP. Notably, a low-frequency absorption peak near 500 cm^{-1} was identified as the Cu-O stretching vibration, providing direct evidence of the metal–ligand coordination bond formation.

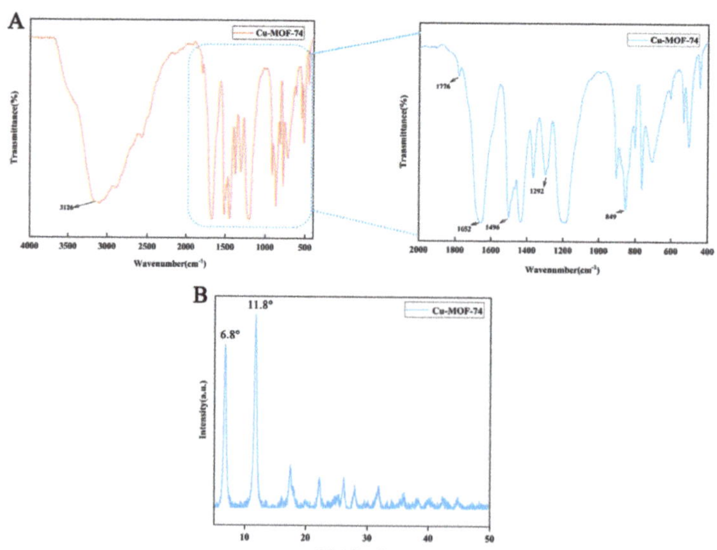

Figure 2. FTIR spectra (**A**) and XRD pattern (**B**) of Cu-MOF-74.

XRD analysis further confirmed the crystalline structure of Cu-MOF-74 (Figure 2B). The diffraction pattern exhibited two distinct and sharp peaks at $2\theta = 6.8°$ and $11.8°$, corresponding to the (001) and (200) crystal planes, respectively. These diffraction features validate the successful synthesis of Cu-MOF-74 with high crystallinity.

3.2. Structure and Physical Characteristics of the 3D-Printed Cu-MOF-74/PCL/HAp Composite Scaffolds

The 3D-printed Cu-MOF-74/PCL/HAp composite scaffolds exhibited a rectangular geometry ($10 \times 10 \times 1$ mm^3) and a progressive color transition from white (0 Cu) to light pink (0.05 Cu) and, ultimately, to brownish-red (1 Cu) with increasing Cu-MOF-74 concentrations, while the macroscopic morphology remained unchanged (Figure 3A,B). The well-defined grid-like architecture showed uniform filament diameters and hierarchically interconnected pores (100–200 μm), as confirmed by SEM (Figure 3C–H). SEM analysis revealed smooth, compact surfaces at lower concentrations of Cu-MOF-74 (<0.2%), suggesting effective encapsulation of Cu-MOF-74 particles by the PCL/HAp matrix. Higher Cu-MOF-74 concentrations (≥0.5%) induced modest surface roughening, potentially enhancing cell adhesion through topographical cues.

EDS mapping (Figure 4) verified the homogeneous elemental distribution of C (54.76 ± 5.18%), O (36.21 ± 3.50%), Ca (6.15 ± 0.32%), P (2.71 ± 0.12%), and Cu (0.17 ± 0.02%) in the 0.2 Cu composite scaffolds, confirming successful composite integration without phase segregation.

The hydrophilicity of the composite scaffolds showed significant Cu-MOF-74 dose-dependence ($p < 0.01$). The water contact angles decreased from 103.3 ± 2.02° (0 Cu) to 63.60 ± 1.93° (1 Cu) (Figure 5A), transitioning from hydrophobic to hydrophilic behavior (threshold at 0.2 Cu). This enhancement in hydrophilicity correlates with the increased surface polarity imparted by the coordinatively unsaturated metal sites of Cu-MOF-74.

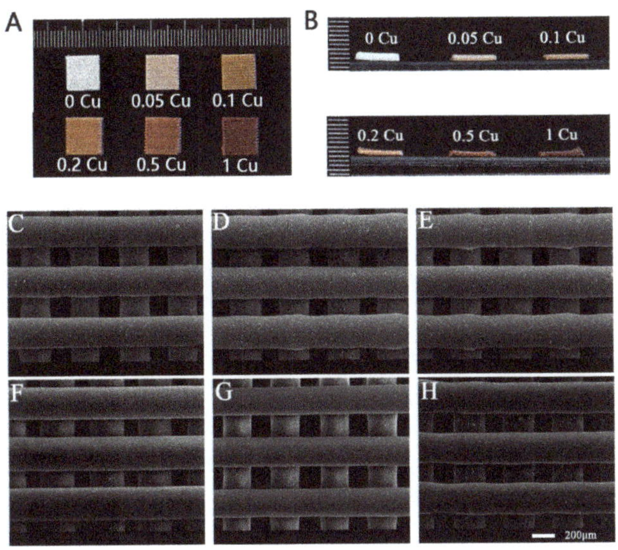

Figure 3. Photographs of the Cu-MOF-74/HAp/PCL composite scaffolds (**A**,**B**) and SEM images of the different Cu-MOF-74/HAp/PCL scaffolds: 0 Cu (**C**), 0.05 Cu (**D**), 0.1 Cu (**E**), 0.2 Cu (**F**), 0.5 Cu (**G**), and 1 Cu (**H**).

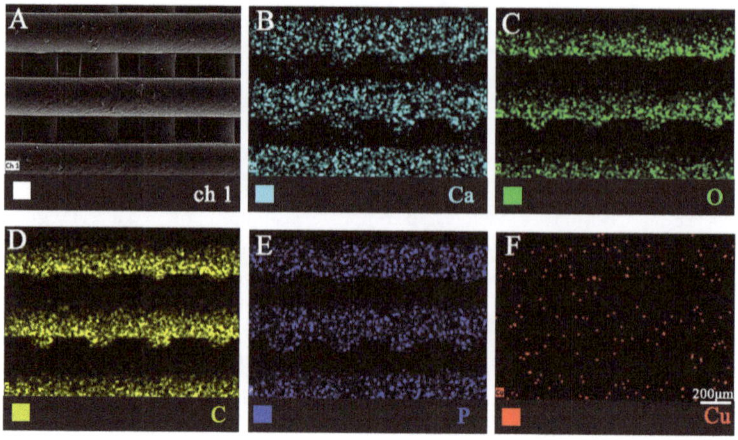

Figure 4. EDS mapping of the 0.2% Cu-MOF-74/HAp/PCL composite scaffolds: ch1 (**A**), Ca (**B**), O (**C**), C (**D**), P (**E**), and Cu (**F**).

The Cu^{2+} release kinetics demonstrated a biphasic pattern: an initial burst release (days 1–7), followed by a sustained release (days 7–28) (Figure 5B). During the first 7 d, rapid dissolution of Cu^{2+} from the scaffold surfaces was observed. From day 7 onward, the release rate decelerated, entering a transient plateau phase. The release persisted until day 28 and potentially beyond. The cumulative release at day 28 reached 28.02 $\pm$ 1.3 ppm (0.05 Cu), 49.33 $\pm$ 2.1 ppm (0.1 Cu), 46.93 $\pm$ 1.8 ppm (0.2 Cu), 68.16 $\pm$ 2.4 ppm (0.5 Cu), and 98.97 $\pm$ 3.1 ppm (1 Cu), showing a strong positive correlation with Cu-MOF-74 loading.

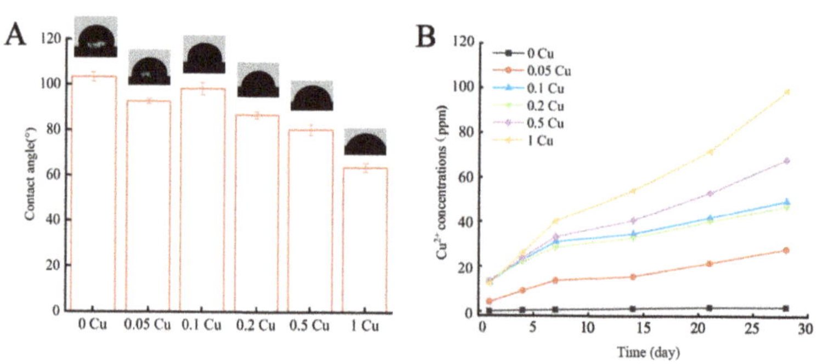

Figure 5. Water absorption process of Cu-MOF-74/HAp/PCL (**A**) and released Cu^{2+} concentration of the Cu-MOF-74/HAp/PCL immersed in PBS solution (**B**).

3.3. Antibacterial Properties of Cu-MOF-74/PCL/HAp Composite Scaffolds

The antimicrobial efficacy of the 3D-printed Cu-MOF-74/PCL/HAp scaffolds was quantitatively assessed against two clinically relevant bacterial strains, *S. aureus* (Gram-positive) and *E. coli* (Gram-negative), using the standardized colony-forming unit counting method. As illustrated in Figure 6, a clear concentration-dependent antibacterial relationship was observed.

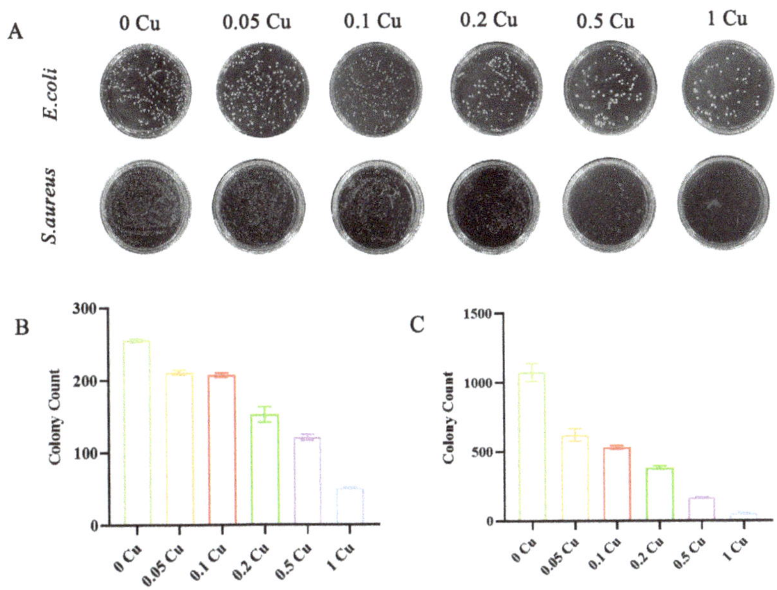

Figure 6. Optical images (**A**) and graphs of the percentage inhibition of colony counts for Cu-MOF-74/HAp/PCL scaffolds against *E. coli* (**B**) and *S. aureus* (**C**) at different concentrations.

Scaffolds containing 0.2% Cu-MOF-74 demonstrated significant antimicrobial activity, achieving inhibition rates of 64.43 ± 3.21% against *S. aureus* and 40.73 ± 2.85% against *E. coli*. Notably, increasing the Cu-MOF-74 loading to 1% substantially enhanced these values to 90.07 ± 1.94% and 80.03 ± 2.17% for *S. aureus* and *E. coli*, respectively ($p < 0.01$). This pronounced enhancement suggests synergistic effects between copper ion release and scaffold surface modification. The corresponding antibacterial rates are summarized in Table 2.

Table 2. List of the bacterial survival rates for 3D-printed scaffolds.

Name	Bacterial Survival Rate (%)	
	E. coli	*S. aureus*
0 Cu	0	0
0.05 Cu	17.36 ± 1.36	42.70 ± 4.17 *
0.1 Cu	18.67 ± 1.13	50.79 ± 1.13 **
0.2 Cu	40.73 ± 2.85 *	64.43 ± 3.21 **
0.5 Cu	52.87 ± 1.63 **	84.56 ± 0.09 **
1 Cu	80.03 ± 2.17 **	90.07 ± 1.94 **

Compared to 0 Cu scaffold, * $p < 0.05$, ** $p < 0.01$.

3.4. Biocompatibility and Proliferation of BMSCs on Cu-MOF-74/HAp/PCL Composite Scaffolds

Primary BMSCs displayed short spindle and polygonal morphologies, exhibited strong adhesion properties, and demonstrated good refractivity (Figure 7A). Confocal microscopy of cytoskeletal staining revealed that BMSCs adhered effectively to the scaffold surfaces, with cells spreading uniformly across the scaffolds. Notably, BMSCs extended pseudopodia to anchor onto the scaffold, suggesting favorable cell–scaffold interactions (Figure 7B).

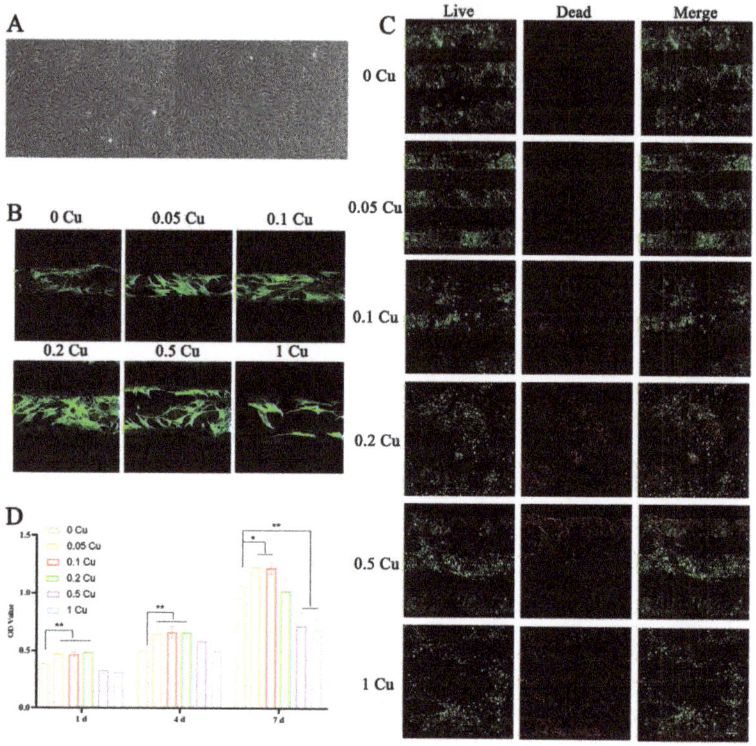

Figure 7. Images of BMSCs extracted from rat bone marrow (**A**). Cytoskeleton staining of BMSCs grown on different scaffolds for 3 d (**B**). Live/dead staining of BMSCs grown on different scaffolds for 3 d (**C**). The proliferation of BMSCs on different composite scaffolds for 1, 4, and 7 d (**D**). * $p < 0.05$, ** $p < 0.01$.

To further assess cell viability, live/dead staining was performed on scaffold-encapsulated BMSCs after three days of culture. As shown in Figure 7C, 80% of the images displayed green fluorescence, indicating high cell viability. The 0 Cu group exhibited negligible red fluorescence, signifying minimal cell death, while the other groups also showed limited red fluorescence, confirming the excellent biocompatibility of the Cu-MOF-74/PCL/HAp composite scaffolds. However, the 0.2 Cu, 0.5 Cu, and 1 Cu groups displayed increased red fluorescence, suggesting higher rates of cell death, which may be attributable to the cytotoxic effects of elevated copper concentrations.

The CCK-8 assay results showed that the 0.05 Cu, 0.1 Cu, and 0.2 Cu groups significantly promoted BMSC proliferation compared to the control group at 1 and 4 days of culture ($p < 0.01$). Over a 7-day culture period, lower copper concentrations ($\leq$0.2 Cu) showed no significant cytotoxicity, with the 0.05 Cu, 0.1 Cu, and 0.2 Cu groups significantly promoting BMSC proliferation ($p < 0.05$). In contrast, the 0.5 Cu and 1 Cu groups exhibited significant inhibition of cell proliferation ($p < 0.01$), likely due to the higher Cu-MOF-74 content in the scaffolds, which may interfere with cell growth (Figure 7D).

3.5. Osteogenic Differentiation and Mineralization of BMSCs on Cu-MOF-74/HAp/PCL Composite Scaffolds

To systematically evaluate the osteogenic potential of the Cu-MOF-74/HAp/PCL composite scaffolds, ALP activity and mineralization capacity were analyzed at multiple timepoints. ALP staining and quantitative analysis revealed distinct temporal and concentration-dependent effects on BMSC differentiation. At day 7, ALP staining showed significantly more intense dark blue coloration in the 0.2 Cu and 1 Cu composite scaffolds compared to the control group (0 Cu), indicating higher ALP activity (Figure 8A). Quantitative analysis further confirmed that the 0.2 Cu and 1 Cu groups exhibited 2.1-fold and 1.8-fold higher ALP activity, respectively, than the control group ($p < 0.01$). Notably, the 0.2 Cu group demonstrated the highest ALP activity at this timepoint (Figure 8B). By day 14, the 0.05 Cu group showed the most pronounced ALP activity, with a 1.4-fold increase compared to the control ($p < 0.01$). However, higher copper concentrations ($\geq$0.2 Cu) did not significantly promote ALP activity ($p > 0.05$).

Mineralization assays were performed at day 21 to evaluate the late-stage osteogenic differentiation of BMSCs. Tetracycline staining revealed stronger fluorescence intensity in the 0.05 Cu group, indicating more abundant calcium deposition compared to the control group (Figure 8C). Similarly, quantitative analysis of Alizarin Red S staining showed higher color intensity in the 0.05 Cu group (Figure 8D). However, there were no significant differences between the Cu-MOF-74-containing groups and the control group.

To further explore the osteogenic potential of Cu-MOF-74/HAp/PCL composite scaffolds, Western blot analysis was performed to assess the expression of osteogenesis-related proteins. BSP levels were significantly upregulated in all Cu-MOF-74-containing composite scaffolds compared to the control ($p < 0.01$) (Figure 9A,D). The 0.05 Cu group showed significantly elevated levels of OPN expression, with a 2.0-fold increase compared to the control ($p < 0.01$) (Figure 9B,E). In contrast, the 0.2 Cu group exhibited the highest expression of OCN. Notably, the 1 Cu group showed significantly decreased OCN expression ($p < 0.01$) (Figure 9C,F).

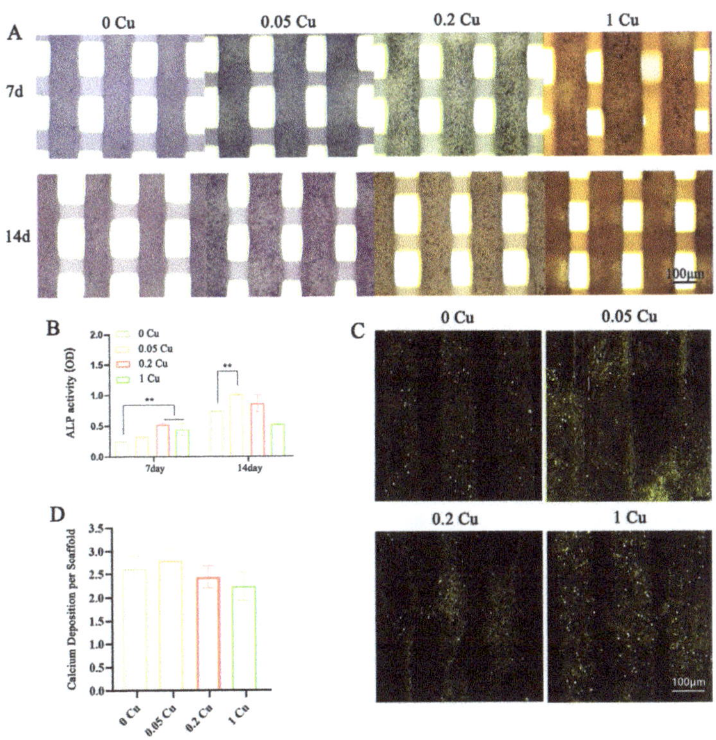

Figure 8. ALP staining (**A**) and ALP quantitative analysis (**B**) at 7 d and 14 d, and tetracycline staining (**C**) and Alizarin Red staining quantitative analysis (**D**) for 21 d; ** $p < 0.01$.

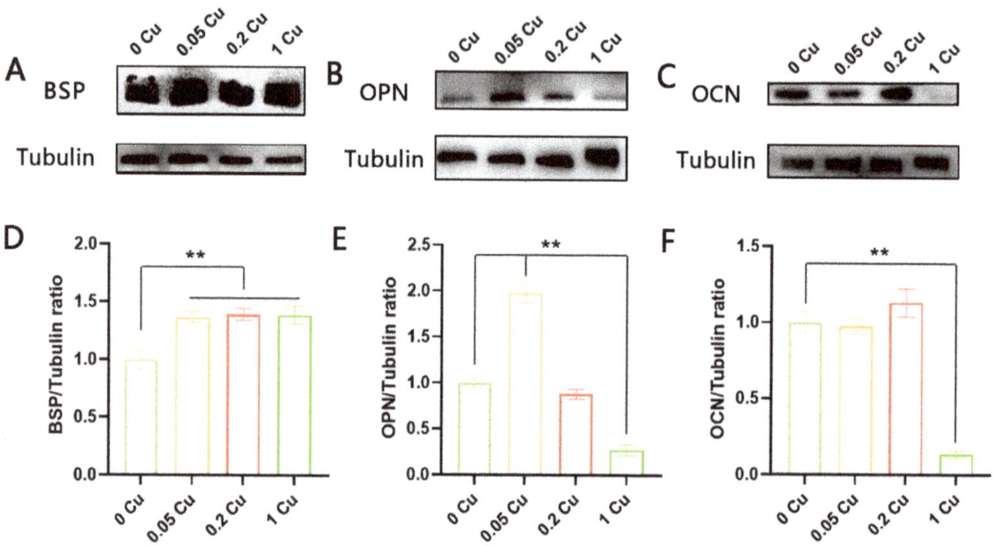

Figure 9. Western blot analysis of osteogenic marker expression (**A–D**) and quantitative analysis (**D–F**) at 21 days; (**A**,**D**) bone sialoprotein (BSP), (**B**,**E**) osteopontin (OPN), (**C**,**F**) osteocalcin (OCN); β-tubulin was used to normalize protein input and quantification; ** $p < 0.01$.

4. Discussion

The development of multifunctional scaffolds that combine antibacterial and osteogenic properties represents a significant advancement in bone tissue engineering. In this study, we successfully fabricated Cu-MOF-74/HAp/PCL composite scaffolds using 3D printing technology and demonstrated their potential for bone defect repair and infection control. The incorporation of Cu-MOF-74 into HAp/PCL scaffolds not only enhanced their antibacterial properties but also promoted the osteogenic differentiation of bone marrow-derived mesenchymal stem cells (BMSCs), making them ideal candidates for clinical applications.

The structural characterization of the scaffolds revealed a uniform porous architecture with interconnected pores ranging from 100 to 200 μm, which is optimal for cell adhesion, nutrient diffusion, and vascularization [28]. The addition of Cu-MOF-74 increased the hydrophilicity of the scaffolds, as evidenced by the reduction in water contact angles, a property known to enhance cell adhesion and proliferation [29]. The sustained release of Cu^{2+} ions from the scaffolds was observed over 28 days, with an initial burst release followed by a gradual release phase. This controlled release mechanism is critical for maintaining therapeutic copper concentrations while minimizing cytotoxicity [23,30]. The release kinetics of Cu^{2+} was influenced by the degradation of the PCL matrix and the hydrolysis of Cu-MOF-74, which is consistent with previous studies on MOF-based drug delivery systems [31].

The antibacterial properties of the scaffolds were evaluated against *S. aureus* and *E. coli*, two common pathogens associated with bone infections. The results demonstrated a concentration-dependent antibacterial effect, with the 1 Cu composite scaffolds showing a 90.07% reduction in *S. aureus* and an 80.03% reduction in *E. coli*. The antibacterial mechanism of Cu^{2+} ions involves the disruption of bacterial cell membranes, generation of reactive oxygen species, and interference with DNA and protein functions [32,33]. These findings are consistent with previous studies demonstrating the potent antibacterial activity of copper-doped biomaterials [19,20,34]. However, it is important to note that excessive copper release can lead to cytotoxicity, as observed in the 0.5% and 1% Cu-MOF-74 groups, which exhibited reduced cell viability and proliferation. Therefore, optimizing the concentration of Cu-MOF-74 is critical for balancing antibacterial activity and biocompatibility.

The osteogenic potential of the scaffolds was evaluated through ALP activity, calcium nodule formation, and the expression of osteogenic markers (BSP, OPN, and OCN). The 0.2% Cu-MOF-74 group exhibited the highest ALP activity and calcium deposition, indicating enhanced osteogenic differentiation of BMSCs. These findings are consistent with those of previous studies demonstrating the osteogenic effects of Cu^{2+}, which are known to promote angiogenesis and bone formation [35,36]. The upregulation of osteogenic markers, particularly BSP and OPN, further supports the role of Cu-MOF-74 in promoting bone regeneration. However, the 1% Cu-MOF-74 group showed reduced osteogenic activity, likely due to the cytotoxic effects of high copper concentrations. This highlights the importance of optimizing copper release to achieve therapeutic efficacy without compromising cell viability.

The multifunctionality of Cu-MOF-74/HAp/PCL scaffolds can be attributed to the synergistic effects of HAp and Cu-MOF-74. HAp provides a bioactive surface that promotes cell adhesion and osteogenic differentiation, while Cu-MOF-74 imparts antibacterial properties and enhances angiogenesis. These findings align with recent reports that Cu^{2+} ions promote angiogenesis [37,38]. The combination of these components in a 3D-printed scaffold offers a promising solution for bone defect repair, particularly in infected or compromised bone sites. However, further studies are needed to evaluate the long-term in vivo performance of these scaffolds, including their degradation kinetics, mechanical stability, and ability to promote vascularized bone regeneration.

In conclusion, our study demonstrates the potential of Cu-MOF-74/HAp/PCL composite scaffolds for bone defect repair and infection control. Optimized concentrations of Cu-MOF-74 (0.05–0.2%) can effectively balance antimicrobial and osteogenic properties, presenting a promising strategy for clinical applications in bone defect repair. While these in vitro assays highlight the osteogenic potential of Cu-MOF-74/HAp/PCL scaffolds and their capacity to control infection and promote bone regeneration, further research is necessary to evaluate their long-term in vivo performance. Future studies should focus on assessing degradation kinetics, mechanical stability, and the ability of these scaffolds to promote vascularized bone regeneration in vivo, alongside continued optimization of Cu-MOF-74 concentrations and comprehensive evaluation in animal models.

Author Contributions: Conceptualization, T.Z. and Q.N.; methodology, T.Z., W.W. and D.G.; software, Q.N.; validation, T.Z. and Q.N.; formal analysis, T.Z. and Y.L.; investigation, T.C.; data curation, D.Z. and X.M.; writing—original draft preparation, T.Z.; writing—review and editing, X.Z.; visualization, T.Z.; supervision, X.Z.; project administration, X.Z.; funding acquisition, X.Z. All authors have read and agreed to the published version of the manuscript.

Funding: This work was supported by the National Natural Science Foundation of China (No. 51377163) and the Research Start-Up Foundation at Northwest University (No. 363042005100).

Institutional Review Board Statement: This study was conducted in accordance with the guidelines of the Declaration of Helsinki and approved by the Institutional Animal Care and Use Committee (IACUC) of Northwest University (No: ACUC2021045).

Informed Consent Statement: Not applicable.

Data Availability Statement: The original contributions presented in this study are included in the article; further inquiries can be directed to the corresponding authors.

Conflicts of Interest: The authors declare no conflicts of interest.

References

1. Xue, N.; Ding, X.; Huang, R.; Jiang, R.; Huang, H.; Pan, X.; Min, W.; Chen, J.; Duan, J.-A.; Liu, P.; et al. Bone Tissue Engineering in the Treatment of Bone Defects. *Pharmaceuticals* **2022**, *15*, 879. [CrossRef] [PubMed]
2. Quarto, R.; Giannoni, P. Bone Tissue Engineering: Past-Present-Future. *Methods Mol. Biol.* **2016**, *1416*, 21–33. [CrossRef] [PubMed]
3. Schmidt, A.H. Autologous Bone Graft: Is It Still the Gold Standard? *Injury* **2021**, *52* (Suppl. S2), S18–S22. [CrossRef]
4. Betz, R.R. Limitations of Autograft and Allograft: New Synthetic Solutions. *Orthopedics* **2002**, *25*, s561–s570. [CrossRef]
5. Ryu, J.-H.; Kang, T.-Y.; Shin, H.; Kim, K.-M.; Hong, M.-H.; Kwon, J.-S. Osteogenic Properties of Novel Methylsulfonylmethane-Coated Hydroxyapatite Scaffold. *Int. J. Mol. Sci.* **2020**, *21*, 8501. [CrossRef]
6. Garcia-Gareta, E.; Coathup, M.J.; Blunn, G.W. Osteoinduction of Bone Grafting Materials for Bone Repair and Regeneration. *Bone* **2015**, *81*, 112–121. [CrossRef]
7. Haugen, H.J.; Lyngstadaas, S.P.; Rossi, F.; Perale, G. Bone Grafts: Which Is the Ideal Biomaterial? *J. Clin. Periodontol.* **2019**, *46*, 92–102. [CrossRef]
8. Luo, Y.; Liu, H.; Zhang, Y.; Liu, Y.; Liu, S.; Liu, X.; Luo, E. Metal Ions: The Unfading Stars of Bone Regeneration-from Bone Metabolism Regulation to Biomaterial Applications. *Biomater. Sci.* **2023**, *11*, 7268–7295. [CrossRef]
9. Todd, E.A.; Mirsky, N.A.; Silva, B.L.G.; Shinde, A.R.; Arakelians, A.R.L.; Nayak, V.V.; Marcantonio, R.A.C.; Gupta, N.; Witek, L.; Coelho, P.G. Functional Scaffolds for Bone Tissue Regeneration: A Comprehensive Review of Materials, Methods, and Future Directions. *J. Funct. Biomater.* **2024**, *15*, 280. [CrossRef]
10. Collins, M.N.; Ren, G.; Young, K.; Pina, S.; Reis, R.L.; Oliveira, J.M. Scaffold Fabrication Technologies and Structure/Function Properties in Bone Tissue Engineering. *Adv. Funct. Mater.* **2021**, *31*, 2010609. [CrossRef]
11. Turnbull, G.; Clarke, J.; Picard, F.; Riches, P.; Jia, L.; Han, F.; Li, B.; Shu, W. 3D Bioactive Composite Scaffolds for Bone Tissue Engineering. *Bioact. Mater.* **2018**, *3*, 278–314. [CrossRef] [PubMed]
12. Yuan, X.; Zhu, W.; Yang, Z.; He, N.; Chen, F.; Han, X.; Zhou, K. Recent Advances in 3D Printing of Smart Scaffolds for Bone Tissue Engineering and Regeneration. *Adv. Mater.* **2024**, *36*, 2403641. [CrossRef] [PubMed]
13. Mirkhalaf, M.; Men, Y.; Wang, R.; No, Y.; Zreiqat, H. Personalized 3D Printed Bone Scaffolds: A Review. *Acta Biomater.* **2023**, *156*, 110–124. [CrossRef]
14. Zhang, Z.; Wang, S.; Zhang, J. 3D-Printed Poly(ε-caprolactone) Scaffold Augmented With Mesenchymal Stem Cells for Total Meniscal Substitution: A 12- and 24-Week Animal Study in a Rabbit Model. *Am. J. Sports Med.* **2017**, *45*, 1497–1511. [CrossRef]
15. Liu, D.; Nie, W.; Li, D.; Wang, W.; Zheng, L.; Zhang, J.; Zhang, J.; Peng, C.; Mo, X.; He, C.L. 3D Printed PCL/SrHA Scaffold for Enhanced Bone Regeneration. *Chem. Eng. J.* **2019**, *362*, 269–279. [CrossRef]
16. Shakir, M.; Jolly, R.; Khan, A.A.; Ahmed, S.S.; Alam, S.; Rauf, M.A.; Owais, M.; Farooqi, M.A. Resol Based Chitosan/Nano-hydroxyapatite Nanoensemble for Effective Bone Tissue Engineering. *Carbohydr. Polym.* **2018**, *179*, 317–327. [CrossRef]
17. Steinmetz, S.; Wernly, D.; Moerenhout, K.; Trampuz, A.; Borens, O. Infection after Fracture Fixation. *EFORT Open Rev.* **2019**, *4*, 468–475. [CrossRef]
18. Wang, Y.; Zhang, W.; Yao, Q. Copper-based Biomaterials for Bone and Cartilage Tissue Engineering. *J. Orthop. Transl.* **2021**, *29*, 60–71. [CrossRef]
19. Lu, Y.; Li, L.; Zhu, Y.; Wang, X.; Li, M.; Lin, Z.; Hu, X.; Zhang, Y.; Yin, Q.; Xia, H.; et al. Multifunctional Copper-Containing Carboxymethyl Chitosan/Alginate Scaffolds for Eradicating Clinical Bacterial Infection and Promoting Bone Formation. *ACS Appl. Mater. Interfaces* **2018**, *10*, 127–138. [CrossRef]
20. Zhang, J.W.; Ye, X.L.; Li, W.H.; Lin, Z.F.; Wang, W.S.; Chen, L.L.; Li, Q.; Xie, X.B.; Xu, X.M.; Lu, Y. Copper-containing Chitosan-based Hydrogels Enabled 3D-printed Scaffolds to Accelerate Bone Repair and Eliminate MRSA-related Infection. *Int. J. Biol. Macromol.* **2023**, *240*, 124463. [CrossRef]
21. Timoshenko, R.V.; Vaneev, A.N.; Savin, N.A.; Klyachko, N.L.; Parkhomenko, Y.N.; Salikhov, S.V.; Majouga, A.G.; Gorelkin, P.V.; Erofeev, A.S. Promising Approaches for Determination of Copper Ions in Biological Systems. *Nanotechnol. Russ.* **2020**, *15*, 121–134. [CrossRef]
22. Pan, C.; Ji, Z.; Wang, Q.; Zhang, Z.; Wang, Z.; Li, C.; Lu, S.; Ge, P. Cuproptosis: Mechanisms, Biological Significance, and Advances in Disease Treatment-A Systematic Review. *CNS Neurosci. Ther.* **2024**, *30*, e70039. [CrossRef] [PubMed]
23. Gwon, K.; Lee, S.; Kim, Y.; Choi, J.; Kim, S.; Kim, S.J.; Hong, H.J.; Hwang, Y.; Mori, M.; Lee, D.N. Construction of a Bioactive Copper-based Metal Organic Framework-embedded Dual-crosslinked Alginate Hydrogel for Antimicrobial Applications. *Int. J. Biol. Macromol.* **2023**, *242*, 124840. [CrossRef]
24. Zhao, T.; Nie, S.; Luo, M.; Xiao, P.; Zou, M.; Chen, Y. Research Progress in Structural Regulation and Applications of HKUST-1 and HKUST-1 Based Materials. *J. Alloys Compd.* **2024**, *974*, 172897. [CrossRef]
25. Liu, J.; Tan, Y.; Shen, E.; Liu, B.; Tian, Y.; Liang, L.; Yan, X.; Wu, H. Highly Water-stable Bimetallic Organic Framework MgCu-MOF74 for Inhibiting Bacterial Infection and Promoting Bone Regeneration. *Biomed. Mater.* **2022**, *17*, 065026. [CrossRef]
26. Zheng, H.; Wang, D.; Sun, X.; Jiang, S.; Liu, Y.; Zhang, D.; Zhang, L. Surface Modified by Green Synthetic of Cu-MOF-74 to Improve the Anti-biofouling Properties of PVDF Membranes. *Chem. Eng. J.* **2021**, *411*, 128524. [CrossRef]

27. Baertl, S.; Rupp, M.; Alt, V. The DAIR-Procedure in Fracture-Related Infection-When and How. *Injury* **2024**, *55*, 111977. [CrossRef]
28. Loh, Q.L.; Choong, C. Three-dimensional Scaffolds for Tissue Engineering Applications: Role of Porosity and Pore Size. *Tissue Eng. Part B Rev.* **2013**, *19*, 485–502. [CrossRef]
29. Lima, A.C.; Mano, J.F. Micro-/Nano-Structured Superhydrophobic Surfaces in the Biomedical Field: Part I: Basic Concepts and Biomimetic Approaches. *Nanomedicine* **2015**, *10*, 103–119. [CrossRef]
30. Shendage, S.S.; Gaikwad, K.; Kachare, K.; Kashte, S.; Ghule, A.V. In Vitro and in Vivo Study of Copper-Doped Bioactive Glass for Bone Regeneration Application. *Mater. Chem. Phys.* **2024**, *313*, 128789. [CrossRef]
31. Cun, J.; Fan, X.; Pan, Q.; Gao, W.; Luo, K.; He, B.; Pu, Y. Copper-based Metal–organic Frameworks for Biomedical Applications. *Adv. Colloid Interface Sci.* **2022**, *305*, 102686. [CrossRef] [PubMed]
32. Chen, L.; Wu, X.; Qi, T.; Feng, K.; Pei, C. Light-Activatable Ionic and Photothermal Therapy by CuAu Nanorods for Treating Wounds and Keratitis Caused by Methicillin-Resistant Staphylococcus aureus. *ACS Appl. Nano Mater.* **2024**, *7*, 15932–15942. [CrossRef]
33. Hong, R.; Kang, T.Y.; Michels, C.A.; Gadura, N. Membrane Lipid Peroxidation in Copper Alloy-mediated Contact Killing of *Escherichia coli*. *Appl. Environ. Microbiol.* **2012**, *78*, 1776–1784. [CrossRef]
34. Zou, F.; Jiang, J.; Lv, F.; Xia, X.; Ma, X. Preparation of antibacterial and osteoconductive 3D-printed PLGA/Cu(I)@ZIF-8 nanocomposite scaffolds for infected bone repair. *J. Nanobiotechnol.* **2020**, *18*, 39. [CrossRef]
35. Lin, Z.; Cao, Y.; Zou, J.; Zhu, F.; Gao, Y.; Zheng, X.; Wang, H.; Zhang, T.; Wu, T. Improved Osteogenesis and Angiogenesis of a Novel Copper Ions Doped Calcium Phosphate Cement. *Mater. Sci. Eng. C* **2020**, *114*, 111032. [CrossRef]
36. Zhang, J.; Wu, H.; He, F.; Wu, T.; Zhou, L.; Ye, J. Concentration-dependent Osteogenic and Angiogenic Biological Performances of Calcium Phosphate Cement Modified with Copper Ions. *Mater. Sci. Eng. C Mater. Biol. Appl.* **2019**, *99*, 1199–1212. [CrossRef]
37. Diao, W.; Li, P.; Jiang, X.; Zhou, J.; Yang, S. Progress in Copper-based Materials for Wound Healing. *Wound Repair Regen.* **2024**, *32*, 314–322. [CrossRef]
38. Hou, X.; Wang, H.; Geng, W.; Yang, L.; Wang, J. The Effect of the Modified Starch with Side Chain on the Morphology of Copper Particles and the Antibacterial Properties of Starch/Copper Composite Material. *Int. J. Biol. Macromol.* **2024**, *283*, 137488. [CrossRef]

Journal of
Functional Biomaterials

Article

Impact of Vitamin D_3 Functionalization on the Osteogenic Capacity of Bioinspired 3D Scaffolds Based on Ce-Doped Bioactive Glass and Spongia Agaricina

Ana-Maria Seciu-Grama [1,†], Sorana Elena Lazăr [2,3,†], Simona Petrescu [2], Oana Cătălina Mocioiu [2], Oana Crăciunescu [1] and Irina Atkinson [2,*]

[1] National Institute of Research and Development for Biological Sciences, 296, Spl. Independentei, 060031 Bucharest, Romania; ana.seciu@yahoo.com (A.-M.S.-G.); oana_craciunescu2009@yahoo.com (O.C.)
[2] "Ilie Murgulescu" Institute of the Physical Chemistry of the Romanian Academy, 202, Spl. Independentei, 060021 Bucharest, Romania; slazar@icf.ro (S.E.L.); simon_pet@yahoo.com (S.P.); omocioiu@icf.ro (O.C.M.)
[3] Faculty of Chemical Engineering and Biotechnologies, National University of Science and Technology Politehnica of Bucharest, 060042 Bucharest, Romania
* Correspondence: irinaatkinson@yahoo.com
† These authors contributed equally to this work.

Abstract: Reconstruction of extensive bone defects due to age, trauma, or post-illness conditions remains challenging. Biomimetic scaffolds with osteogenic capabilities have been proposed as an alternative to the classical autograft and allograft implants. Three-dimensional scaffolds were obtained based on Ce-doped mesoporous bioactive glass (MBG) and *Spongia agaricina* (SA) as sacrificial templates functionalized with vitamin D_3. The study aimed to investigate the effect of vitamin D_3 functionalization on the optimal variant of a 3D scaffold doped with 3 mol% ceria, selected in our previous work based on its biological and physicochemical properties. Scanning electron microscopy (SEM) images of the non-functionalized/functionalized scaffolds revealed a porous structure with interconnected pores ranging from 100 to 350 μm. Fourier transform infrared spectroscopy (FTIR) and SEM analysis confirmed the surface functionalization. Cytotoxicity evaluation showed that all investigated scaffolds do not exhibit cytotoxicity and genotoxicity toward the Saos-2 osteosarcoma cell line. Moreover, the study demonstrated that functionalization with vitamin D_3 enhanced osteogenic activity in dental pulp stem cells (DPSCs) by increasing calcium deposition and osteocalcin secretion, as determined by Alizarin red stain and a colorimetric ELISA kit, as a result of its synergistic action with cerium ions. The results showed that the Ce-doped MBG scaffold functionalized with vitamin D_3 had the potential for applications in bone regeneration.

Keywords: mesoporous bioactive glass; tissue scaffolds; cholecalciferol; ceria

1. Introduction

Tissue engineering aims to develop novel materials and methods for repairing and regenerating traumatic, injured, diseased, or aging bone. While autografts remain the gold standard for bone healing, their clinical use is limited by donor site morbidity and restricted availability. Bone allografts come with high costs, potential disease transmission, and the risk of adverse immune responses [1].

Metallic alloys are primarily used as implants for large bone defects due to their mechanical strength. However, these materials can have fixation issues and, unlike natural bone, cannot self-repair or adapt to changes in physiological conditions [2]. The drawbacks of current treatments have driven scientific efforts to design biomaterials with enhanced osteogenesis, a critical area of research.

Bioactive glasses (BGs) have emerged as a promising solution to address the limitations of current bone graft materials since the discovery of 45S5-BG by Prof. Hench [3], which is considered a scaffold for bone repair [4,5]. BGs are well known for their capacity to promote bone cell growth [6,7] and form strong bonds with both hard and soft tissues [8,9]. Following implantation, these glasses undergo specific processes that result in the formation of either amorphous calcium phosphate (ACP) or crystalline hydroxyapatite (HAP) on their surface, which is crucial for their integration with the surrounding tissue [10]. Furthermore, BG are reported to release ions that trigger the expression of osteogenic genes [11,12] and stimulate angiogenesis [13,14].

However, BGs can be doped with various therapeutic elements to improve their biological properties. Rare earth elements, particularly Cerium, have been proven to promote the osteogenic differentiation of mesenchymal stem cells. Furthermore, the addition of cerium to BGs can provide anti-inflammatory, antibacterial, and antioxidant properties [15,16]. H. Mostajeran et al. [17] reported that cerium-doped BG within a polymeric matrix of an alginate-gel scaffold provided an adequate environment for the osteogenic differentiation of human bone marrow-derived mesenchymal stem cells. The paper [18] reported that the scaffold obtained from cerium-doped hydroxyapatite and cashew gum exhibits anti-inflammatory and healing properties, promising features for replacing small bone parts. Our previous study [19] produced cerium-containing MBG scaffolds (0, 1, 3 mol% ceria) using SA as a sacrificial template. Cerium-doped MBG scaffolds, especially those doped with 3 mol% ceria, demonstrated antibacterial activity, osteogenic differentiation, and non-cytotoxicity.

In addition to therapeutic ions, BG-based scaffolds can release other agents from their structure, which can stimulate tissue regeneration. Recent research has focused on surface functionalization of these scaffolds using biomolecules (e.g., growth factors (bone morphogenetic protein-2 (BMP-2), fibroblast growth factors), proteins (fibronectin derived peptides, osteonectin), sterols (liposomes, phytosterols), and hormones (calcitonin, Vitamin D_3) [20] to create multifunctional materials that combine the properties of both the scaffolds and grafted molecules.

Vitamin D_3 (cholecalciferol), a fat-soluble vitamin, plays a crucial role in maintaining bone structure and promoting calcium absorption, which is essential for the mineralization process, thus enhancing the overall strength and density of the regenerated bone [21]. Furthermore, vitamin D_3 influences the expression of genes involved in osteoblast differentiation, such as Runx2 and Osterix, thereby accelerating the maturation of osteoblasts and promoting faster bone formation. It also stimulates osteoclast resorption, thereby balancing bone remodeling [22]. It can modulate the immune response at the implantation site, reducing inflammation and promoting a favorable environment for bone healing [23]. Deficiency in vitamin D_3 is associated with various bone disorders, including osteoporosis, rickets, and impaired fracture healing [24].

Due to its crucial role in bone metabolism, integrating vitamin D_3 into biomaterials for bone regeneration has been proposed as a method to enhance the osteogenic potential of biomaterials, such as BGs [25].

Few studies in the research literature have reported the functionalization of BG-based scaffolds with vitamin D_3. M.E. Abad-Javier et al. reported [26] the positive effect of vitamin D_3 on the bioactivity of the BG scaffold by accelerating the hydroxyapatite condensation on the surface. I. Negut et al. [27] demonstrated the feasibility of using thin films obtained based on BG and vitamin D_3 to enhance osseointegration and provide corrosion protection for Ti-like implant surfaces. A. A. Gupta [28] studied a biodegradable chitosan scaffold for the sustained release of vitamin D_3 and demonstrated its ability to deliver functional vitamin D_3 over 3–5 days without significant cytotoxic effects. Many studies emphasize in vitro analysis, leaving a gap for in vivo data regarding biological responses, immune reactions, and bone healing, which are essential for clinical applications.

Building on the mineralization and osteogenic differentiation properties of cerium and the crucial role of vitamin D_3 in calcium and phosphate metabolism, this research objective is to functionalize with vitamin D_3 the optimal three-dimensional scaffold variant doped with 3 mol% ceria investigated in our earlier work and to explore the synergistic effect of cerium and vitamin D_3 on the osteogenic differentiation capacity of bioactive scaffolds, an underexplored strategy in tissue engineering.

2. Materials and Methods

2.1. Materials

Tetraethylorthosilicate (Merck, Darmstadt, Germany), Pluronic® P-123 (Sigma Aldrich, St. Louis, MO, USA), triethyl phosphate (Sigma Aldrich, St. Louis, MO, USA), calcium nitrate tetrahydrate (Carl Roth, Karlsruhe, Germany), and cerium nitrate hexahydrate (Sigma Aldrich, St. Louis, MO, USA) were used as starting materials for the synthesis of MBGs. A marine sponge (SA) (PureSponges.co.uk, Solihull, UK) was used as a sacrificial template for scaffold preparation.

2.2. Scaffold Preparation

The template replica method was employed for 3D scaffold preparation based on Ce-doped MBGs in the $70SiO_2$-(26-y)CaO-$4P_2O_5$-$yCeO_2$ system (y denotes 3 mol% ceria) and SA, serving as template, as described in the paper [19]. Marine sponges are often used as templates because they closely resemble the architecture needed for bone scaffolds, in addition to other advantages such as biocompatibility, adjustable shape and size, and controlled degradation. The selection of 3 mol% ceria doping was based on our previous study [19].

In brief, SA was immersed in the prepared MBG solutions [19] for 15 min. Excess solution was removed by squeezing the sponge, and this process was repeated four times The green scaffold was then left at room temperature for 24 h followed by thermal treatment in two stages: I stage, heating to 700 °C, for 1 h, at a rate of 2 °C/min, and II stage up to a maximum temperature of 1200 °C at a rate of 5 °C/min.

2.3. Scaffold Functionalization

The scaffold surface was functionalized by immersion in a 0.03 mg/mL vitamin D_3 [26] (Sigma-Aldrich, St. Louis, MO, USA) solution in phosphate buffer (PBS) at pH 7, subjected to continuous agitation for 20 min, at 150 rpm and 25 °C in light-protected tubes, and subsequently washed with distilled water to remove inadequately adhered vitamin D_3. The scaffold was labeled as follows: S3Ce for the non-functionalized scaffold and S3CeD3 for the functionalized scaffold with vitamin D_3.

2.4. Characterization of Functionalized/Non-Functionalized Scaffolds

Non-functionalized and functionalized Ce-doped MBG-based scaffolds were evaluated using an FEI Quanta 3D FEG microscope (FEI, Brno, Czech Republic) operating in the 5–30 kV range in high-vacuum mode coupled with energy-dispersive X-ray spectroscopy (EDS) and Fourier transform infrared spectroscopy (FTIR) conducted using a 6700 Nicolet Spectrometer (Thermo Fisher Scientific, Waltham, MA, USA).

2.5. Evaluation of Biological Properties

The influence of vitamin D_3 functionalization on the biological properties of the scaffold was assessed through in vitro assays.

2.5.1. In Vitro Cytocompatibility Testing

The cytocompatibility of non-functionalized and functionalized Ce-doped MBG-based scaffolds was evaluated by the extract method in the human Saos-2 osteosarcoma cell culture from the European Collection of Cell Cultures (ECACC).

Sample preparation. Before testing, the samples were washed with 75% ethanol and sterilized under UV light for 2 h. The scaffold extracts were prepared by incubating the samples in DMEM: F12 culture medium for 72 h at 37 °C. The extraction medium (25 mg/mL) was centrifuged for 5 min at 1500 rpm. The supernatant was collected, and 6 concentrations (namely 1.25, 2.5, 6.25, 12.5, 18.75, and 25 mg/mL) were tested for each sample.

Cell culture and treatment. SaOS-2 cells were cultured in T75 flasks with McCoy 5A culture medium until they achieved 80% confluence. Then, the trypsinized cells were transferred to 96-well culture plates at a density of 5×10^4 cells/mL. Cells were then cultivated in McCoy 5A culture medium, maintaining a standard environment of 5% CO_2 at 37 °C for 24 h to ensure optimal cell adhesion. After this period, the culture medium was replaced with a fresh medium containing extract concentrations ranging from 1.25 to 25 mg/mL. The plates were subsequently incubated under standard conditions for 48 and 96 h.

Cell viability assessment by MTT assay. After replacing the culture medium, the plates were incubated with 100 µL of 0.25 mg/mL MTT solution at 37 °C for 3 h. After incubation, 100 µL of isopropanol was added to the plates and the mixture was stirred for 15 min. A Sunrise microplate reader (Tecan, Grödig, Austria) was used to determine the optical density (OD) at 570 nm. A 100 µM H_2O_2 solution was used as a positive control. The obtained results were presented as the percentage of viable cells relative to the control sample, which consisted of cells incubated without extract and was considered to have 100% viability. The number of metabolically active cells was directly correlated with the results obtained.

Live/Dead cell viability assay. The viability of the cells was also evaluated employing the Live/Dead® viability/cytotoxicity kit (Molecular Probes, Thermo Fisher Scientific, Waltham, MA, USA), as described in [29]. This method allows for dual staining of live cells with calcein (green fluorescence) and dead cells with ethidium bromide (red fluorescence). The cells were washed with PBS and treated with a solution of PBS containing 20 µM calcein-AM and 5 µM ethidium homodimer-1, following the manufacturer's instructions after each incubation. Under standard conditions, the plates were incubated for 20 min at 37 °C. An inverted fluorescence microscope (Axio Observer D1, Carl Zeiss, Oberkochen, Germany) with a digital camera was used to capture the images. Calcein-AM staining suggests intracellular esterase activity, whereas ethidium homodimer-1 staining is linked to plasma membrane integrity.

2.5.2. Genotoxicity Assessment

Genotoxicity testing was performed on the murine fibroblast cell line NCTC clone L929, cultured in MEM culture medium supplemented with 10% fetal bovine serum and 1% antibiotics, under standard conditions (37 °C temperature and 5% CO_2 atmosphere) [30]. L929 cells were seeded at a density of 5×10^4 cells/mL and allowed to adhere overnight. At 24 h after seeding, the cells were treated with the test samples at 18.75 mg/mL concentration. Following another 24 h of treatment, the cells were trypsinized and re-suspended in PBS.

The positive control was incubated with 100 μM H_2O_2. The resulting cell suspension was mixed with a 1% LMA (low melting point agarose) solution and added onto a slide previously coated with an agarose gel. A final layer of 1% LMA agarose was added, and the prepared slides were immersed in a lysis buffer at 4 °C. The slides were deposited in an electrophoresis tank and incubated in migration buffer (1 mM EDTA, 0.3 M NaOH) to denature DNA in an alkaline environment. Migration was carried out at 0.74 V/cm for 20 min, followed by washing with Tris neutralization buffer (pH 7.5). The slides were stained with DAPI and analyzed using an Axio Observer D1 Carl Zeiss fluorescence microscope (Oberkochen, Germany) with visualization performed using a 20× objective. The median comet score, expressed as a percentage of DNA tail for each sample, was determined using Comet Score Pro v.1.5 software.

2.5.3. In Vitro Osteogenic Capacity Testing

For the in vitro evaluation of osteogenic capacity, a cryopreserved culture of dental pulp stem cells (DPSCs) at passage 9 was used. DPSCs were isolated from third molars obtained from specialized clinics during the mandatory surgical or orthodontic treatment of healthy participants, with their informed consent, in accordance with the ethical regulations. The cells were preserved in liquid nitrogen until use. For the experiments, the DPSCs were cultivated in T75 flasks with α-MEM medium, 10% fetal bovine serum (FBS), 100 μg/mL streptomycin, and 100 U/mL penicillin. Until they reached 80% confluence, they were incubated in a humidified atmosphere with 5% CO_2 and 95% air at 37 °C. These cells exhibited stem-like properties, including self-renewal, surface antigen expression, and the ability to differentiate into multiple lineages, such as osteogenic, chondrogenic, and adipogenic [31].

DPSCs were placed in a 24-well plate cell culture at a density of 5×10^4 cells/mL and cultured in α-MEM with 10% fetal bovine serum (FBS) until confluence. Subsequently, the culture medium was changed with a certain osteogenic medium (Gibco, Thermo Fisher Scientific, Waltham, MA, USA) containing extract concentrations ranging from 12.5 to 18.75 mg/mL, followed by incubation under standard conditions for 14 and 21 days at 37 °C. The osteogenic marker was quantified at specific times over 21 days to evaluate the metabolic activity of the differentiated cells. A negative control group was cultured in osteogenic conditions in the absence of the extract.

Alizarin Red S staining. The formation of mineralized matrix nodules was evaluated using Alizarin Red S staining, following the protocol described in the paper, after 21 days of incubation [31]. Images were captured using an Axio Observer D1 optical microscope (Carl Zeiss, Oberkochen, Germany). Alizarin Red S-stained cells were rinsed with 2 M NaCl and incubated with 10% N-cetyl pyridinium chloride under gentle stirring for 10 min. The OD at 562 nm of the dye extract was measured using a Sunrise microplate reader (Tecan, Grödig, Austria).

Calcium quantification. The calcium colorimetric test kit (Sigma-Aldrich) was used to measure the concentration of calcium ions. In brief, 50 μL of conditioned medium collected at specific intervals was combined with 90 μL of o-cresolphthalein reagent and 60 μL of buffer to create a chromogenic complex. Afterward, the resulting mixture was incubated in the dark at room temperature for 10 min. The OD at 575 nm was measured using a Sunrise microplate reader (Tecan, Grödig, Austria). The concentration of calcium was determined based on a standard curve prepared with a 500 mM standard calcium solution.

Osteocalcin quantification. The osteocalcin concentration was measured using an ELISA kit (R&D Systems, Minneapolis, MN, USA). The concentration was ascertained using the filtrate collected following 14 and 21 days of cell treatment. OD at 450 nm was performed using a Sunrise microplate reader (Tecan, Grödig, Austria).

2.6. Statistical Analysis

Cell culture experiments were conducted in triplicate for each tested sample. Data are presented as mean ± standard deviation (SD) (n = 3). Statistical analysis was performed using a two-tailed, two-sample equal variance Student's t-test (Excel v. 10). Statistically significant differences were considered at $p < 0.05$.

3. Results and Discussion

3.1. Surface Characterization of Functionalized/Non-Functionalized Ce-Containing MBG-Based Scaffolds

The FTIR spectra of vitamin D_3 along with those of the 3Ce MBG-based scaffold, before and after functionalization, are presented in Figure 1. In the vitamin D_3 spectrum, several characteristic absorption bands were observed, indicating the presence of the functional groups: C=O stretching vibration of the aldehyde group at 1723 cm^{-1}, while the C–O stretching was evident at 2352 cm^{-1}. Additionally, alkyl C–H stretching vibrations were observed at 2944 cm^{-1} and 2853 cm^{-1}, corresponding to the aliphatic hydrocarbon chains of vitamin D_3. The broad band at 3420 cm^{-1} was attributed to hydrogen-bonded O–H stretching, suggesting the presence of hydroxyl groups [32].

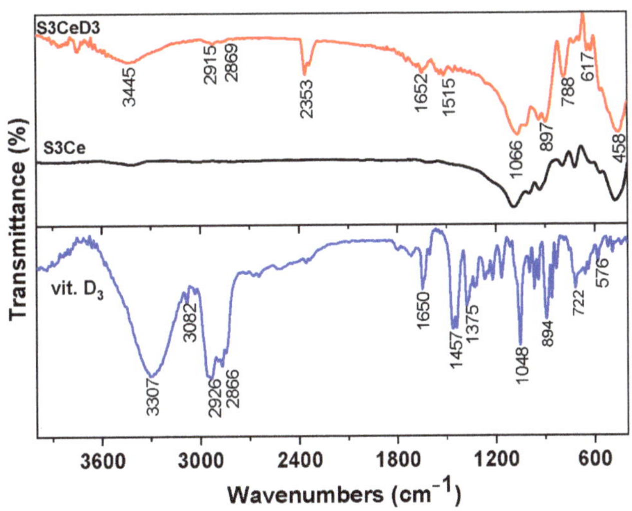

Figure 1. FTIR analysis of vitamin D_3, S3Ce and S3CeD3 scaffolds.

Furthermore, the spectral region between 1645 cm^{-1} and 900 cm^{-1} exhibited multiple absorption peaks associated with C–H bending vibrations, indicative of the complex molecular structure of vitamin D_3 [33]. In comparison, the FTIR spectra of the S3Ce scaffold before functionalization primarily displayed characteristic silicate-related vibration bands, including Si–O–Si asymmetric stretching at around 1080 cm^{-1} and Si-OH vibrations at 950 cm^{-1}, as well as a phosphate group [34]. However, after vitamin D_3 functionalization, the scaffold exhibited additional bands corresponding to the molecular fingerprint of vitamin D_3, confirming the successful surface modification of the scaffold. The band at 2353 cm^{-1} is attributed to CO_2 included in the vitamin D_3 film after its exposure to air for some time, as confirmed by [35]. Characteristics bands of phosphate and silicate groups remained unchanged in the S3CeD3 scaffold, indicating that the structural integrity of the scaffold was preserved after functionalization.

SEM images and EDX analysis further validated the surface modification (Figure 2). The SEM image of the S3Ce scaffold (Figure 2a) revealed a highly porous architecture with interconnected pores ranging from 100 to 350 µm. The scaffold demonstrated an open porosity of 87.67%, as revealed by micro-CT analysis in our previous study [19]. Porosity, interconnectivity, and pore size play a crucial role in tissue regeneration due to their direct impact on extracellular matrix formation and vascularization. Three-dimensional scaffolds with pore sizes ranging from 50 to 700 µm are commonly used in bone tissue engineering [36]. No significant changes in porosity were observed following the functionalization with vitamin D_3.

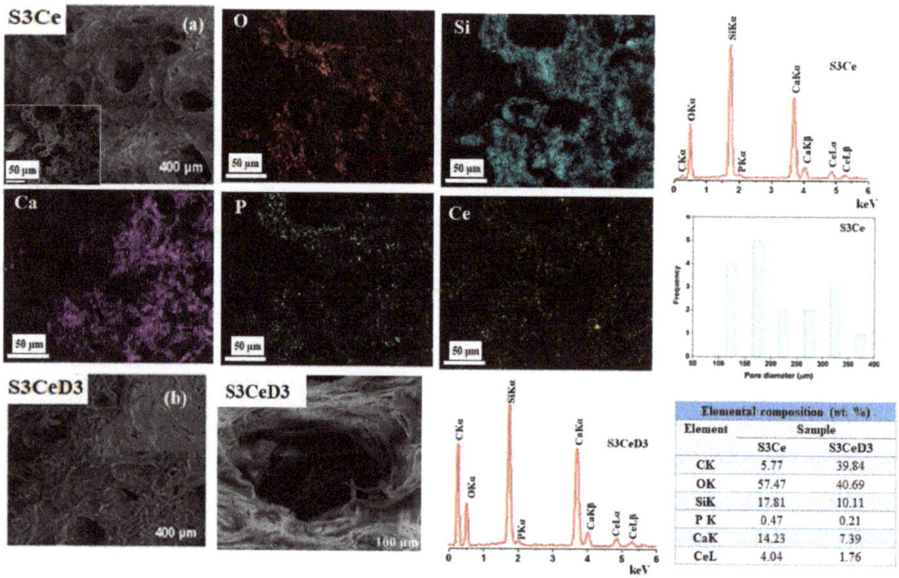

Element	S3Ce	S3CeD3
CK	5.77	39.84
OK	57.47	40.69
SiK	17.81	10.11
PK	0.47	0.21
CaK	14.23	7.39
CeL	4.04	1.76

Figure 2. SEM image of the S3Ce scaffold, EDX mapping for O, Si, Ca, P, and Ce, EDX analysis of the S3Ce scaffold surface and pore size distribution of S3Ce scaffold (**a**); SEM image at different magnifications, EDX analysis of the S3CeD3 scaffold and elemental composition of the investigated scaffolds (**b**).

The primary components of the S3Ce scaffold, as observed by EDX elemental mapping and analysis (Figure 2a), include calcium, phosphorus, silicon, and cerium. Notably, the EDX spectrum of the S3CeD3 scaffold (Figure 2b) showed an increased carbon signal generated by the carbon structure within the vitamin D_3 molecule, confirming the functionalization of the scaffold surface.

3.2. *Cytocompatibility Testing*

The cytocompatibility of MBG-based scaffolds was analyzed using the MTT assay at concentrations ranging from 1.25 to 25 mg/mL, with cultivation periods of 48 and 96 h. The obtained results are presented in Figure 3.

The data show that treatment with the S3Ce scaffold within the concentration range of 1.25–12.5 mg/mL resulted in high cell viability of SaOS-2 osteoblasts, with values exceeding that of the untreated culture (control, 100%), at both 48 (Figure 3a) and 96 h (Figure 3b) of incubation. A slight decrease in cell viability was observed at 96 h, but the values remained above 80%, indicating good cytocompatibility at concentrations of 18.75 mg/mL and 25 mg/mL.

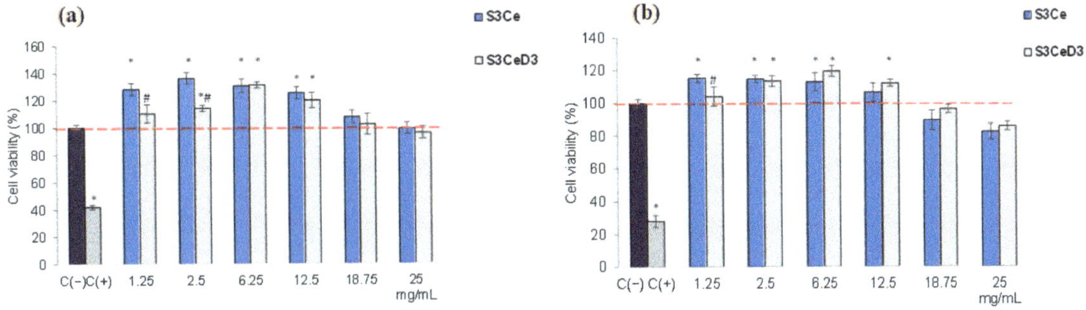

Figure 3. SaOS-2 osteoblast viability assessed by MTT assay in different concentrations of S3Ce and S3CeD3 scaffold extracts after 48 h (**a**) and 96 h (**b**) of cultivation. The results were obtained as the mean of three determinations ± SD and presented relative to the control culture (cells cultivated in the absence of the sample), which was considered 100% viable; * $p < 0.05$, compared to the control. # $p < 0.05$, compared to S3Ce.

The incubation of SaOS-2 osteoblasts with a D_3 functionalized scaffold resulted in high cell viability, with values exceeding that of control (100%), in the concentration range of 1.25 to 12.5 mg/mL, at 48 and 96 h (Figure 3). At higher concentrations, the cell viability remained above 80%, with values of 96.1% for 18.75 mg/mL and 85.6% for 25 mg/mL. Additionally, treatment with the S3CeD3 scaffold yielded the highest cell viability values across the concentration range of 6.25–18.75 mg/mL. Specifically, after 96 h of treatment, the viability values were 119.46% (6.25 mg/mL), 112.06% (12.5 mg/mL), and 96.1% (18.75 mg/mL) for the S3CeD3 functionalized scaffold, higher than those recorded for the non-functionalized S3Ce scaffold of 112.94% (6.25 mg/mL), 106.37% (12.5 mg/mL) and 89.24% (18.75 mg/mL). However, the values were not statistically different ($p > 0.05$).

The data obtained in the present study are consistent with other findings on vitamin D_3-functionalized biomaterials. Bose et al. [37] reported that 3D-printed calcium phosphate scaffolds functionalized with vitamin D_3 and quercetin showed a 1.3- and 1.6-fold increase in osteoblast cell proliferation and differentiation and reduced osteoclast activity, respectively.

The viability and morphology of SaOS-2 osteoblasts cultured with 3SCe and 3SceD3 scaffolds were analyzed using the Live/Dead assay after 96 h of incubation. The results presented in Figure 4 indicated that the cells treated with the tested scaffolds exhibited high viability, as evidenced by the green fluorescence of the calcein-stained live cells. The cells were uniformly distributed throughout the cultivation period and maintained a morphology similar to that of the control group. However, the cell density decreased in all analyzed scaffolds at concentrations of 18.75 mg/mL and 25 mg/mL. Despite this decrease, the cells preserved a normal morphology and uniform distribution, with only a small number of dead cells (indicated by red staining) detected.

3.3. Genotoxicity Testing

Considering the reported research that excessive amounts of vitamin D_3 can lead to harmful effects on the body [38], such as alterations in cell cycle regulatory pathways [39] and increased pro-oxidation [40], the genotoxicity analysis of the scaffolds was conducted using the Comet assay (alkaline single-cell gel electrophoresis). This method is based on the property of denatured DNA fragments to migrate out of the cell under the influence of an electric field, forming the so-called "comet tails". The larger the comet size, the greater the number of lesions in the DNA macromolecules.

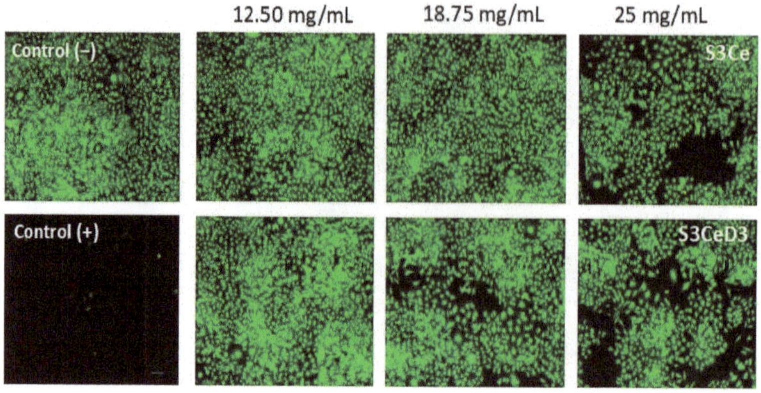

Figure 4. Fluorescence microscopy images indicating the SaOS-2 osteoblast viability after 96 h of cultivation in the presence of S3Ce and S3CeD3 scaffolds obtained using the Live and Dead assay.

Figure 5 presents the results of the genotoxicity analysis. Specific comet formations were observed only in the positive control group treated with 100 μM H_2O_2. In contrast, no comet formations were detected in the cells treated with the investigated scaffolds, and the results were comparable to those of the untreated culture control.

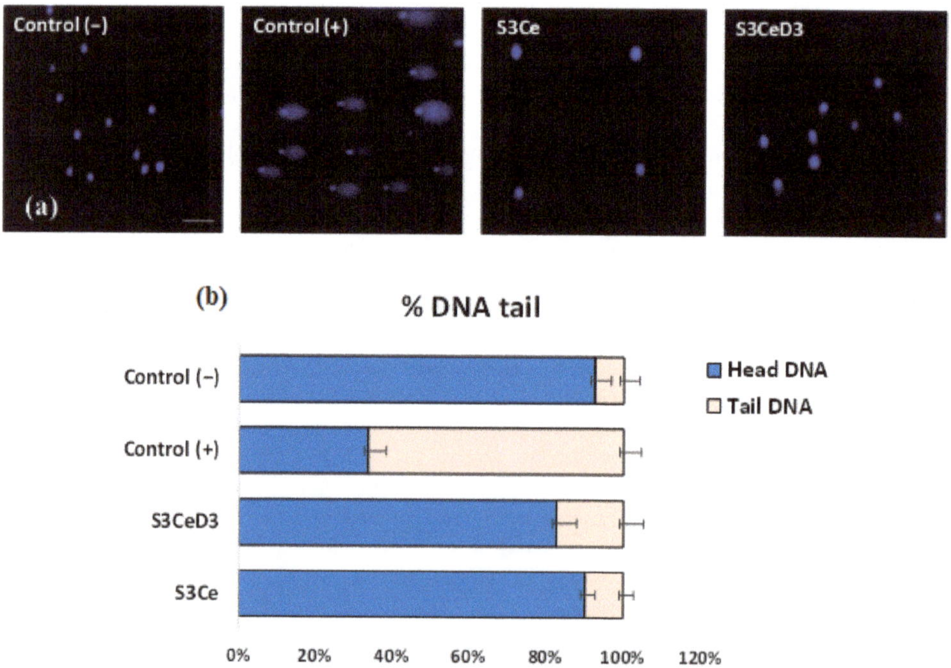

Figure 5. Fluorescence micrographs of DAPI-stained SaOS-2 osteoblasts cultivated in the presence of S3Ce and S3CeD3 scaffolds for 24 h (**a**) and histogram of the percentage of DNA tail in comet cells determined using Comet Score Pro v.1.5 software (**b**). The results are presented as the mean ± SD (n = 3). Untreated cells were used as a negative control (C (−)). Cells treated with 100 μM H_2O_2 were used as a positive control (C (+)). Scale bar = 100 μm.

Using the Comet Score v. 1.5 software, the size of the comets obtained after electrophoresis in an alkaline medium was quantified. Figure 5b presents the parameter %

DNA tail, representing the amount of damaged cellular DNA present in the comet's "tail". The results confirmed the qualitative observations and showed that the treatment with the scaffolds did not affect the treated cells, as values not significantly ($p > 0.05$) different from those of the untreated culture control were obtained.

3.4. Osteogenic Capacity of Functionalized Scaffolds

This study on the osteogenic capacity of functionalized scaffolds aimed to provide insights into the mechanisms of biomaterial-mediated bone regeneration in a culture of osteogenic, differentiated DPSCs. Thus, the influence of the S3Ce and S3CeD3 scaffolds on the extracellular matrix mineralization of DPSCs was analyzed after 21 days of incubation. The calcium deposits in treated cultures were qualitatively and quantitatively evaluated using Alizarin Red S staining. In the untreated culture (control), Alizarin Red S staining resulted in partial red coloration of the extracellular matrix of the DPSCs, indicating the presence of small calcium or mineralized deposits. In contrast, the treated culture exhibited an intense red coloration of the extracellular matrix, particularly at 18.75 mg/mL for both non-functionalized and functionalized scaffolds (Figure 6a).

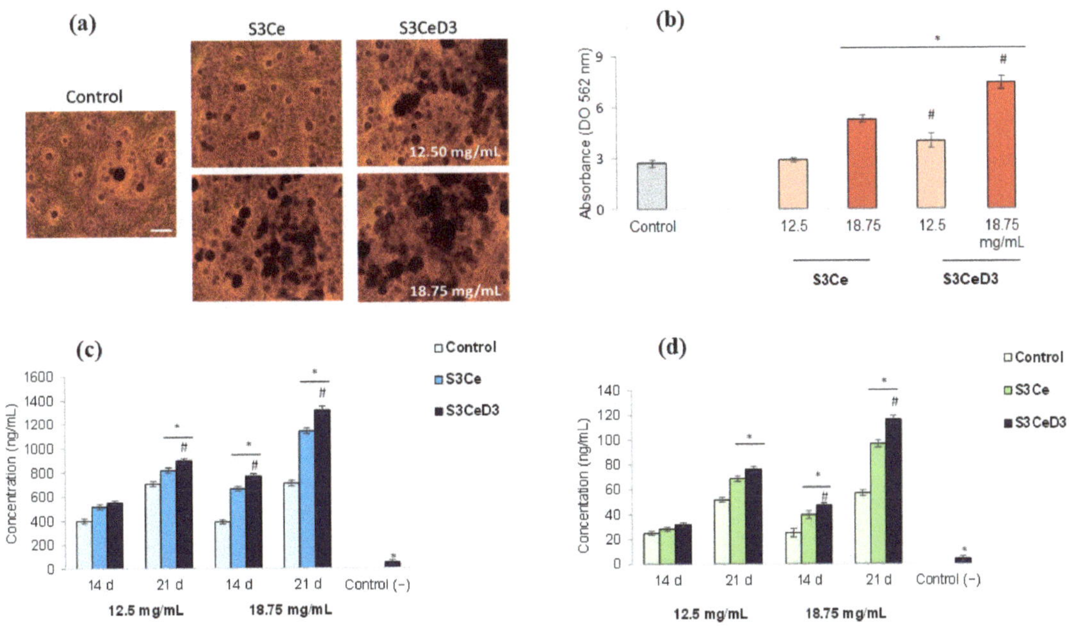

Figure 6. Micrographs of DPSC cultivated in osteogenic medium for 21 days in the presence of S3Ce and S3CeD3 scaffolds showing (**a**) mineralization nodules stained with Alizarin Red S and (**b**) quantitative data based on absorbance at 562 nm. * $p < 0.05$ compared to control, # $p < 0.05$ compared to S3Ce. Scale bar = 100 µm. Calcium (**c**) and osteocalcin (**d**) concentrations secreted by cells osteogenically differentiated in the presence of non-functionalized scaffolds for 14 and 21 days, estimated with a colorimetric kit and ELISA kit, respectively. Negative control: cells cultivated only in osteogenic medium. Values are expressed as the mean ± SD ($n = 9$), * $p < 0.05$ compared to control, # $p < 0.05$ compared to S3Ce.

Additionally, the quantitative analysis of calcium deposit formation in the presence of functionalized MBG scaffolds indicated their increase in a dose-dependent manner, confirming the stimulation of osteogenic differentiation of stem cells. Thus, the obtained results showed that cells that osteogenically differentiated in the presence of both functionalized (S3CeD3) and non-functionalized (S3Ce) scaffolds formed significantly ($p < 0.05$) higher amounts of calcium deposits, compared to the untreated culture of stem cells, except for 12.5 mg/mL S3Ce (Figure 6b). At a concentration of 12.5 mg/mL in the presence of a S3CeD3 functionalized scaffold, an increase in calcium deposit formation 1.49 times higher than that of the untreated control was observed. At a concentration of 18.75 mg/mL, calcium deposit formation was 1.98 times higher for the S3Ce scaffold and even higher for the functionalized S3CeD3 scaffold, with values 2.77 times higher compared to the untreated control. Moreover, it was observed that at the concentration of 18.75 mg/mL, the S3CeD3 functionalized scaffold exhibited a value 1.4 times higher compared to the S3Ce scaffold. These results suggest a synergistic effect of cerium content and vitamin D_3 functionalization, which enhanced the osteogenic capacity.

The late stage of the osteogenic differentiation of DPSCs (14–21 days) is characterized by the high expression of calcium, phosphate, and osteocalcin deposition [19]. The in vitro investigation of these markers could predict the osteogenic potential of the biomaterials after in vivo implantation. In the present study, the values obtained for the quantification of calcium ions in the culture medium of osteogenically differentiated cells in the presence of the S3CeD3 scaffold are shown in Figure 6c. After 21 days of cultivation, the S3CeD3 scaffold significantly increased calcium secretion by 1.15 times compared to the S3Ce scaffold at a concentration of 18.75 mg/mL ($p < 0.05$). This result indicates that vitamin D_3 significantly stimulated the osteogenic capacity of Ce-doped MBG scaffolds.

Osteocalcin is an osteogenic marker synthesized by mature osteoblasts. The concentration values of osteocalcin secretion by cells osteogenically differentiated in the presence of non-functionalized and functionalized scaffolds are presented in Figure 6d. The results show a similar concentration variation to that observed for calcium ion secretion at 14 and 21 days of cultivation. After 21 days, osteocalcin secretion was 1.2 times higher at a concentration of 18.75 mg/mL when using the functionalized S3CeD3 scaffold compared to the non-functionalized S3Ce scaffold.

The obtained results, an increase in calcium ion secretion by 1.15 times and osteocalcin levels by 1.2 times by the S3CeD3 functionalized scaffold, compared to the S3Ce scaffold after 21 days of incubation, align with other work showing that vitamin D_3-functionalized scaffolds enhance osteocalcin expression and mineralization via integrin-mediated signaling and MAPK/ERK pathway activation [41]. The osteogenic properties of this latter scaffold were positively modulated by the polyurethane foam templating, which provided a grain-like microtopography, upregulating the expression of osteogenic genes through the MAPK/ERK signaling pathway activated by the adsorption of fibronectin and the expression of integrin $\alpha5\beta1$ and focal adhesion kinase (p-FAK) [41]. Another study has revealed an increase in mineral deposition, as observed using Alizarin red staining, in human adipose-derived stem cells incubated with a polycaprolactone/gelatin scaffold on day 21 [42].

Osteogenesis is a complex process involving several transcription factors that govern the differentiation of mesenchymal cells into osteoblasts. The key transcription factor responsible for initiating the osteodensification of mesenchymal cells is runt-related transcription factor 2 (Runx2). Once these cells differentiate into pre-osteoblasts, Runx2, osterix, and β-catenin direct their development into immature osteoblasts that produce specific proteins [43]. The Runx2 expression must be decreased for bone maturation to continue because it inhibits osteoblast maturation by maintaining its immature state. Osteocalcin is the most abundant non-collagenous protein found in bones. It is considered a suitable marker for osteogenic maturation, and it is regarded as a late indicator of osteodifferentiation. Mature osteoblasts express high levels of osteocalcin when they are finally embedded in the bone matrix and transform into osteocytes [44].

The functionalization of the scaffold creates a biomimetic microenvironment that mimics the native extracellular matrix (ECM) of bone, fostering cell-matrix interactions crucial for osteogenesis.

The contribution of Ce doping to osteogenesis lies in its antioxidant properties, resulting from the presence of Ce^{3+}/Ce^{4+} at the scaffold surface, which can catalyze the dismutation reaction of hydrogen peroxide, thereby mitigating reactive oxygen species (ROS) and creating a favorable environment for osteoblast activity. Our previous study determined a Ce^{4+}/Ce^{3+} ratio of ~0.5 for the S3Ce scaffold using UV-Vis spectroscopy [19]. Jinhua Li et al. [45] reported a close relationship between the Ce^{4+}/Ce^{3+} ratio and bone formation, indicating that a higher Ce^{4+}/Ce^{3+} ratio promotes better osseointegration. Ce^{3+} scavenges ROS, protecting cells from apoptosis and enhancing their responsiveness to osteogenic stimuli like vitamin D_3.

Vitamin D_3 enhances osteogenic differentiation through multiple mechanisms, including the activation of key osteogenic markers such as RUNX2, osteocalcin, and alkaline phosphatase, regulation of calcium–phosphate homeostasis, stimulation of osteoblast differentiation, and modulation of signaling pathways, such as the β-catenin signaling pathway [46].

This synergistic interplay between Ce and vitamin D_3 may explain why the S3CeD3 scaffold exhibited superior performance: Ce scavenged reactive oxygen species (ROS), while vitamin D_3 directly stimulated osteogenic differentiation. Combining the antioxidant benefits of Ce with the osteogenic signaling of vitamin D_3 represents a promising approach. While our findings highlighted their potential to enhance the osteogenic differentiation of the scaffold, further studies are necessary to elucidate the mechanisms involved in osteogenesis and ultimately translate these findings into clinical applications.

4. Conclusions

Three-dimensional scaffolds were obtained by doping MBGs with 3 mol% ceria and SA and then functionalized with vitamin D_3.

SEM and FTIR investigations confirmed the successful functionalization of the S3CeD3 scaffold. In vitro testing confirmed the cytocompatibility of the functionalized scaffolds, promoting the proliferation of SaOS-2 osteoblasts in a higher proportion than the non-functionalized variants.

The cytotoxicity and genotoxicity evaluation confirmed the safe use of the investigated scaffolds and their potential for biomedical applications. An increase in calcium concentration by 1.15 times and osteocalcin levels by 1.2 times, as well as the formation of mineralization nodules, were observed in experiments on the osteogenic differentiation of DPSCs using the vitamin D_3-functionalized scaffold compared to a non-functionalized one. These results confirmed the synergistic effect of cerium doping and vitamin D_3 functionalization on MBG scaffolds, improving their osteogenic potential.

Future studies, particularly those utilizing in vivo animal models, should evaluate osteogenesis differentiation and confirm its potential clinical applications.

Author Contributions: Conceptualization I.A. and A.-M.S.-G.; investigation, formal analysis S.E.L., A.-M.S.-G., O.C.M., S.P. and O.C.; writing—original draft preparation, I.A. and A.-M.S.-G.; writing—review and editing, I.A. and A.-M.S.-G.; supervision, I.A. and A.-M.S.-G. All authors have read and agreed to the published version of the manuscript.

Funding: This research was partly funded by the Program Nucleu within the National Plan for Research, Development, and Innovation 2022–2027, implemented with the support of the Ministry of Education and Research, project no. 23020201 (contract 7N/2023). This research was also funded by Executive Agency for Higher Education, Research, Development and Innovation Funding (UEFISCDI), grant number PN-III-P2-2.1-PED-2019-0598, no. 258 PED/2020.

Institutional Review Board Statement: The study was conducted in accordance with the Declaration of Helsinki, and approved by the Ethics Committee of NIRDBS (no.756/02.10.2019).

Informed Consent Statement: Informed consent was obtained from all subjects involved in the study.

Data Availability Statement: The data presented in this study are contained within the article.

Conflicts of Interest: The authors declare no conflict of interest. The funders had no role in the design of the study; in the collection, analyses, or interpretation of data; in the writing of the manuscript, or in the decision to publish the results.

References

1. Sohn, H.S.; Oh, J.K. Review of bone graft and bone substitutes with an emphasis on fracture surgeries. *Biomater. Res.* **2019**, *23*, 9. [CrossRef] [PubMed]
2. Agarwal, R.; Garcia, A.J. Biomaterial strategies for engineering implants for enhanced osseointegration and bone repair. *Adv. Drug Deliv. Rev.* **2015**, *94*, 53–62. [CrossRef]
3. Hench, L.L. The story of Bioglass®. *J. Mater. Sci. Mater. Res.* **2006**, *17*, 967–973. [CrossRef]
4. Fiume, E.; Barberi, J.; Verné, E.; Baino, F. Bioactive Glasses: From Parent 45S5 Composition to Scaffold-Assisted Tissue-Healing Therapies. *J. Funct. Biomater.* **2018**, *9*, 24. [CrossRef] [PubMed]
5. Biano, F. Bioactive glasses-When glass science and technology meet regenerative medicine. *Ceram. Int.* **2018**, *44*, 14953–14966. [CrossRef]
6. Ciraldo, F.E.; Boccardi, E.; Melli, V.; Westhauser, F.; Boccaccini, A.R. Tackling bioactive glass excessive in vitro bioreactivity: Preconditioning approaches for cell culture tests. *Acta Biomater.* **2018**, *75*, 3–10. [CrossRef]
7. Piatti, E.; Miola, M.; Verné, E. Tailoring of bioactive glass and glass-ceramics properties for in vitro and in vivo response optimization: A review. *Biomater. Sci.* **2024**, *12*, 4546–4589. [CrossRef]
8. Ranjbar, F.E.; Farzad-Mohajeri, S.; Samani, S.; Saremi, J.; Khademi, R.; Dehghan, M.M.; Azami, M. Kaempferol-loaded bioactive glass-based scaffold for bone tissue engineering: In vitro and in vivo evaluation. *Sci. Rep.* **2023**, *13*, 12375. [CrossRef]
9. Krishnan, L.; Chakrabarty, P.; Govarthanan, K.; Rao, S.; Santra, T.S. Bioglass and nano bioglass: A next-generation biomaterial for therapeutic and regenerative medicine applications. *Int. J. Biol. Macromol.* **2024**, *277*, 133073. [CrossRef]
10. Kowalska, K.J.; Czechowska, J.P.; Yousef, E.S.; Zima, A. Novel phosphate bioglasses and bioglass-ceramics for bone regeneration. *Ceram. Int.* **2024**, *50 Pt A*, 45976–45985. [CrossRef]
11. Meng, L.; Zhao, P.; Jiang, Y.; You, J.; Xu, Z.; Yu, K.; Boccaccini, A.R.; Ma, J.; Zheng, K. Extracellular and intracellular effects of bioactive glass nanoparticles on osteogenic differentiation of bone marrow mesenchymal stem cells and bone regeneration in zebrafish osteoporosis model. *Acta Biomater.* **2024**, *174*, 412–427. [CrossRef] [PubMed]
12. Krishnamoorthy, E.; Subramanian, B. Synergistic effects of silica-enriched bioactive glass and tri-calcium phosphate nanocomposites on BMP2 gene expression for bone repair and regeneration applications. *Int. J. Pharm.* **2025**, *669*, 125026. [CrossRef] [PubMed]
13. Chen, X.; Ran, X.; Wei, X.; Zhu, L.; Chen, S.; Liao, Z.; Xu, K.; Xia, W. Bioactive glass 1393 promotes angiogenesis and accelerates wound healing through ROS/P53/MMP9 signaling pathway. *Regen. Ther.* **2024**, *26*, 132–144. [CrossRef]
14. Moll, M.; Scheurle, A.; Nawaz, Q.; Walker, T.; Kunisch, E.; Renkawitz, T.; Boccaccini, A.R.; Westhauser, F. Osteogenic and angiogenic potential of molybdenum-containing mesoporous bioactive glass nanoparticles: An ionic approach to bone tissue engineering. *J. Trace Elem. Med. Biol.* **2024**, *86*, 127518. [CrossRef]
15. Zheng, K.; Torre, E.; Bari, A.; Taccardi, N.; Cassinelli, C.; Morra, M.; Fiorilli, S.; Vitale-Brovarone, C.; Iviglia, G.; Boccaccini, A.R. Antioxidant mesoporous Ce-doped bioactive glass nanoparticles with anti-inflammatory and pro-osteogenic activities. *Mater. Today Bio* **2020**, *5*, 100041. [CrossRef] [PubMed]
16. Chen, Z.; Zhou, X.; Mo, M.; Hu, X.; Liu, J.; Chen, L. Systematic review of the osteogenic effect of rare earth nanomaterials and the underlying mechanisms. *J. Nanobiotechnol.* **2024**, *22*, 185. [CrossRef]
17. Mostajeran, H.; Baheiraei, N.; Bagheri, H. Effects of cerium-doped bioactive glass incorporation on an alginate/gelatin scaffold for bone tissue engineering: In vitro characterizations. *Int. J. Biol. Macromol.* **2024**, *255*, 128094. [CrossRef]
18. Beserra dos Santos, M.V.; Bastos Nogueira Rocha, L.; Gomes Vieira, E.; Oliveira, A.L.; Oliveira Lobo, A.; Martins de Carvalho, M.A.; Osajima, J.A.; Silva-Filho, E.C. Development of Composite Scaffolds Based on Cerium Doped-Hydroxyapatite and Natural Gums—Biological and Mechanical Properties. *Materials* **2019**, *12*, 2389. [CrossRef]
19. Atkinson, I.; Seciu-Grama, A.M.; Serafim, A.; Petrescu, S.; Voicescu, M.; Anghel, E.M.; Marinescu, C.; Mitran, R.A.; Mocioiu, O.C.; Cusu, J.P.; et al. Bioinspired 3D scaffolds with antimicrobial, drug delivery, and osteogenic functions for bone regeneration. *Drug Deliv. Transl. Res.* **2024**, *14*, 1028–1047. [CrossRef]

20. Szwed-Georgiou, A.; Płociński, P.; Kupikowska-Stobba, B.; Urbaniak, M.M.; Rusek-Wala, P.; Szustakiewicz, K.; Piszko, P.; Krupa, A.; Biernat, M.; Gazińska, M.; et al. Bioactive Materials for Bone Regeneration: Biomolecules and Delivery Systems. *ACS Biomater. Sci. Eng.* **2023**, *9*, 5222–5254. [CrossRef]
21. Wacker, M.; Holick, M.F. Vitamin D-Effects on Skeletal and Extraskeletal Health and the Need for Supplementation. *Nutrients* **2013**, *5*, 111–148. [CrossRef] [PubMed]
22. Van de Peppel, J.; Van Leeuwen, J.P.T.M. Vitamin D and gene networks in human osteoblasts. *Front. Physiol.* **2014**, *9*, 137. [CrossRef] [PubMed]
23. Ao, T.; Kikuta, J.; Ishii, M. The Effects of Vitamin D on Immune System and Inflammatory Diseases. *Biomolecules* **2021**, *11*, 1624. [CrossRef] [PubMed]
24. Christodoulou, S.; Goula, T.; Ververidis, A.; Drosos, G. Vitamin D and Bone Disease. *BioMed Res. Int.* **2012**, *27*, 396541. [CrossRef]
25. Sundar, R.; Rai, A.B.; Naveen Kumar, J.; Devang Divakar, D. The role of Vitamin D as an adjunct for bone regeneration: A systematic review of literature. *Saudi Dent. J.* **2023**, *35*, 220–232. [CrossRef]
26. Abad-Javier, M.E.; Cajero-Juárez, M.; Nuñez-Anita, R.E.; Contreras-García, M.E. Effect of collagen type I and vitamin D_3 functionalization of biomimetic bioglass scaffolds on hydroxyapatite condensation. *J. Eur. Ceram. Soc.* **2019**, *39*, 3505–3512. [CrossRef]
27. Negut, I.; Gradisteanu-Pircalabioru, G.; Dinu, M.; Bita, B.; Constantina Parau, A.; Grumezescu, V.; Ristoscu, C.; Chifiriuc, C.M. Bioglass and Vitamin D3 Coatings for Titanium Implants: Osseointegration and Corrosion Protection. *Biomedicines* **2023**, *11*, 2772. [CrossRef]
28. Gupta, A.A.; Kheur, S.; Badhe, R.V.; Raj, A.T.; Bhonde, R.; Jaisinghani, A.; Vyas, N.; Patil, V.R.; Alhazmi, Y.A.; Parveen, S.; et al. Assessing the potential use of chitosan scaffolds for the sustained localized delivery of vitamin D. *Saudi J. Biol. Sci.* **2021**, *4*, 2210–2215. [CrossRef]
29. Gaspar-Pintiliescu, A.; Oancea, A.; Cotarlet, M.; Vasile, A.M.; Bahrim, G.E.; Shaposhnikov, S.; Craciunescu, O.; Oprita, E.I. Angiotensin-converting enzyme inhibition, antioxidant activity and cytotoxicity of bioactive peptides from fermented bovine colostrum. *Int. J. Dairy Technol.* **2020**, *73*, 108–116. [CrossRef]
30. Craciunescu, O.; Seciu, A.M.; Zarnescu, O. In vitro and in vivo evaluation of a biomimetic scaffold embedding silver nanoparticles for improved treatment of oral lesions. *Mater. Sci. Eng. C* **2021**, *123*, 112015. [CrossRef]
31. Seciu, A.M.; Craciunescu, O.; Zarnescu, O. Advanced regenerative techniques based on dental pulp stem cells for the treatment of periodontal disease. In *Periodontology and Dental Implantology*; Manakil, J., Ed.; InTechOpen Ltd.: London, UK, 2020; pp. 129–148. [CrossRef]
32. Jan, Y.; Al-Keridis, L.A.; Malik, M.; Haq, A.; Ahmad, S.; Kaur, J.; Adnan, M.; Alshammari, N.; Ashraf, S.A.; Panda, B.P. Preparation, modelling, characterization and release profile of vitamin D3 nanoemulsion. *LWT* **2022**, *169*, 113980. [CrossRef]
33. Torrisi, A.; Cutroneo, M.; Torrisi, L.; Lavalle, S.; Forzina, A.; Pegreffi, F. Unveiling the Potential of Vitamin D3 Orodispersible Films: A Comprehensive FTIR and UV–Vis Spectroscopic Study. *Molecules* **2024**, *29*, 3762. [CrossRef] [PubMed]
34. Atkinson, I.; Seciu-Grama, A.M.; Petrescu, S.; Culita, D.; Mocioiu, O.C.; Voicescu, M.; Mitran, R.A.; Lincu, D.; Prelipcean, A.M.; Craciunescu, O. Cerium-Containing Mesoporous Bioactive Glasses (MBGs)-Derived Scaffolds with Drug Delivery Capability for Potential Tissue Engineering Applications. *Pharmaceutics* **2022**, *14*, 1169. [CrossRef] [PubMed]
35. Torrisi, L.; Roszkowska, A.M.; Silipigni, L.; Cutroneo, M.; Torrisi, A. Effects of 365nm UV lamp irradiation of polymethyl methacrylate (PMMA). *Radiat. Eff. Defects Solids* **2024**, *179*, 264–274. [CrossRef]
36. Alkentar, R.; Kladovasilakis, N.; Tzetzis, D.; Mankovits, T. Effects of pore size parameters of titanium additively manufactured lattice structures on the osseointegration process in orthopedic applications: A comprehensive review. *Crystals* **2022**, *13*, 113. [CrossRef]
37. Bose, S.; Chaudhari, V.C.; Kushram, P. 3D printed scaffolds with quercetin and vitamin D3 nanocarriers: In vitro cellular evaluation. *J. Biomed. Mater. Res. A* **2024**, *112*, 2110–2123. [CrossRef]
38. Owens, D.; Tang, J.C.Y.; Bradley, W.J.; Sparks, A.S.; Fraser, W.D.; Morton, J.P.; Close, G.L. Efficacy of high-dose vitamin D supplements for elite athletes. *Med. Sci. Sports Exerc.* **2017**, *49*, 349–356. [CrossRef]
39. Irving, A.A.; Plum, L.A.; Blaser, W.J.; Ford, M.R.; Weng, C.; Clipson, L.; DeLuca, H.F.; Dove, W.F. Cholecalciferol or 25-Hydroxycholecalciferol neither prevents nor treats adenomas in a rat model of familial colon cancer. *J. Nutr.* **2015**, *145*, 291–298. [CrossRef]
40. Koren, R.; Hadari-Naor, I.; Zuck, E.; Rotem, C.; Liberman, U.A.; Ravid, A. Vitamin D is a prooxidant in breast cancer cells. *Cancer Res.* **2001**, *61*, 1439–1444.
41. Liu, Y.; Ma, Y.; Zhang, J.; Xie, Q.; Wang, Z.; Yu, S.; Yuan, Y.; Liu, C. MBG-Modified β-TCP Scaffold Promotes Mesenchymal Stem Cells Adhesion and Osteogenic Differentiation via a FAK/MAPK Signaling Pathway. *ACS Appl. Mater. Interfaces* **2017**, *9*, 30283–30296. [CrossRef]

42. Sattary, M.; Rafienia, M.; Kazemi, M.; Salehi, H.; Mahmoudzade, M. Promoting effect of nano hydroxyapatite and vitamin D3 on the osteogenic differentiation of human adipose-derived stem cells in polycaprolactone/gelatin scaffold for bone tissue engineering. *Mater. Sci. Eng. C* **2019**, *19*, 141–155. [CrossRef] [PubMed]
43. Dvorakova, J.; Wiesnerova1, L.; Chocholata, P.; Kulda, V.; Landsmann, L.; Cedikova, M.; Kripnerova, M.; Eberlova, L.; Babuska, V. Human cells with osteogenic potential in bone tissue research. *BioMed. Eng. OnLine* **2023**, *22*, 33. [CrossRef] [PubMed]
44. Komori, T. Regulation of osteoblast differentiation by transcription factors. *J. Cell. Biochem.* **2006**, *99*, 1233–1239. [CrossRef] [PubMed]
45. Li, J.; Wen, J.; Li, B.; Li, W.; Qiao, W.; Shen, J.; Jin, W.; Jiang, X.; Yeung, K.W.K.; Chu, P.K. Valence State Manipulation of Cerium Oxide Nanoparticles on a Titanium Surface for Modulating Cell Fate and Bone Formation. *Adv. Sci* **2017**, *5*, 1700678. [CrossRef]
46. Lou, Y.R.; Toh, T.C.; Tee, Y.H.; Yu, H. 25-Hydroxyvitamin D3 induces osteogenic differentiation of human mesenchymal stem cells. *Sci. Rep.* **2017**, *7*, 42816. [CrossRef]

Journal of
Functional Biomaterials

MDPI

Article

Combined Effects of Dual-Scale Modified Surface with Micro- and Nanostructures on the Cellular Biocompatibility, Osteoinduction, and Antibacterial Properties of Titanium Implants

Shaheer Maher [1], Nenad L. Ignjatović [2,*], Miloš Lazarević [3], Sanja Petrović [3], Andrijana Žekić [4] and Dusan Losic [5,*]

1 Faculty of Pharmacy, Assiut University, Assiut 71526, Egypt; shaheer.maher@aun.edu.eg
2 Institute of Technical Sciences of the Serbian Academy of Sciences and Arts, Knez Mihailova 35/4, PAK 104105, 11000 Belgrade, Serbia
3 School of Dental Medicine, University of Belgrade, Dr Subotica 8, 11000 Belgrade, Serbia; milos.lazarevic@stomf.bg.ac.rs (M.L.); sanja.petrovic@stomf.bg.ac.rs (S.P.)
4 Faculty of Physics, University of Belgrade, Studentski trg 12, 11000 Belgrade, Serbia; andrijana@ff.bg.ac.rs
5 School of Chemical Engineering, The University of Adelaide, Adelaide, SA 5005, Australia
* Correspondence: nenad.ignjatovic@itn.sanu.ac.rs (N.L.I.); dusan.losic@adelaide.edu.au (D.L.)

Abstract: Titanium implants are widely used in biomedical applications due to their excellent mechanical properties and biocompatibility. However, implant-associated bacterial infections and suboptimal osseointegration remain significant challenges. Recent studies have demonstrated that the interplay between micro- and nanostructures can enhance both biocompatibility and antibacterial properties. This study explores the synergistic effects of hierarchical and dual surface topography on Ti surfaces with micro- and nanostructures to demonstrate their ability to promote cellular biocompatibility and osteoinduction while simultaneously inhibiting bacterial colonization. The combination of selective laser melting (SLM) to create micro-structured surfaces and hydrothermal processes is used to generate distinctive nanopillar structures. By integrating nanoscale features that mimic the extracellular matrix with microscale topographies that influence cellular responses, we achieve a balance between enhanced osseointegration and antimicrobial performance. The physicochemical properties of these dual-scale topographies are characterized through cellular assays using dental pulp stem cells (DPSCs), demonstrating sustained support for long-term cell viability (above 78% in MTT and NR assays ($p < 0.05$), low levels of LDH release, and high levels of cellular migration) and osteoinduction (statistically significant ($p < 0.0001$) ALP activity increase and higher levels of calcified matrix deposition, up-regulation of *ALP* and *OCN* genes compared with smooth surface topographies). Their antibacterial properties against *S. aureus* and *E. coli* showed a significant reduction ($p < 0.05$) in bacterial attachment and biofilm formation. Our findings highlight the potential of multi-scale surface modifications as a promising strategy for next-generation titanium implants, paving the way for improved clinical outcomes in orthopedic and dental applications.

Keywords: titanium; titanium implants; osseointegration; nanostructures; microstructures; additive manufacturing; selective laser melting; hydrothermal process

1. Introduction

Numerous studies have been performed aimed at eliminating the inherent limitations and improving the performance of current titanium (Ti) implants with enhanced antibacterial properties and osseointegration capabilities for dental and orthopedic applications [1–5]. These studies indicate that modifying the physicochemical properties of implant surfaces, such as surface charge, hydrophobicity, roughness, and structures, has shown promise in improving both cell attachment and bacterial resistance.

One notable approach involves the electrochemical anodization of titanium to form titania nanotubes (TNTs) to make these unique nanostructures on the Ti surface [6,7]. These nanotubes not only exhibit antibacterial properties but also facilitate the localized delivery of antimicrobial agents [3,8,9]. Another strategy introduced by Ivanova et al. involves mimicking the surface features of insect wings (e.g., cicadas or dragonflies) with nanoscale sharp structures capable of mechanically disrupting bacterial cells by inducing membrane rupture, effectively eliminating bacterial attachment and biofilm formation without the need for antibiotics [10–12]. This approach offers a cost-effective and scalable method for generating antibacterial surfaces on titanium implants. From these studies, hierarchical topographies on the implant surface, which combine microscale roughness with nanoscale features, have been introduced in many studies showing great promise in improving biocompatibility and antibacterial properties. Microscale structures contribute to mechanical interlocking with bone tissue, promoting osseointegration, while nanoscale features mimic the extracellular matrix, facilitating cellular interactions and reducing bacterial adhesion [1,3,5,13–16].

These dual-scale modifications offer a novel approach to achieving a balance between enhanced osteogenic potential and antibacterial efficacy. However, the impact of various surface topographies—such as micro- and nano-features—on bone tissue formation remains a topic of ongoing debate, largely due to variations in fabrication methods and structural inconsistencies. While micro-topographical surfaces have been shown to improve bone fixation and in vivo pullout strength, recent studies suggest that nanoscale features, which mimic the hierarchical structure of bone, may further enhance osteoblast activity, increase cell attachment, and promote bone formation [17–20]. Nano-textured surfaces have demonstrated excellent biocompatibility by facilitating protein adsorption, promoting apatite formation, and providing an optimal environment for cell proliferation and anchoring [17,19–21]. The processes of osseointegration and bacterial infection are intricately linked. The "race to the surface" hypothesis suggests that bacteria often colonize the implant surface before human cells can attach, due to factors such as the surgical environment, implant handling, and the presence of contaminants [21,22]. Therefore, developing an implant surface that balances the promotion of bone cell attachment with resistance to bacterial colonization and biofilm formation is essential for improving implant longevity and functionality.

Over the past decade, additive manufacturing (AM) technologies, such as 3D printing, have gained considerable attention for their potential to revolutionize the production of titanium-based medical implants [23]. AM offers unique advantages, including design flexibility and the ability to fabricate implants with complex geometries that closely resemble bone structures [24]. Moreover, AM allows for patient-specific implants that can be produced rapidly and tailored to individual anatomical needs, reducing the need for mass production and excess inventory. Among various AM techniques, selective laser melting (SLM), also known as direct metal printing (DMP), stands out as a highly effective method for manufacturing titanium implants. SLM enables the fabrication of titanium and titanium-alloy implants with precisely controlled properties, reduces material waste, and lowers overall production costs [2,24–26]. In our previous work, we explored several approaches to enhance the performance of 3D-printed titanium for biomedical applications, focusing on surface modifications via electrochemical treatments, chemical etching, and hydrothermal processes [13–15,27–29]. Our findings demonstrated that nanopillar structures significantly improved antibacterial properties by mechanically disrupting bacterial cell membranes while supporting osteoblastic cell growth [13,27]. AM of Ti implants presents a promising solution for developing the next generation of low-cost, customizable, and efficient Ti alloy implants for dental and orthopedic applications, and their further advancements in surface engineering are essential to fully realize their potential, enhancing their performance and biocompatibility.

Building upon this promising development, the present study aims to further investigate the biocompatibility and antibacterial properties of titanium alloys with dual micro- and nano-topography fabricated via AM (SLM). Specifically, we examine how different surface features at varying scales, such as micro-smooth, micro-rough, and nanopillar structures, can affect both cellular behavior (osteoblast attachment and proliferation) and bacterial attachment, biofilm formation, and bactericidal activity. This comprehensive study on the interaction of stem cells and bacteria on fabricated nano- and microstructures on Ti surfaces fabricated by a combined SLM and hydrothermal method is illustrated in Figure 1. Dental pulp stem cells (DPSCs) are employed as model cells to assess the integration potential of these surfaces, which are particularly relevant to the emerging field of 3D-printed dental implants [30,31]. Additionally, we use Escherichia coli and Staphylococcus aureus to evaluate the antibacterial efficacy of the implants. The findings of this study provide valuable insights into the interplay between implant surface topography, cell behavior, and bacterial colonization, contributing to the development of advanced medical implants with optimized biocompatibility and antibacterial properties using AM technologies.

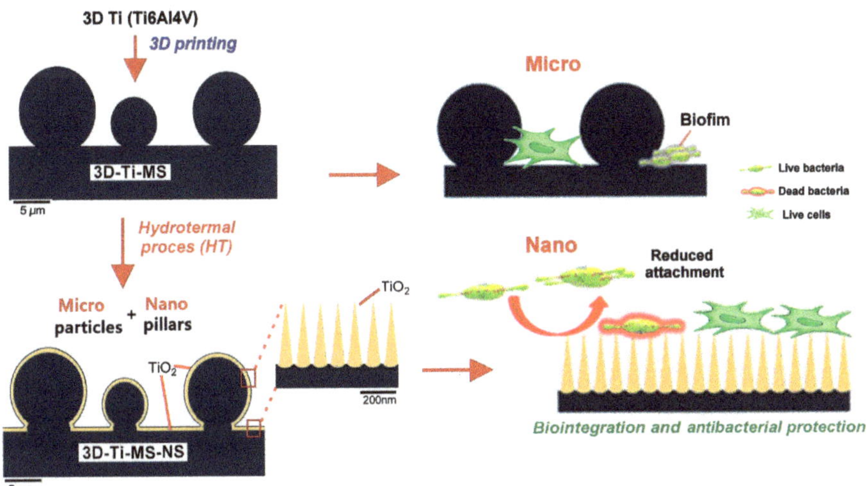

Figure 1. Schematic diagram of the cellular biocompatibility and antibacterial protection of titanium alloys with dual micro- and nano-topography.

2. Materials and Methods

2.1. Fabrication of 3D-Ti Implants

Titanium implants (termed 3D-Ti-MS) were fabricated using titanium alloy (Ti6Al4V) powder (Titanium grade 5, TLS Technik GmbH & Co. Spezialpulver, Bitterfeld, Germany) in the forms of square discs (1.5×1.5 cm^2) via selective laser melting (SLM) using a ProX 200 Production 3D Printer (Phenix Systems PXM, Alpharetta, GA, USA). The SLM system was equipped with a 300 W laser (wavelength 1070 nm, operating at 50% power) and processed under an inert argon atmosphere. Detailed information on the fabrication process and the average particle diameter of the powder can be found in our previous work [28]. Following fabrication, any residual powder particles were removed from the wafer surfaces by sonication in acetone for 10 min. The resulting implants, featuring a micro-smooth surface, will be referred to as "Ti" throughout the paper.

2.2. Surface Modification

To make a nanostructured surface on a 3D-Ti-MS, hydrothermal surface modification was performed, adapting previously described HT procedures [32–34]. Briefly, 3D-Ti-MS discs were submerged in 50 mL of 1 M NaOH (Chem-Supply, Adelaide, Australia) in a Teflon-lined vessel, which was then placed inside a stainless-steel autoclave. The samples were hydrothermally (HT) treated for 4 h at 160 °C. After the HT process, the autoclave was removed from the oven and allowed to cool to room temperature. The discs were rinsed three times with Milli-Q water to remove any remaining NaOH and were then air-dried at room temperature overnight. The samples were subsequently annealed in a tube furnace at 300 °C for 3 h under atmospheric conditions. These HT-treated samples are referred to as 3D-Ti-MS-NS substrates.

Prior to cell and bacterial studies, the prepared 3D-Ti-MS and 3D-Ti-MS-NS samples were immersed in Milli-Q water for 2 weeks to ensure complete removal of any residual chemicals from the fabrication process.

2.3. Materials Characterizations

High-resolution imaging of various surfaces was performed using a Focused Ion Beam (FIB)-Scanning Electron Microscope (SEM, FEI Helios Nanolab 600 Dual Beam, Thermo Fisher Scientific, Scoresby, Australia), operated at 10 kV. Surface chemical characterization was carried out via Energy Dispersive X-ray Spectroscopy (EDX, Oxford Ultim Max Large Area SDD EDS detector, Oxford Instruments, Concord, MA, USA). Prior to SEM imaging, all wafers were coated with a 5 nm layer of platinum. The crystallinity of the surfaces was analyzed by X-ray diffraction (XRD, Rigaku MiniFlex 600, Osaka, Japan), operating at 40 kV and 15 mA, using Cu Kα radiation. XRD spectra were recorded at an angular scan rate of 10° per minute over the 30°–80° range under ambient conditions.

Water contact angle (WCA) measurements were conducted using the sessile drop method at room temperature, employing a Tension Theta optical tensiometer (KSV Instruments, Helsinki, Finland) equipped with an automated stage, droplet dispenser, and digital camera. A 2 µL drop of Milli-Q water (resistivity 18.2 MΩ·cm) was used, and measurements were taken at three distinct locations on each sample, which was prepared in triplicate—three separate wafers. The images were analyzed with OneAttension software (version 3.2, Biolin Scientific, Gothenburg, Sweden). To minimize the influence of environmental factors on WCA measurements, prepared samples were stored in a clean, airtight desiccator immediately after fabrication. This approach prevents contamination and moisture adsorption, which can significantly alter surface properties and affect CA values. Furthermore, CA measurements were conducted promptly after removing the samples from the desiccator, thereby reducing exposure to ambient conditions such as humidity and airborne contaminants that could impact the results.

2.4. In Vitro Studies

2.4.1. Cell Cultures

Dental pulp stem cells (DPSCs) were isolated using an explant culture method from dental pulp tissues of immature permanent third molars obtained from healthy donors (n = 3, aged between 18 and 21 years) who provided written informed consent. The study was approved by the Ethical Committee (No. 36/26, 19 November 2021) at the School of Dental Medicine, University of Belgrade. The cells were cultured in complete growth medium consisting of DMEM/F12, supplemented with 10% fetal bovine serum (FBS) and

1% antibiotic/antimycotic (all from Gibco, ThermoFisher, Waltham, MA, USA). Cultures were maintained at 37 °C in a humidified 5% CO_2 atmosphere and passaged routinely when they reached 80% confluence. The medium was refreshed every 2–3 days. The cells were characterized as DPSCs based on criteria outlined in a previous study [35]. Only cells between the 3rd and 5th passage were used for the experiments. Before cell seeding, alloy wafers were sterilized by immersion in 70% (v/v) ethanol for 5 min, followed by air-drying in a Class II biohazard hood for 3 h and three rinses in sterile water. Pure titanium (Ti) wafers were used as controls in all experiments. Triplicate substrates were placed in a single well of a 6-well tissue culture plate.

2.4.2. MTT and Neutral Red Assays

Two distinct methods, MTT and Neutral Red (NR) assays, were used to evaluate cytotoxicity: indirect cytotoxicity assay and direct cytotoxicity assay.

For indirect cytotoxicity assays, Ti-based wafers (Ti, 3D-Ti-MS, and 3D-Ti-MS-NS) were first immersed in complete growth medium for 7 days. Subsequently, DPSCs were seeded at a density of 5×10^3 cells/mL into 96-well plates (Corning, New York, NY, USA) and exposed to the supernatants from the respective wafers (Ti, 3D-Ti-MS, or 3D-Ti-MS-NS) for 1, 3, or 7 days. After the designated incubation periods, 100 µL of MTT solution (0.5 mg/mL) was added to each well and incubated for 3 h at 37 °C. For the NR assay, 150 µL of 1× NR Staining Solution (Sigma-Aldrich, St. Louis, MO, USA) was added to each well and incubated for 4 h at 37 °C. Following incubation, the medium was aspirated, and 100 µL of dimethyl sulfoxide (DMSO, Sigma-Aldrich) for MTT or 150 µL of NR eluent (96% ethanol: dH_2O: CH_3COOH, 50:49:1) for NR assay was added per well. The plates were then shaken at 250 RPM for 15 min. Optical density (OD) was measured at 570 nm using an automated microplate reader (RT-2100c, Microplate Reader, Rayto, Shenzhen, China). Cells without treatment served as the positive control. The percentage of viable cells was calculated using the following formula:

$$\text{Viable cells} = (\text{OD (sample)} - \text{OD (blank)})/(\text{OD (control)} - \text{OD (blank)}) \times 100.$$

For the direct cytotoxicity assays, Ti wafers were placed in individual wells of a 12-well tissue culture plate (Corning, New York, NY, USA), and DPSCs were seeded directly onto the wafer surfaces at a density of 5×10^3 cells/mL. The cells were incubated for 1, 3, or 7 days, with cell viability assessed using both MTT and NR assays, as described above. For the MTT assay, 500 µL of MTT solution was added, while for the NR assay, 500 µL of 1× NR Staining Solution was used. After the incubation period, the culture medium was removed, and either DMSO or NR eluent (depending on the assay) was added. The OD of the plates and the percentage of viable cells were calculated as described above.

2.4.3. Lactate Dehydrogenase (LDH) Assay

Cytotoxicity was also assessed by measuring the release of extracellular lactate dehydrogenase (LDH) using the CyQUANT™ LDH Cytotoxicity Assay (Catalog no. C20300, ThermoFisher, Waltham, MA, USA). DPSCs were seeded at a density of 2×10^4 cells/well in 12-well plates with Ti, 3D-Ti-MS, or 3D-Ti-MS-NS substrates and incubated for 1, 3, or 7 days. At each time point, the LDH release was quantified according to the manufacturer's protocol. Absorbance was measured at 490 nm using a microplate reader (RT-2100c Microplate Reader, Rayto, Shenzhen, China).

2.4.4. Wound Healing Assay

DPSCs were seeded at a density of 2×10^4 cells/well in 12-well plates, with titanium samples placed in each well. Once the cells reached confluence, a scratch was made using a 200 μL pipette tip along the edge of the Ti sample (Figure S1). The wells were then washed with phosphate-buffered saline (PBS, Gibco, ThermoFisher, Waltham, MA, USA) to remove cell debris. The closure of the scratched area was monitored over 48 h at 5 min intervals using a CytoSMART™ Lux2 camera (CytoSmart Technologies, Amsterdam, The Netherlands) [32]. CytoSMART software (version 1.0) was used to quantify the healing area (μm^2). Representative images were extracted at 0, 12, 24, 36, and 48 h to track the migration process.

2.4.5. Alkaline Phosphatase (ALP) Activity Assay

To evaluate functional alkaline phosphatase (ALP) activity, 4×10^5 DPSCs were seeded in 12-well plates (1 mL per well) with different titanium wafers and cultured in human StemMACSOsteoDiff Media (osteogenic medium, OM, Miltenyi Biotec, Tokyo, Japan) for 7 days. After the incubation period, ALP activity was assessed using a commercially available pNPP Alkaline Phosphatase assay kit (Sigma-Aldrich, St. Louis, MO, USA). The OD was measured at 405 nm using a microplate reader (RT-2100C Microplate Reader, Rayto, Shenzhen, China).

2.4.6. Osteogenic Differentiation and Alizarin Red Staining

One day after seeding DPSCs (5×10^4 cells/well) on Ti, 3D-Ti-MS, or 3D-Ti-MS-NS substrates in 12-well plates, OM was added. The cells were incubated for 7, 14, and 21 days in the osteogenic media and wafers, with OM changes every 3 days. The effect of 3D-Ti-MS and 3D-Ti-MS-NS on mineralized nodule formation was evaluated by Alizarin Red S (ARS) staining. At each time point (7, 14, and 21 days), the cultures were fixed with 4% neutral formalin for 15 min and stained with 2% ARS (Sigma-Aldrich). To quantify mineralization, ARS-bound dye was extracted by incubating the cultures with 250 μL of 1% hydrochloric acid in 70% ethanol for 20 min. The absorbance of the extracted dye was measured at 450 nm using a microplate reader (RT-2100c, Microplate Reader, Rayto, Shenzhen, China) [36]. Cells cultured in OM only served as the control.

2.4.7. Real-Time Quantitative Polymerase Chain Reaction (qPCR)

DPSCs were seeded on wafers and cultured for 21 days to allow osteogenic differentiation. Total RNA was then extracted from the cells using Trizol reagent (Invitrogen, Thermo Fisher Scientific, Waltham, MA, USA). Complementary DNA (cDNA) was synthesized using the RevertAid First Strand cDNA Synthesis Kit (Thermo Fisher Scientific), following the manufacturer's instructions. RT-qPCR analysis was performed on the cDNA using the same RevertAid kit to assess the expression of osteogenic markers *ALP* and *OCN*. Specific primers for these markers are listed in Table S1. GAPDH was used as the reference housekeeping gene. Fold induction values were calculated using the $2^{-\Delta\Delta Ct}$ method [34].

2.4.8. Scanning Electron Microscope (SEM) Imaging of Cells

DPSCs (5×10^4 cells per well) were seeded onto different wafer substrates and incubated for 4 days in complete growth medium in 12-well plates. The morphology of the cells attached to the substrates was assessed using scanning electron microscopy (SEM) [37]. After the 4-day incubation, the samples were gently rinsed with Milli-Q water and fixed with 2.5% (v/v) glutaraldehyde (Sigma). Following fixation, the samples were rinsed twice with Milli-Q water. The cells were then dehydrated through a graded series of ethanol solutions (30%, 50%, 70%, 90%, and 100% v/v) for 10 min each, with an additional 10 min treatment in 100% ethanol. After dehydration, the samples were immersed in hexamethyl disilazane (HMDS) (Sigma) for 10 min. The samples were left to air-dry and then sputter-coated with gold using a JFC 1100 ion sputter for imaging. High-resolution images were captured using a JEOL JSM-840A SEM instrument at an acceleration voltage of 30 kV.

2.5. Antibacterial Activity Studies

2.5.1. Bacterial Growth

Escherichia coli ATCC 25922 and *Staphylococcus aureus* ATCC 25923 (Microbiologics KWIK-STIK, Manassas, VA, USA) were revived from frozen stocks stored at −80 °C by incubation at 37 °C for 24 h, as outlined in Table S2. After activation of the reference strains, 3–4 colonies of *S. aureus* and *E. coli* were transferred to separate Brain Heart Infusion (BHI) broth (HIMEDIA, Maharashtra, India) and incubated for an additional 24 h at 37 °C in aerobic conditions. The bacterial suspensions were centrifuged (3000 RPM for 5 min). The supernatant was discarded, and the pellet of each suspension was resuspended in sterile PBS to achieve a turbidity corresponding to the 0.5 McFarland standard ($\approx 10^8$ bacterial cells/mL). Each suspension was further diluted in enriched BHI broth to obtain two suspensions (one of *S. aureus* and the other of *E. coli*) with a final concentration of approximately 10^6 bacterial cells/mL.

2.5.2. Biofilm Formation

Monomicrobial bacterial biofilms were formed by adding 1 mL of each described bacterial suspension ($\approx 10^6$ bacteria cells/mL) to the surface of prepared Ti (control), 3D-Ti-MS, and 3D-Ti-MS-NS samples placed in 24-well microplates. The plates were incubated statically under aerobic conditions at 37 °C. For each microorganism, monomicrobial biofilm was formed on three groups of materials: Ti ($n = 3$), 3D-Ti-MS ($n = 3$), and 3D-Ti-MS-NS ($n = 3$).

SEM Visualization of Monomicrobial Biofilms

For visualization of the formed monomicrobial biofilm of *S. aureus* and *E. coli* on 3D-Ti-MS and 3D-Ti-MS-NS samples, scanning electron microscopy was performed. After 24 h of biofilm formation, wafers were removed from the medium, gently washed with sterile PBS to remove unattached bacterial cells, and fixed in 2.5% glutaraldehyde (Sigma, Aldrich) for 48 h. Subsequently, wafers were dehydrated by applying a series of solutions of 3% acetic acid, 3% acetic acid and 25% ethanol, 3% acetic acid and 50% ethanol, and 70% ethanol, according to a previously described fixation method [38].

Determination of Colony Forming Units (CFU) from Medium Around the Samples

After biofilm formation on the samples, twelve 10-fold serial dilutions of the medium surrounding the samples were seeded on solid media: Columbia agar with 5% sheep blood for *S. aureus* and Endo agar for *E. coli*. The plates were incubated at 37 °C under aerobic conditions, and CFUs were counted after 24 h.

MTT Assay of Medium Around the Samples

After the removal of the wafers for MTT analysis, 50 µL of the medium from each well was transferred to eight wells of a new 96-well microplate. The same volume (50 µL) of MTT solution was added to each well containing the medium. The plates were incubated in lightproof conditions at 37 °C for 3.5 h under static aerobic conditions. After incubation, 100 µL of dimethyl sulfoxide (DMSO) was added to each well, and the plates were shaken at 250 rpm for 15 min at 37 °C in the dark. The absorbance of the resulting colored product, which reflects bacterial viability, was measured at 540 nm using a spectrophotometer.

2.6. Statistical Analysis

Statistical analysis and generation of graphical images were conducted using GraphPad Prism version 9 (GraphPad Software, Inc., San Diego, CA, USA). All results in this study are presented as the mean ± standard deviation (SD) of at least three independent experiments. Statistical comparisons were made using one-way analysis of variance

(ANOVA) followed by Tukey's multiple comparisons test, with a single pooled variance. The normality of data distribution was assessed using the Kolmogorov–Smirnov test. Statistical significance was set at $p < 0.05$ for all comparisons. All in vitro experiments were performed in triplicate and repeated at least twice.

3. Results

3.1. Surface Morphology and Physico-Chemical Properties of Fabricated Combined Micro- and Nanostructures

The topographic morphology and structural characteristics of fabricated titanium (Ti) wafers with micro- and nanostructures obtained by SEM imaging are summarized in Figure 2. The surface of the 3D-Ti-MS fabricated using SLM equipment exhibits randomly distributed microspheres, which result from the partial melting of Ti alloy powder during the SLM process (Figure 2a,b). During the SLM process, layers of Ti alloy powders with spherical microparticles are melted by the laser beam to generate microparticles inside the bulk structure on the top surface. These microspheres, which have a diameter ranging from 5 μm to 20 μm, are randomly distributed on the surface, with different inter-distances and very strong adhesion on the surface with an average surface roughness (Ra) of 17 ± 3 μm.

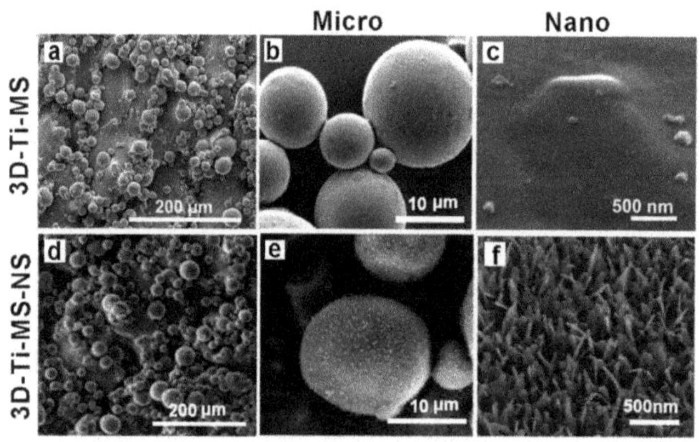

Figure 2. SEM images showing the top surface of **(a–c)** a 3D-Ti-MS substrate with spherical microparticle topography fabricated by the SLM process showing typical structures of microparticles and a flat surface between these particles; **(d–f)** a 3D-Ti-MS-NS substrate that retains microstructure topography, the surface of which is covered by nanopillar structures generated by the hydrothermal process with 1 M NaOH.

Apart from the microparticles, the peak and valley structures of the surface appeared smooth, exhibiting no nanostructures. After fabrication, prepared 3D-Ti-MS samples were sonicated in acetone to remove any loosely adhered particles an ensure the stability of the topographic features during subsequent processing. Apart from the microspheres, the surface exhibits a smooth topography with peaks and valleys but lacks nanostructures (Figure 2c). Following the fabrication of 3D-Ti-MS, hydrothermal treatment was applied to generate 3D-Ti-MS-NS nanostructures, which have typical topography presented in Figure 2d–f. The images show that these nanostructures are composed of arrays of nanopillars with wide bases and sharp ends with a length in the range of 400–500 nm and inter-distances between 150–250 nm.

The physical and chemical properties of 3D-Ti-MS and 3D-Ti-MS-NS were characterized using EDX, XRD, and WCA measurements, the results of which are presented in Figure 3. EDX spectra of all samples displayed the typical peaks of Ti, Al, and V that are expected, as they were fabricated from titanium alloy (Ti_6Al_4V) powders by the SLM printing process (Figure 3a,b). In addition, an Na peak is detected for 3D-Ti-MS-NS, which could correspond to sodium titanate ($Na_2Ti_3O_7$) that is created during the HT process. The crystal structure of the prepared 3D-Ti-MS and 3D-Ti-MS-NS samples assessed by XRD showed the distinct diffraction peaks presented in Figure 3c,d. The high crystalline structure of all samples was confirmed through sharp, high-intensity diffraction peaks. All samples showed Ti peaks (JCPDS 44-1294), while anatase TiO_2 peaks (JCPDS 21-1272) and ($Na_2Ti_3O_7$) (JCPDS 31-1329) were evident after hydrothermal processing.

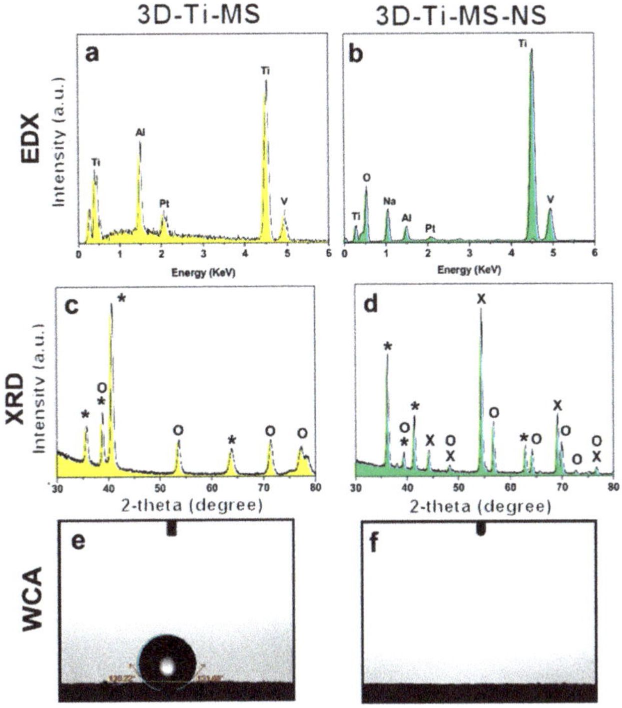

Figure 3. Chemical and physical characterization and surface properties of prepared 3D-Ti-MS and 3D-Ti-MS-NS substrates demonstrating (**a**,**b**) EDX spectra showing surface chemical composition, (**c**,**d**) XRD patterns with peaks corresponding to Ti, anatase TiO_2, and $Na_2Ti_3O_7$ (*—Ti, O—TiO_2, X—titanate), and (**e**,**f**) wetting properties by water contact angle (WCA).

The measurement of the water contact angle (WCA) on the prepared substrates presented in Figure 3e,f revealed distinct differences in surface wettability based on the surface modifications. The 3D-Ti-MS exhibited hydrophobic characteristics, with WCAs of 133 ± 1 degrees, respectively, consistent with the typical hydrophobic behavior of metal surfaces. On the other hand, the 3D-Ti-MS-NS surface displayed superhydrophilic behavior, where the water droplet spread instantly, preventing accurate WCA measurement. This dramatic increase in wettability can be primarily attributed to the formation of sodium titanate ($Na_2Ti_3O_7$) during the hydrothermal treatment. Sodium titanate is known to enhance surface hydrophilicity, likely by increasing the surface energy and creating a more porous surface structure.

3.2. DPSCs Viability and Integration Studies on Combined Micro- and Nanostructures

3.2.1. MTT Assay

The results of the MTT assay, showing DPSCs viability on prepared samples including Ti, 3D-Ti-MS, and 3D-Ti-MS-NS with microspheres and combined microspheres and nanopillar topography over 7 days, are summarized in Figure 4. Results indicate that cells in direct contact with the different Ti materials did not exhibit a significant loss in viability (1st day: Ti group 96.1% (±5.9); 3D-Ti-MS group 92.9% (±10.5); 3D-Ti-MS-NS group 78.1% (±5.9); 3rd day: Ti group 106.6% (±5.5); 3D-Ti-MS group 99.3% (±5.9); 3D-Ti-MS-NS group 87.6% (±6.3); 7th day: Ti group 109.6% (±2.5); 3D-Ti-MS group 86.9% (±6.8); 3D-Ti-MS-NS group 97.2% (±5.3)), with cell viability remaining above 70%, indicating good support for cell growth. On day 1, the cell viability of both Ti and 3D-Ti-MS surfaces was similar (Figure 4a). However, by day 3, 3D-Ti-MS showed a reduction in cell viability compared with Ti (Figure 4b), and by day 7, the cell viability of 3D-Ti-MS was lower compared with 3D-Ti-MS-NS (Figure 4c).

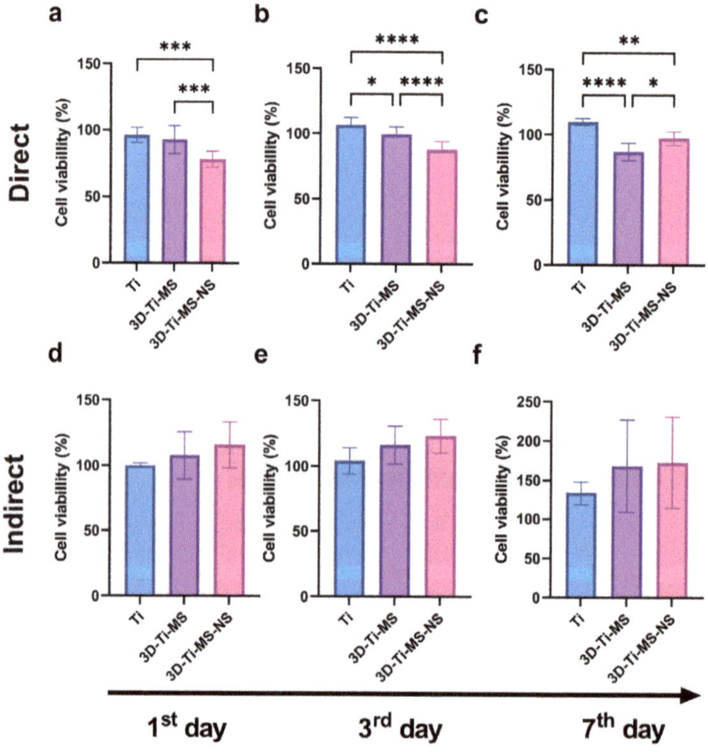

Figure 4. MTT assay of dental pulp stem cells (DPSCs) on prepared samples including Ti, 3D-Ti-MS, and 3D-Ti-MS-NS showing (**a**) direct MTT assay 1-day treatment; (**b**) direct MTT assay 3-day treatment; (**c**) direct MTT assay 7-day treatment; (**d**) indirect MTT assay 1-day treatment; (**e**) indirect MTT assay 3-day treatment; (**f**) indirect MTT assay 7-day treatment; * $p \leq 0.05$; ** $p \leq 0.01$; *** $p \leq 0.001$; ****$p \leq 0.0001$.

This trend confirms that surface topography influences cell viability, as previously reported in many other studies [3,5,7,15,17–19,28,29,39,40]. The highest cell viability was observed in the Ti group (smooth surface), where it exceeded 100% after 3 and 7 days of culture. Interestingly, the 3D-Ti materials (3D-Ti-MS and 3D-Ti-MS-NS) demonstrated a significantly better effect on cell viability (1st day: Ti group 100.2% (±1.9); 3D-Ti-MS group

107.8% (±18.1), 3D-Ti-MS-NS group 116.0% (±17.8); 3rd day: Ti group 104.2% (±10.1); 3D-Ti-MS group 116.3% (±14.3), 3D-Ti-MS-NS group 123.2% (±12.9); 7th day: Ti group 113.4% (±14.9); 3D-Ti-MS group 168.2% (±59.1), 3D-Ti-MS-NS group 172.8% (±58.3)) after indirect treatment (indirect MTT, Figure 4d–f), although this difference was not statistically significant when compared with the Ti group.

3.2.2. Neutral Red Assay

The results of the Neutral Red (NR) assay of DPSCs on prepared substrates, including Ti, 3D-Ti-MS, and 3D-Ti-MS-NS, are presented in Figure 5. Cells in direct contact with all these surfaces showed no significant loss in viability (1st day: Ti group 94.4% (±3.8); 3D-Ti-MS group 88.8% (±5.6); 3D-Ti-MS-NS group 88.4% (±4.4); 3rd day: Ti group 101.7% (±15.4); 3D-Ti-MS group 107.4% (±15.9); 3D-Ti-MS-NS group 98.2% (±15.1); 7th day: Ti group 113.1% (±10.1); 3D-Ti-MS group 113.2% (±16.5); 3D-Ti-MS-NS group 107.7% (±14.8)) and did not indicate cytotoxicity for any of the materials tested. Additionally, no significant differences in viability were observed between the different substrates. Similar results were obtained using the indirect method to assess cell viability (1st day: Ti group 101.2% (±5.1); 3D-Ti-MS group 102.4% (±18.8); 3D-Ti-MS-NS group 103.6% (±14.3); 3rd day: Ti group 102.0% (±8.5); 3D-Ti-MS group 130.2% (±49.1); 3D-Ti-MS-NS group 174.2% (±92.8); 7th day: Ti group 120.3% (±8.1); 3D-Ti-MS group 111.1% (±7.8); 3D-Ti-MS-NS group 116.2% (±13.9)), where no significant differences in viability were noted between the substrates after 7 days of incubation with DPSCs. Therefore, the NR assay aligns with MTT results and confirms that all tested surfaces are biocompatible and can support DPSCs growth after implantation into the body.

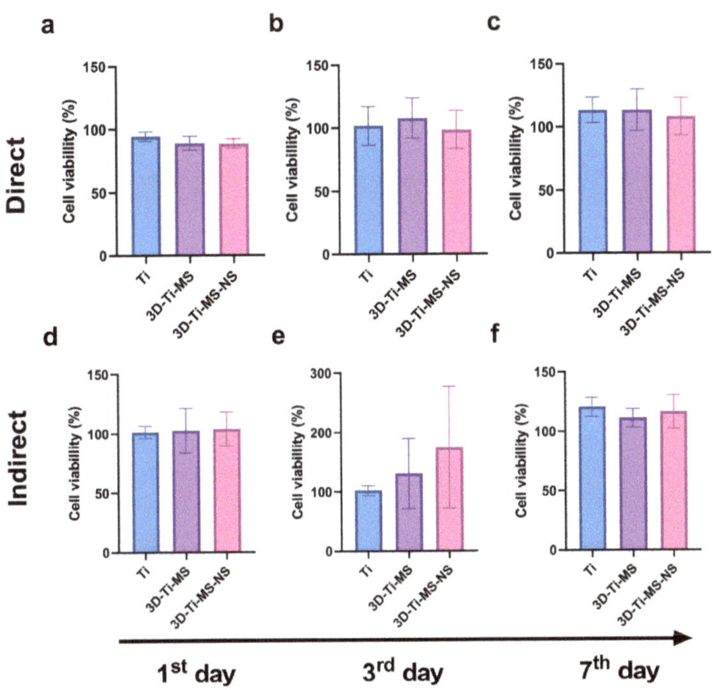

Figure 5. NR assay with dental pulp stem cells (DPSCs) on prepared samples including Ti (control), 3D-Ti-MS, and 3D-Ti-MS-NS showing (**a**) direct NR assay 1-day treatment; (**b**) direct NR assay 3-day treatment; (**c**) direct NR assay 7-day treatment; (**d**) indirect NR assay 1-day treatment; (**e**) indirect NR assay 3-day treatment; (**f**) indirect NR assay 7-day treatment.

3.2.3. LDH Assay

LDH activity in the DPSC culture media was measured as an indicator of cell membrane integrity during 7 days of interactions with Ti (control), 3D-Ti-MS, and 3D-Ti-MS-NS substrates and is presented in Figure 6. The 3D-surfaced materials showed significantly lower levels of LDH after one and three days of culturing with DPSCs compared with the Ti control (Figure 6a,b). These results suggest that Ti substrates caused considerable cell membrane damage, whereas 3D-Ti-MS-NS supported the maintenance of DPSCs membrane integrity. Based on the cell viability and LDH assay results, 3D-Ti-MS-NS emerges as a promising candidate for implant surfaces, particularly in the critical initial days following implant insertion, a period that plays a pivotal role in determining the success of the implant [41].

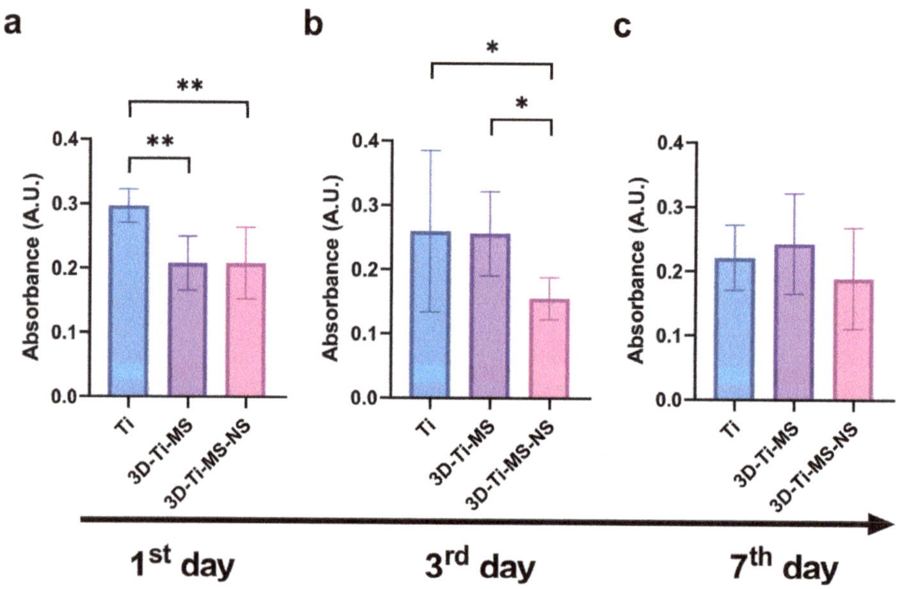

Figure 6. LDH activity of dental pulp stem cells (DPSCs) cultured on Ti (control), 3D-Ti-MS, and 3D-Ti-MS-NS substrates measured at (**a**) the 1st day; (**b**) the 3rd day; and (**c**) the 7th day of direct contact with materials; * $p \leq 0.05$; ** $p \leq 0.01$.

The interaction of DPSCs with 3D-Ti-MS and 3D-Ti-MS-NS substrates was evaluated using SEM imaging, with the results presented in Figure 7. These images reveal differences in cellular interactions between the two substrate types. On the 3D-Ti-MS surfaces, most cell propagation occurs at the interface between the microspheres and the surface, with cells attempting to bridge these gaps (Figure 7a–c, arrows). On the 3D-Ti-MS-NS substrates, which feature both micro- and nanopillar topographies, such interconnections are also present; however, the majority of cells are observed lying directly on the microsphere surfaces or in the spaces between them (Figure 7d–f). These findings suggest a stronger interaction of DPSCs with the 3D-Ti-MS-NS substrate, indicating improved biointegration properties as a result of synergetic interaction between micro- and nanostructures that are proposed by other studies [13–15,29].

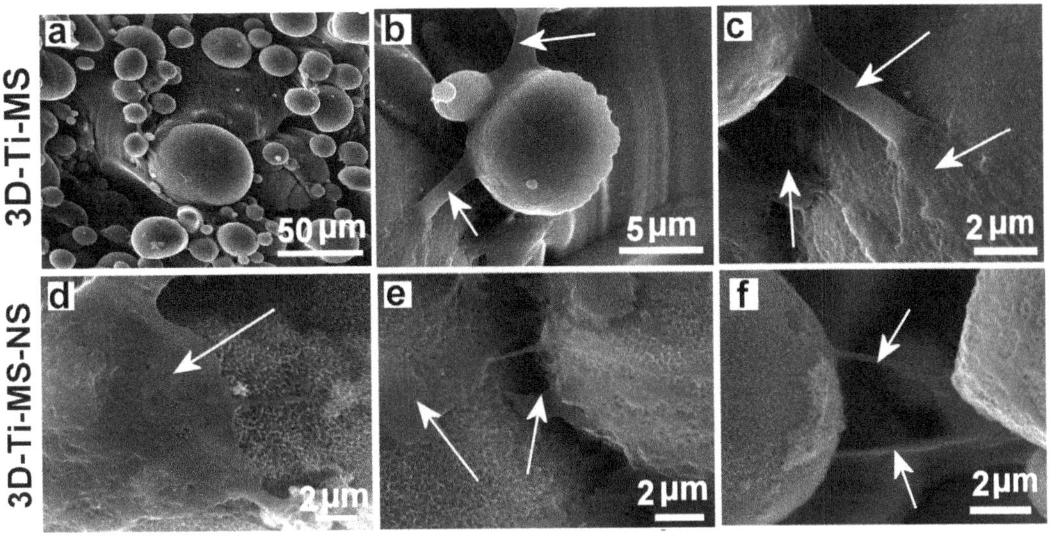

Figure 7. SEM images showing the dental pulp stem cells (DPSCs) interacting with (**a–c**) 3D-Ti-MS with spherical microparticle topography and (**d–f**) 3D-Ti-MS-NS substrate that combined microstructure topography covered but nanopillar structures generated by hydrothermal process. Arrows show the filopodia projection of the cells in the direction of bridging structures on the surface.

3.2.4. Wound Scratch Healing Assay

To assess the impact of the prepared samples, including Ti (control), 3D-Ti-MS, and 3D-Ti-MS-NS substrates, on cell migration, the cell monolayers were "wounded" near the material borders and allowed to heal over 48 h, the results of which are summarized in Figure 8. Results show that interaction with the 3D-printed materials resulted in a significantly faster cell migration rate when compared with Ti (0 h) (Figure 8a,b and Supplementary Videos S1–S3), indicating that 3D-Ti-MS and 3D-Ti-MS-NS surfaces promoted rapid cell healing. These results suggest that both 3D-Ti-MS and 3D-Ti-MS-NS could better support cell growth and provide the long-term attachment necessary for successful permanent implants compared with smooth Ti surfaces. This enhanced performance can be attributed to the ability of the micro- and combined micro- and nano-textured surfaces to promote cell adhesion and proliferation, as previously confirmed in our and other studies [13–15,29].Results showed that both bone cells and fibroblasts exhibited strong anchoring through the formation of extended pseudopodia and focal adhesions, which can contribute to more effective osseointegration.

3.2.5. Alkaline Phosphatase (ALP) Activity Assay

The interaction of DPSCs cells cultured in the osteogenic medium over 7 days with Ti (control), 3D-Ti-MS, and 3D-Ti-MS-NS substrates is presented in Figure 9. Notably, the interaction of DPSCs with the 3D materials with micro and combined micro and nanostructures resulted in significantly higher ALP activity ($p \leq 0.0001$) compared with Ti control samples. The highest ALP activity was observed in the cells in contact with 3D-Ti-MS, with microstructures that indicate that these structures effectively promote mineralization in DPSCs, a process that is closely associated with cellular differentiation. This process is more intensive compared with 3D-Ti-MS-NS with combined micro- and nanostructures.

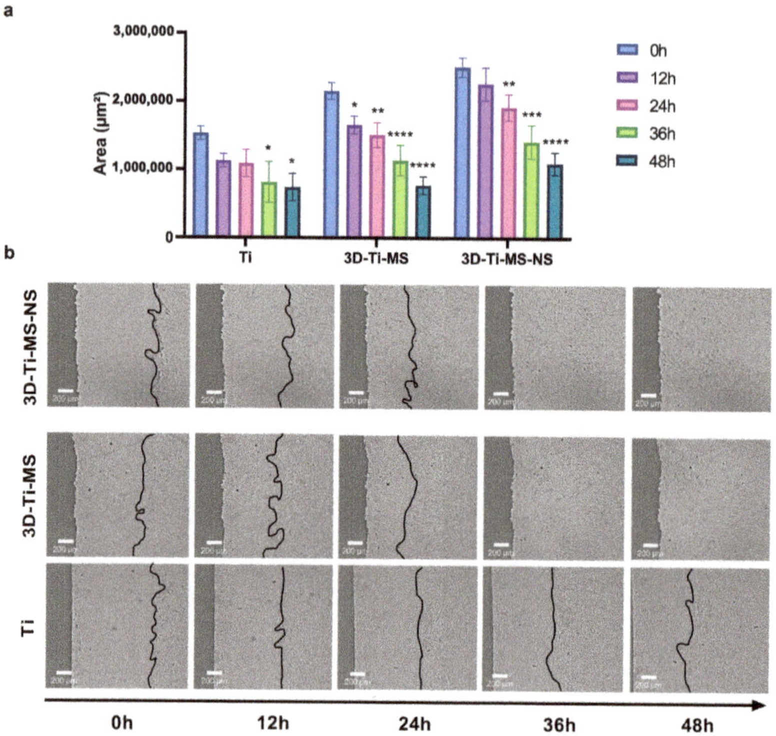

Figure 8. Wound scratch healing assay on Ti (control), 3D-Ti-MS, and 3D-Ti-MS-NS substrates on cell migration showing (**a**) the graph of closing of the scratch area during 48 h; (**b**) representative micrographs at 0, 12, 24, 36, and 48 h after the scratch; * $p \leq 0.05$; ** $p \leq 0.01$; *** $p \leq 0.001$; **** $p \leq 0.0001$ (vs. 0 h group). (the real-time of the cells movements are presented in Videos S1–S3).

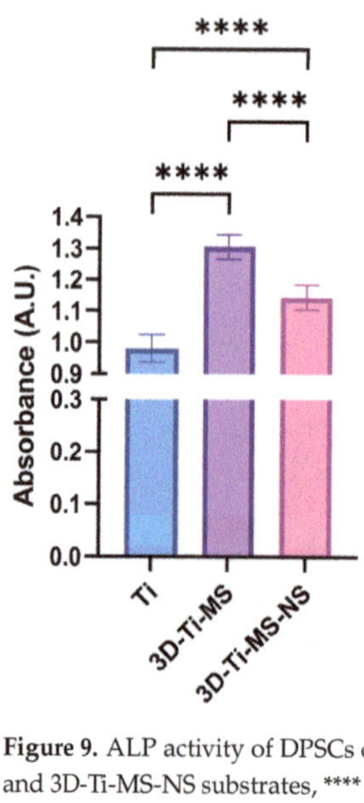

Figure 9. ALP activity of DPSCs cultured in osteogenic media measured on Ti (control), 3D-Ti-MS, and 3D-Ti-MS-NS substrates, **** $p \leq 0.0001$.

3.2.6. Alizarin Red S Quantification

Alizarin Red S staining of cells upon 7, 14, and 21 days of osteogenic differentiation on different Ti surfaces, including Ti, 3D-Ti-MS, 3D-Ti-MS-NS, and control (cells cultured only with osteogenic medium—OM), are presented in Figure 10. Cells cultured on flat and non-modified Ti surfaces and 3D-Ti-MS-NS exhibited a higher level of calcified matrix deposition compared with the control group after just 7 days of osteoinduction. These findings are consistent with our previous studies and align with the ALP activity results [29].

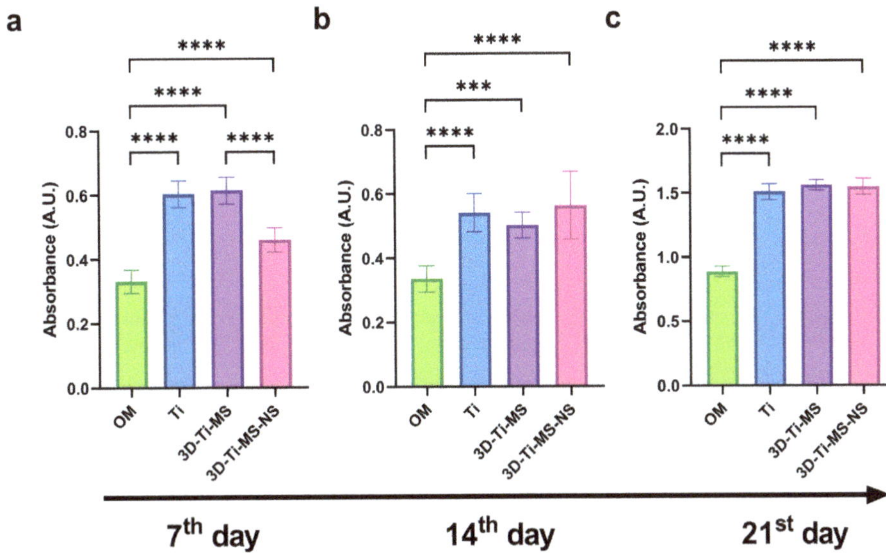

Figure 10. Alizarin Red S staining quantification of DPSCs upon osteogenic differentiation on Ti, 3D-Ti-MS, and 3D-Ti-MS-NS substrates after (**a**) 7, (**b**) 14, and (**c**) 21 days of osteoinduction; OM—cells cultured only in osteogenic medium. *** $p \leq 0.001$; **** $p \leq 0.0001$.

Alizarin Red staining revealed enhanced mineralization in cells exposed to the materials compared with those cultured with osteogenic media alone over periods of 7, 14, and 21 days, with a significant difference ($p \leq 0.001$) compared with the control. This increased mineralization is indicative of cellular differentiation. During this process, an extracellular matrix (ECM) rich in collagen is initially formed, followed by the deposition of hydroxyapatite (HAP)-like mineralized protein matrix, which is in agreement with previous studies [42,43].

3.2.7. qPCR

qPCR analysis and gene expression after 21 days of osteoinduction of cells in contact with Ti (control), 3D-Ti-MS, and 3D-Ti-MS-NS substrates are presented in Figure 11. Gene expression plays a crucial role in DPSC differentiation into osteoblasts, and this study showed that cells cultured on titanium materials with micro- and combined micro- and nanostructures exhibited significantly ($p \leq 0.05$) higher expression of key osteogenic genes, such as alkaline phosphatase (*ALP*) and osteocalcin (*OCN*), after 21 days of osteoinduction.

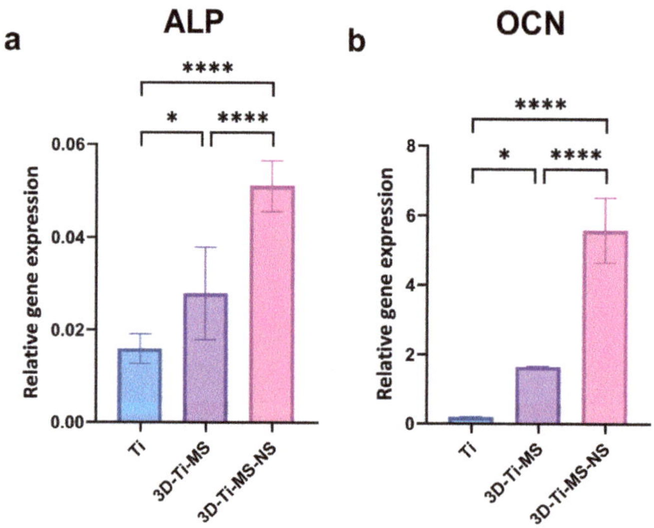

Figure 11. qPCR analysis of gene expression after 21 days of osteoinduction of cells in contact with Ti (control), 3D-Ti-MS, and 3D-Ti-MS-NS substrates showing (**a**) *ALP* and (**b**) *OCN* osteogenic markers; * $p \leq 0.05$; **** $p \leq 0.0001$.

3.3. Bacterial Interaction Studies on Combined Micro- and Nanostructures

The number of bacterial cells in the medium around Ti (control), 3D-Ti-MS, and 3D-Ti-MS-NS substrates assessed using CFU and MTT assays after 24 h of incubation with *E. coli* and *S. aureus* are presented in Figure 12. CFUs of both bacteria in the medium around the samples were lower for 3D-Ti-MS than around control samples. The number of *S. aureus* around samples was lower for both types of 3D-Ti samples compared with the control, presenting that nanopillars showed better results against Gram-positive bacteria (Figure 12c,d). In addition, the MTT assay results, as a measure of bacterial metabolism, were consistent with the CFU findings. Bacterial viability was significantly lower ($p \leq 0.0001$) for *E. coli* around the 3D-Ti-MS and 3D-Ti-MS-NS plates compared with the smooth Ti (Figure 12a). The same trend was noted in the *S. aureus* group, however, without reaching a statistically significant difference (Figure 12b).

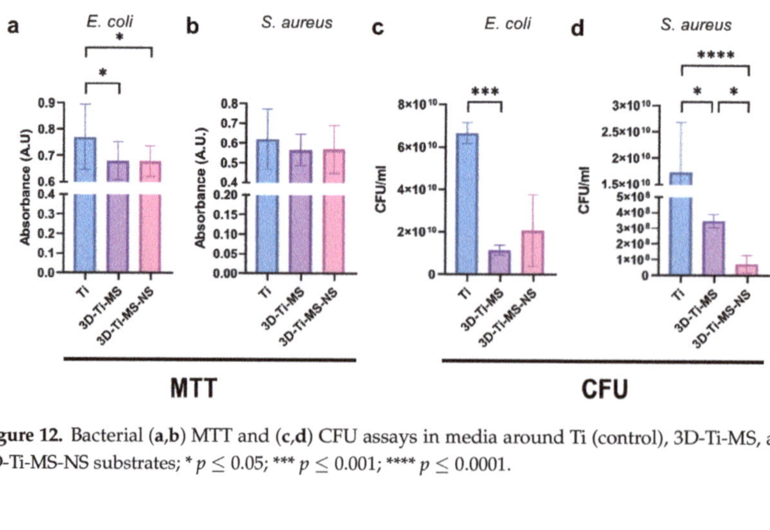

Figure 12. Bacterial (**a**,**b**) MTT and (**c**,**d**) CFU assays in media around Ti (control), 3D-Ti-MS, and 3D-Ti-MS-NS substrates; * $p \leq 0.05$; *** $p \leq 0.001$; **** $p \leq 0.0001$.

The interaction of 3D-Ti-MS and 3D-Ti-MS-NS substrates with micro- and combined micro- and nano-topography and two bacteria, *E. coli* and *S. aureus*, is visualized by SEM imaging showing a series of SEM images summarized in Figure 13. These results show that more *E. coli* and *S. aureus* bacteria cells are observed attached to 3D-Ti-MS compared with 3D-Ti-MS-NS. In both cases, the formation of biofilms is observed on the crevices of microspheres where large numbers of bacteria have settled down and agglomerated. In the case of 3D-Ti-MS-NS, a lower number of bacteria was seen on the microsphere surface and the surface between microspheres, indicating the ability of nanopillar structures to reduce *E. coli* and *S. aureus* viability compared with 3D-Ti-MS substrates. Moreover, SEM images confirmed the nanopillar structures induced membrane damage inflicted on attached *E. coli* bacterial cells, as seen in Figure 13g, showing the destroyed bacteria body. Overall, it can be concluded that 3D-Ti-MS-NS could significantly reduce the occurrence of bacterial attachment and eradicate those bacterial cells that do attach to the surface compared with smooth Ti substrates.

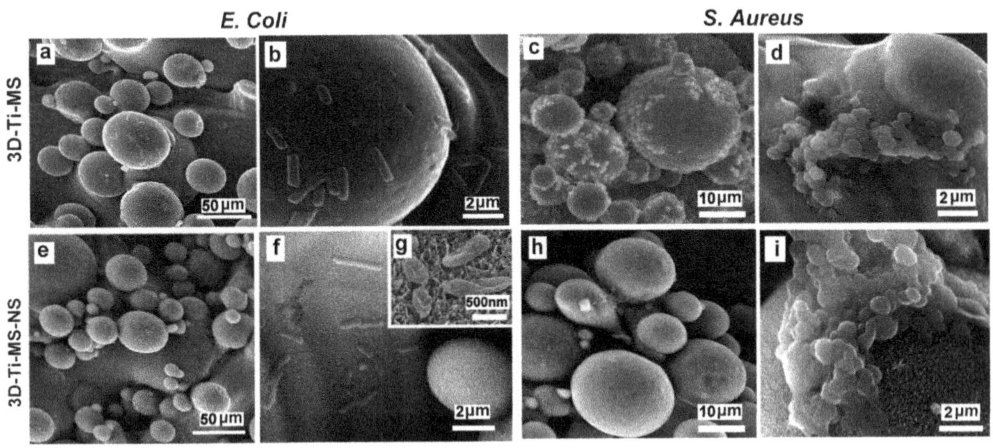

Figure 13. SEM images showing the interaction of 3D-Ti-MS substrate with spherical microparticle topography with (**a**,**b**) *E. coli* bacteria and (**c**,**d**) *S. aureus* and 3D-Ti-MS-NS substrate that combines microstructure topography covered with nanopillar structures with (**e**–**g**) *E. coli* and (**h**,**i**) *S. aureus*.

4. Discussion

4.1. Projected Design of Micro- and Nano-Topography

To investigate the biocompatibility and antibacterial properties of titanium alloys with dual micro- and nano-topography fabricated via AM (SLM) in this study, we successfully fabricated these model implant substrates that have different surface features at varying scales, such as micro-smooth and micro-rough with spherical particle morphology, which are additionally modified with nanostructured surfaces with distinctive sharp nanopillar structures. Our morphological analysis confirmed that these 3D-printed Ti alloy substrates have characteristic microparticle topographies. Previous studies showed many advantages of these microstructures, which improve the biointegration of these emerging 3D-printed implants in orthopedic medicine and dentistry [44]. The sharp nanostructures in the form of nanopillars are successfully generated on our 3D-Ti-MS substrates (Figure 2). The HT process, involving oxidation and etching of the 3D-Ti-MS wafers followed by the growth of sodium titanate ($Na_2Ti_3O_7$) crystals on the surface, gradually made the distinct nanopillar structures. One of our hypotheses was that these nanopillar structures demonstrate strong antibacterial properties, primarily attributed to their ability to mechanically disrupt the membranes of bacterial cells and eventually kill bacteria [45,46].The second hypothesis posited that, despite the antibacterial properties of these surfaces, they would maintain cellular biocompatibility.

The ability to produce wafers with varying hydrophobic and hydrophilic properties is of great importance, as the interaction between implants, cells, and bacteria is significantly affected by surface hydrophilicity. Previous studies have indicated that hydrophilic surfaces enhance the repulsion between bacterial cells and the implant surface while simultaneously reducing the hydrophobic interactions between the bacterial cell membrane and the implant, leading to a decrease in bacterial attachment on the surface [47].The nanostructures formed on the 3D-Ti-MS-NS surface, particularly the nanopillars, contributed to the superhydrophilic nature of the surface (Figure 3f). The wettability of the surface is a critical feature influencing the interaction between cells or bacteria and the implant surface. Previous studies showed that hydrophilicity causes a significant reduction in bacterial cell attachment and film formation on the surface of implants due to inhibition of hydrophobic interactions, promoting the repulsion between the surface and bacteria [15,30]. This effect is particularly important for implants, as bacterial colonization can lead to infection or Ti implant failure. In addition, combining nanostructures with hydrophilicity can enhance protein adsorption and improve osseointegration [42,47].

4.2. Influence of Surface Topography on Cellular Biocompatibility

The MTT, NR, and LDH assay results provided insights into the biocompatibility of titanium-based materials with different surface topographies, namely Ti, 3D-Ti-MS, and 3D-Ti-MS-NS. The findings indicate that all tested substrates supported DPSCs viability, with no significant decrease in cell viability, suggesting that these materials are generally biocompatible and provide a favorable environment for cell growth. The viability of cells in direct contact with the various surfaces remained high, with no essential differences in DPSCs viability between the Ti, 3D-Ti-MS, and 3D-Ti-MS-NS groups (Figure 4a–c, Figures 5a–c and 6). This is a promising outcome, indicating that all these surface modifications support cell growth effectively. Additionally, the indirect MTT and NR viability test, which simulates the conditions after implantation where the cells are not directly in contact with the surface but exposed to the materials microenvironment, showed no significant differences in DPSCs viability across the substrates over the 7-day incubation period. The 3D-Ti-MS-NS surface, which features both microspheres and nanopillars, showed a more favorable outcome in terms of cell membrane integrity compared with 3D-Ti-MS (LDH assay, Figure 6). The presence of nanopillars, which potentially mimic natural extracellular matrix features and likely contributed to enhanced cell attachment and proliferation, possibly by providing additional surface area for cell interaction. Previous studies have indicated that surface roughness and microstructure can affect cell adhesion and proliferation, which might explain the observed statistical differences in cell viability over time.

Cell morphology analysis by SEM (Figure 7) revealed that the surface roughness and topography of the 3D-Ti-MS and 3D-Ti-MS-NS materials facilitate better contact between cells and the implant surface, leading to stronger focal adhesions and more effective cell spreading. In our previous studies, both osteoblasts and fibroblasts showed improved anchorage to roughened titanium surfaces, forming multiple, longer pseudopodia and focal cell-matrix adhesions [28,29]. These interactions are key to the successful integration of the implant into the bone, contributing to more effective osseointegration over time.

Our study additionally assessed the material biocompatibility using a wound scratch healing assay (Figure 8). This method involved the mechanical disruption of cell monolayers at the material interface, resulting in a standardized "wound," which was then monitored for migration and closure. In our study, this assay indicated whether the materials hinder or promote cell movement, which is essential for implant-tissue integration. The results indicated that while cell migration on 3D-Ti (3D-Ti-MS and 3D-Ti-MS-NS) materials was slightly higher than on the smooth Ti surface during the first 24 h, the difference was not statistically significant. However, between 24 and 48 h, cells in contact with the 3D surfaces showed a significantly greater migration rate. This enhanced cell migration is a critical factor in the initial stages of osseointegration, as it supports the healing process and encourages the attachment and proliferation of bone cells around the implant. The rapid migration observed on 3D-Ti surfaces suggests that these materials may promote better early-stage cell attachment and proliferation, which is important for the long-term success of dental implants. Overall, based on cell viability, cell morphology, and migration, 3D-Ti-MS-NS demonstrates the potential to be a suitable candidate for implantable surfaces, as it supports cellular attachment, growth, and migration, minimizing cellular damage.

4.3. Influence of Surface Topography on Parameters of Osseointegration

The ultimate goal of dental implants is to provide long-term functionality and promote rapid bone healing, which begins with osseointegration as the result of the interaction between osteoblasts and the implant surface. This is achieved through osseointegration, a critical process where bone cells interact with and attach to the implant surface. For this reason, we investigated the impact of different titanium surface topographies on the osteogenic differentiation of DPSCs, measured by ALP activity, ARS staining, and qPCR analysis of osteogenesis-related genes. Our results demonstrated that treatment with 3D-Ti materials significantly enhanced ALP activity in DPSCs (Figure 9), with the most pronounced increase observed in cells cultured on 3D-Ti-MS. This suggests that the 3D-Ti surface topography is highly effective in promoting osteoblastic differentiation and mineralization, processes essential for bone formation. The activity of osteoblasts and their ability to form new bone is influenced by the enzyme alkaline phosphatase. In mineralized tissues, ALP plays a crucial role in promoting the formation of hydroxyapatite crystals by increasing the local concentration of inorganic phosphate while simultaneously reducing the extracellular levels of pyrophosphate, an inhibitor of mineral formation [5]. ALP is a well-established biomarker of osteoblastic activity and plays a central role in bone formation by facilitating the mineralization process.

The enhanced ALP activity observed in the 3D-Ti-MS group supports our third hypothesis that micro/nanoscale surface modifications can improve the osteogenic potential of titanium materials. This increased ALP activity is indicative of improved cellular differentiation, which is critical for the success of dental implants and other bone-related applications. The results from this study are consistent with previous research showing that microstructured surfaces can enhance osteoblastic differentiation and mineralization [1,5,7,17,40,48].

The enhanced mineralization observed in cells cultured on the 3D titanium surfaces was further confirmed by Alizarin Red S quantification (Figure 10), which highlighted a greater deposition of mineralized matrix compared with cells treated only with osteogenic media. This difference was statistically significant across all time points (7, 14, and 21 days), suggesting that the surface topography of titanium plays an important role in accelerating osteo-differentiation and matrix mineralization. These findings are consistent with our previous studies, where we demonstrated that surface features such as roughness and micro/nanostructures can improve osteogenic differentiation and mineralization [1,5,7,17,40,42,48,49].

However, the ALP activity assay and ARS staining assessed osteogenic differentiation at a functional level. For that reason, we performed qPCR analysis (Figure 11) of *ALP* and *OCN* to provide additional mechanistic insights. Upregulated *ALP* and *OCN* gene expression confirmed that DPSCs were undergoing robust osteogenic differentiation at both the functional and transcriptional levels. Upregulated *ALP* and *OCN* gene expression suggests that differentiation is transcriptionally regulated, ensuring that high enzymatic activity and mineralization are not just transient effects but are driven by osteogenic gene activation. ALP is a marker of early osteoblast differentiation [50], while OCN is a late-stage marker primarily involved in the mineralization process [49]. The upregulation of both these genes indicates that the titanium materials, particularly the 3D surface structures, are not only biocompatible but also enhance osteoinduction of DPSCs. This is consistent with our earlier findings, where we demonstrated that micro/nanoscale surface modifications on titanium enhanced osteogenic differentiation and mineralization in vitro [29,42,46]. Further research, including in vivo studies, would be beneficial to confirm these findings and explore the long-term effects of these materials on osteointegration.

4.4. Influence of Surface Topography on Biofilm Formation and Antibacterial Properties

The ability of biomaterial surfaces to influence bacterial colonization and biofilm formation is a crucial factor in determining the success of implants, particularly in preventing infection and promoting long-term integration with the host tissue [51–53]. In this study, we evaluated the bacterial colonization on titanium surfaces with different topographies, such as 3D-Ti-MS and 3D-Ti-MS-NS substrates, by assessing bacterial cell counts through CFU and MTT assays and following SEM imaging. Interestingly, the bacterial load was higher on the smooth Ti surface than on 3D-Ti-MS and 3D-Ti-MS-NS substrates. This suggests that while the 3D surfaces may promote bacterial attachment in direct contact, they may also have a greater propensity for bacterial killing or inhibition of bacterial growth in the surrounding area. The indirect bacterial killing observed in the 3D-Ti-MS-NS substrates by SEM imaging showing many bacteria with destroyed membranes (Figure 13g,i), indicated by the lower viable bacteria counts in the surrounding area, could be attributed to the enhanced surface characteristics of the nanopillar structures. These results are in agreement with previous studies showing exceptional antibacterial properties of nanostructured surfaces fabricated by the HT process [13,14,27]. The observed reduction in bacterial adhesion and biofilm formation on the dual-scale modified titanium surfaces may have clinical implications, particularly in the prevention of peri-implantitis—a multifactorial inflammatory condition primarily driven by biofilm accumulation and a leading cause of late-stage implant failure. Our findings align with previous reports that highlight the significance of surface modifications in controlling peri-implant infections [54–56]. These studies have demonstrated that *S. aureus*, due to its strong biofilm-forming ability, plays an important role in the pathogenesis of peri-implantitis [56]. In addition, *E. coli* has been identified in early-stage peri-implant mucositis. Similarly to our study, it was reported that Poly[2-(methacryloyloxy)ethyl choline phosphate]-modified Ti surfaces could disrupt early *E. coli* and *S. aureus* colonization [54].

These nanopillar structures appear to support bacterial killing in adjacent areas, offering a complex but potentially beneficial interaction that could be harnessed in the design of implant surfaces to reduce infection rates and improve implant success in clinical applications. In contrast with the control smooth Ti surfaces, bacteria seem to be able to survive more readily in the surrounding areas, possibly due to a lack of surface modifications that could otherwise disrupt bacterial activity. Overall, these results of the CFU, MTT assays, and SEM imaging indicate that while the 3D-Ti-MS-NS materials with combined micro- and nanostructures support bacterial attachment, they show properties for inducing bacterial killing areas, which are desirable features for implant materials to reduce the risk of infection. Further investigations into the specific mechanisms underlying these effects, such as antimicrobial surface properties or the influence of surface nanostructures on bacterial behavior, are needed to fully understand the interaction between titanium implants and bacterial biofilms. While *S. aureus* and *E. coli* were selected as representative

Gram-positive and Gram-negative strains, respectively, we acknowledge that the use of only two bacterial species limits the extent of our antimicrobial evaluation. Peri-implant infections are often polymicrobial in nature, involving complex biofilms that include anaerobes and opportunistic pathogens (e.g., *Candida albicans*). Future studies could expand the microbial panel and incorporate multispecies biofilm models to provide a more comprehensive understanding of the antimicrobial efficacy under conditions that more closely mimic the in vivo peri-implant environment.

5. Conclusions

This study demonstrated that 3D surface modifications on titanium substrates with combined micro- and nanostructures significantly enhance cell viability, osteointegration, and antimicrobial properties compared with smooth Ti surfaces. SEM, EDX, and XRD analyses confirmed the formation of microspheres on 3D-Ti-MS and nanopillars on 3D-Ti-MS-NS surfaces, with the latter exhibiting increased hydrophilicity that is favorable for better cell interactions that are confirmed by the presented results. Dental pulp stem cell viability assays showed that 3D-Ti-MS-NS promoted higher cell growth and membrane integrity, while osteogenic assays confirmed its ability to enhance osteodifferentiation. Additionally, both 3D-Ti-MS and 3D-Ti-MS-NS demonstrated reduced bacterial adhesion and biofilm formation, whereas 3D-Ti-MS-NS showed the ability to mechanically destroy attached bacterial cells, highlighting their potential for preventing implant infections.

In conclusion, 3D-fabricated titanium implants with combined micro- and nanostructures offer significant advantages compared with traditional Ti implants, providing low-cost, fast, and custom-made on-site implants when they are needed. Their additional advancements in terms of biological performance and antimicrobial properties make them promising candidates for the next generation of Ti implants for various applications, specifically for dental implants. Future studies should focus on optimizing surface topography, further enhancing antimicrobial effects through functional coatings, and conducting long-term in vivo studies to assess the clinical efficacy of these materials. Exploring the molecular mechanisms of osteogenesis on 3D-Ti-MS-NS surfaces and validating these findings in clinical trials will be key steps toward ensuring their successful translation to implant applications.

Supplementary Materials: The following supporting information can be downloaded at: https://www.mdpi.com/article/10.3390/jfb16050157/s1, Figure S1: Schematic diagram of wound healing assay; Figure S2. Representative SEM image of Ti (Control). Bar scale is 10 μm, Table S1: The list of specific primers used in the study; Table S2. Growth conditions for activation of bacteria strains used for monomicrobial biofilm formation. Manufacturer of growth medium: *HIMEDIA (India); **ProReady (Serbia). Video S1: Wound scratch healing assay on Ti substrate; Video S2: Wound scratch healing assay on 3D-Ti-MS substrate; Video S3: Wound scratch healing assay on 3D-Ti-MS-NS substrate.

Author Contributions: Conceptualization, D.L., N.L.I. and M.L.; methodology, D.L., N.L.I. and M.L.; validation, D.L., N.L.I. and M.L. investigation, S.M., M.L., S.P. and A.Ž.; resources, D.L., N.L.I. and M.L., data curation, D.L. and N.L.I., writing—original draft preparation, S.M, D.L., M.L. and N.L.I.; writing—review and editing, D.L., M.L. and N.L.I.; visualization, S.M., D.L. and M.L.; supervision, D.L., M.L. and N.L.I.; project administration, D.L. and N.L.I.; funding acquisition, N.L.I. and M.L. All authors have read and agreed to the published version of the manuscript.

Funding: This research was funded by the Ministry of Science, Technological Development and Innovation of the Republic of Serbia on the research programs: Grant No. 451-03-136/2025-03/200129 (University of Belgrade, School of Dental Medicine) and Grant No. 451-03-136/2025-03/200175 (Institute of Technical Sciences of SASA).

Institutional Review Board Statement: The study was approved by the Ethical Committee (No. 36/26, 19 November 2021) at the School of Dental Medicine, University of Belgrade.

Informed Consent Statement: Informed consent was obtained from all subjects involved in the study.

Data Availability Statement: Data underlying the results presented in this paper may be obtained from the authors upon reasonable request.

Conflicts of Interest: The authors declare no conflicts of interest.

References

1. Subramani, K.; Mathew, R.T.; Pachauri, P. Titanium Surface Modification Techniques for Dental Implants—From Microscale to Nanoscale. In *Emerging Nanotechnologies in Dentistry*; Elsevier: Amsterdam, The Netherlands, 2018; pp. 99–124, ISBN 978-0-12-812291-4.
2. Pantović Pavlović, M.R.; Stanojević, B.P.; Pavlović, M.M.; Mihailović, M.D.; Stevanović, J.S.; Panić, V.V.; Ignjatović, N.L. Anodizing/Anaphoretic Electrodeposition of Nano-Calcium Phosphate/Chitosan Lactate Multifunctional Coatings on Titanium with Advanced Corrosion Resistance, Bioactivity, and Antibacterial Properties. *ACS Biomater. Sci. Eng.* **2021**, *7*, 3088–3102. [CrossRef]
3. Chopra, D.; Guo, T.; Jayasree, A.; Gulati, K.; Ivanovski, S. Bioinspired, Bioactive, and Bactericidal: Anodized Nanotextured Dental Implants. *Adv. Funct. Mater.* **2024**, *34*, 2314031. [CrossRef]
4. Li, X.; Wu, B.; Chen, H.; Nan, K.; Jin, Y.; Sun, L.; Wang, B. Recent Developments in Smart Antibacterial Surfaces to Inhibit Biofilm Formation and Bacterial Infections. *J. Mater. Chem. B* **2018**, *6*, 4274–4292. [CrossRef]
5. Zamparini, F.; Prati, C.; Generali, L.; Spinelli, A.; Taddei, P.; Gandolfi, M.G. Micro-Nano Surface Characterization and Bioactivity of a Calcium Phosphate-Incorporated Titanium Implant Surface. *J. Funct. Biomater.* **2021**, *12*, 3. [CrossRef]
6. Roy, P.; Berger, S.; Schmuki, P. TiO_2 Nanotubes: Synthesis and Applications. *Angew. Chem. Int. Ed.* **2011**, *50*, 2904–2939. [CrossRef]
7. Gulati, K.; Maher, S.; Findlay, D.M.; Losic, D. Titania Nanotubes for Orchestrating Osteogenesis at the Bone–Implant Interface. *Nanomedicine* **2016**, *11*, 1847–1864. [CrossRef]
8. Losic, D. Advancing of Titanium Medical Implants by Surface Engineering: Recent Progress and Challenges. *Expert. Opin. Drug Deliv.* **2021**, *18*, 1355–1378. [CrossRef]
9. Maher, S.; Mazinani, A.; Barati, M.R.; Losic, D. Engineered Titanium Implants for Localized Drug Delivery: Recent Advances and Perspectives of Titania Nanotubes Arrays. *Expert. Opin. Drug Deliv.* **2018**, *15*, 1021–1037. [CrossRef]
10. Linklater, D.P.; Baulin, V.A.; Juodkazis, S.; Crawford, R.J.; Stoodley, P.; Ivanova, E.P. Mechano-Bactericidal Actions of Nanostructured Surfaces. *Nat. Rev. Microbiol.* **2021**, *19*, 8–22. [CrossRef]
11. Ivanova, E.P.; Linklater, D.P.; Werner, M.; Baulin, V.A.; Xu, X.; Vrancken, N.; Rubanov, S.; Hanssen, E.; Wandiyanto, J.; Truong, V.K.; et al. The Multi-Faceted Mechano-Bactericidal Mechanism of Nanostructured Surfaces. *Proc. Natl. Acad. Sci. USA* **2020**, *117*, 12598–12605. [CrossRef]
12. Linklater, D.P.; Juodkazis, S.; Ivanova, E.P. Nanofabrication of Mechano-Bactericidal Surfaces. *Nanoscale* **2017**, *9*, 16564–16585. [CrossRef]
13. Maher, S.; Linklater, D.; Rastin, H.; Liao, S.T.-Y.; Martins De Sousa, K.; Lima-Marques, L.; Kingshott, P.; Thissen, H.; Ivanova, E.P.; Losic, D. Advancing of 3D-Printed Titanium Implants with Combined Antibacterial Protection Using Ultrasharp Nanostructured Surface and Gallium-Releasing Agents. *ACS Biomater. Sci. Eng.* **2022**, *8*, 314–327. [CrossRef]
14. Maher, S.; Wijenayaka, A.R.; Lima-Marques, L.; Yang, D.; Atkins, G.J.; Losic, D. Advancing of Additive-Manufactured Titanium Implants with Bioinspired Micro- to Nanotopographies. *ACS Biomater. Sci. Eng.* **2021**, *7*, 441–450. [CrossRef]
15. Maher, S.; Kaur, G.; Lima-Marques, L.; Evdokiou, A.; Losic, D. Engineering of Micro- to Nanostructured 3D-Printed Drug-Releasing Titanium Implants for Enhanced Osseointegration and Localized Delivery of Anticancer Drugs. *ACS Appl. Mater. Interfaces* **2017**, *9*, 29562–29570. [CrossRef]
16. Aparicio, C.; Padrós, A.; Gil, F.-J. In Vivo Evaluation of Micro-Rough and Bioactive Titanium Dental Implants Using Histometry and Pull-out Tests. *J. Mech. Behav. Biomed. Mater.* **2011**, *4*, 1672–1682. [CrossRef]
17. Lyons, J.G.; Plantz, M.A.; Hsu, W.K.; Hsu, E.L.; Minardi, S. Nanostructured Biomaterials for Bone Regeneration. *Front. Bioeng. Biotechnol.* **2020**, *8*, 922. [CrossRef]
18. Civantos, A.; Martínez-Campos, E.; Ramos, V.; Elvira, C.; Gallardo, A.; Abarrategi, A. Titanium Coatings and Surface Modifications: Toward Clinically Useful Bioactive Implants. *ACS Biomater. Sci. Eng.* **2017**, *3*, 1245–1261. [CrossRef]
19. Wang, H.; Lai, Y.-K.; Zheng, R.-Y.; Bian, Y.; Lin, C.-J.; Zhang, K.-Q. Tuning the Surface Microstructure of Titanate Coatings on Titanium Implants for Enhancing Bioactivity of Implants. *Int. J. Nanomed.* **2015**, 3887–3896. [CrossRef]
20. Kunze, J.; Müller, L.; Macak, J.M.; Greil, P.; Schmuki, P.; Müller, F.A. Time-Dependent Growth of Biomimetic Apatite on Anodic TiO_2 Nanotubes. *Electrochim. Acta* **2008**, *53*, 6995–7003. [CrossRef]

21. Gristina, A.G. Biomaterial-Centered Infection: Microbial Adhesion Versus Tissue Integration. *Science* **1987**, *237*, 1588–1595. [CrossRef] [PubMed]
22. Ayliffe, G.A.J. Role of the Environment of the Operating Suite in Surgical Wound Infection. *Clin. Infect. Dis.* **1991**, *13*, S800–S804. [CrossRef]
23. Ni, J.; Ling, H.; Zhang, S.; Wang, Z.; Peng, Z.; Benyshek, C.; Zan, R.; Miri, A.K.; Li, Z.; Zhang, X.; et al. Three-Dimensional Printing of Metals for Biomedical Applications. *Mater. Today Bio* **2019**, *3*, 100024. [CrossRef]
24. Tshephe, T.S.; Akinwamide, S.O.; Olevsky, E.; Olubambi, P.A. Additive Manufacturing of Titanium-Based Alloys- A Review of Methods, Properties, Challenges, and Prospects. *Heliyon* **2022**, *8*, e09041. [CrossRef]
25. Tofail, S.A.M.; Koumoulos, E.P.; Bandyopadhyay, A.; Bose, S.; O'Donoghue, L.; Charitidis, C. Additive Manufacturing: Scientific and Technological Challenges, Market Uptake and Opportunities. *Mater. Today* **2018**, *21*, 22–37. [CrossRef]
26. Qu, M.; Wang, C.; Zhou, X.; Libanori, A.; Jiang, X.; Xu, W.; Zhu, S.; Chen, Q.; Sun, W.; Khademhosseini, A. Multi-Dimensional Printing for Bone Tissue Engineering. *Adv. Healthc. Mater.* **2021**, *10*, 2001986. [CrossRef]
27. Maher, S.; Linklater, D.; Rastin, H.; Le Yap, P.; Ivanova, E.P.; Losic, D. Tailoring Additively Manufactured Titanium Implants for Short-Time Pediatric Implantations with Enhanced Bactericidal Activity. *ChemMedChem* **2022**, *17*, e202100580. [CrossRef]
28. Gulati, K.; Prideaux, M.; Kogawa, M.; Lima-Marques, L.; Atkins, G.J.; Findlay, D.M.; Losic, D. Anodized 3D-Printed Titanium Implants with Dual Micro- and Nano-Scale Topography Promote Interaction with Human Osteoblasts and Osteocyte-like Cells: 3D Printed Titanium Implants with Dual Micro- and Nano-Scale Topography. *J. Tissue Eng. Regen. Med.* **2017**, *11*, 3313–3325. [CrossRef] [PubMed]
29. Qin, J.; Yang, D.; Maher, S.; Lima-Marques, L.; Zhou, Y.; Chen, Y.; Atkins, G.J.; Losic, D. Micro- and Nano-Structured 3D Printed Titanium Implants with a Hydroxyapatite Coating for Improved Osseointegration. *J. Mater. Chem. B* **2018**, *6*, 3136–3144. [CrossRef]
30. Iezzi, G.; Zavan, B.; Petrini, M.; Ferroni, L.; Pierfelice, T.V.; D'Amora, U.; Ronca, A.; D'Amico, E.; Mangano, C. 3D Printed Dental Implants with a Porous Structure: The in Vitro Response of Osteoblasts, Fibroblasts, Mesenchymal Stem Cells, and Monocytes. *J. Dent.* **2024**, *140*, 104778. [CrossRef]
31. Lee, U.-L.; Yun, S.; Lee, H.; Cao, H.-L.; Woo, S.-H.; Jeong, Y.-H.; Jung, T.-G.; Kim, C.M.; Choung, P.-H. Osseointegration of 3D-Printed Titanium Implants with Surface and Structure Modifications. *Dent. Mater.* **2022**, *38*, 1648–1660. [CrossRef]
32. Mohandas, A.; Krishnan, A.G.; Biswas, R.; Menon, D.; Nair, M.B. Antibacterial and Cytocompatible Nanotextured Ti Surface Incorporating Silver via Single Step Hydrothermal Processing. *Mater. Sci. Eng. C* **2017**, *75*, 115–124. [CrossRef] [PubMed]
33. Anitha, V.C.; Banerjee, A.N.; Joo, S.W.; Min, B.K. Morphology-Dependent Low Macroscopic Field Emission Properties of Titania/Titanate Nanorods Synthesized by Alkali-Controlled Hydrothermal Treatment of a Metallic Ti Surface. *Nanotechnology* **2015**, *26*, 355705. [CrossRef]
34. Cao, Y.; Su, B.; Chinnaraj, S.; Jana, S.; Bowen, L.; Charlton, S.; Duan, P.; Jakubovics, N.S.; Chen, J. Nanostructured Titanium Surfaces Exhibit Recalcitrance towards Staphylococcus Epidermidis Biofilm Formation. *Sci. Rep.* **2018**, *8*, 1071. [CrossRef] [PubMed]
35. Vukovic, M.; Lazarevic, M.; Mitic, D.; Jaksic Karisik, M.; Ilic, B.; Andric, M.; Jevtic, B.; Roganovic, J.; Milasin, J. Acetylsalicylic-Acid (ASA) Regulation of Osteo/Odontogenic Differentiation and Proliferation of Human Dental Pulp Stem Cells (DPSCs) in Vitro. *Arch. Oral Biol.* **2022**, *144*, 105564. [CrossRef]
36. Karisik, M.J.; Lazarevic, M.; Mitic, D.; Nikolic, N.; Markovic, M.M.; Jelovac, D.; Milasin, J. Osteogenic and Adipogenic Differentiation Potential of Oral Cancer Stem Cells May Offer New Treatment Modalities. *Int. J. Mol. Sci.* **2023**, *24*, 4704. [CrossRef]
37. Mathar, M.I. Surface Morphology and Chemical Composition Analysis of Titanium Dental Implants Using SEM and EDX. *Bioinformation* **2024**, *20*, 926–932. [CrossRef]
38. Stojkovska, J.; Kostić, D.; Jovanović, Ž.; Vukašinović-Sekulić, M.; Mišković-Stanković, V.; Obradović, B. A Comprehensive Approach to in Vitro Functional Evaluation of Ag/Alginate Nanocomposite Hydrogels. *Carbohydr. Polym.* **2014**, *111*, 305–314. [CrossRef]
39. Wennerberg, A.; Albrektsson, T. Effects of Titanium Surface Topography on Bone Integration: A Systematic Review. *Clin. Oral Implant. Res.* **2009**, *20*, 172–184. [CrossRef]
40. Neoh, K.G.; Hu, X.; Zheng, D.; Kang, E.T. Balancing Osteoblast Functions and Bacterial Adhesion on Functionalized Titanium Surfaces. *Biomaterials* **2012**, *33*, 2813–2822. [CrossRef] [PubMed]
41. Stoilov, M.; Stoilov, L.; Enkling, N.; Stark, H.; Winter, J.; Marder, M.; Kraus, D. Effects of Different Titanium Surface Treatments on Adhesion, Proliferation and Differentiation of Bone Cells: An In Vitro Study. *J. Funct. Biomater.* **2022**, *13*, 143. [CrossRef]
42. Chen, M. Recent Advances and Perspective of Nanotechnology-Based Implants for Orthopedic Applications. *Front. Bioeng. Biotechnol.* **2022**, *10*, 878257. [CrossRef]
43. Bernar, A.; Gebetsberger, J.V.; Bauer, M.; Streif, W.; Schirmer, M. Optimization of the Alizarin Red S Assay by Enhancing Mineralization of Osteoblasts. *Int. J. Mol. Sci.* **2022**, *24*, 723. [CrossRef]
44. Meng, M.; Wang, J.; Huang, H.; Liu, X.; Zhang, J.; Li, Z. 3D Printing Metal Implants in Orthopedic Surgery: Methods, Applications and Future Prospects. *J. Orthop. Transl.* **2023**, *42*, 94–112. [CrossRef]

45. Ren, B.; Wan, Y.; Liu, C.; Wang, H.; Yu, M.; Zhang, X.; Huang, Y. Improved Osseointegration of 3D Printed Ti-6Al-4V Implant with a Hierarchical Micro/Nano Surface Topography: An in Vitro and in Vivo Study. *Mater. Sci. Eng. C* **2021**, *118*, 111505. [CrossRef]
46. Yang, S.; Jiang, W.; Ma, X.; Wang, Z.; Sah, R.L.; Wang, J.; Sun, Y. Nanoscale Morphologies on the Surface of 3D-Printed Titanium Implants for Improved Osseointegration: A Systematic Review of the Literature. *IJN* **2023**, *18*, 4171–4191. [CrossRef]
47. Yuan, Y.; Hays, M.P.; Hardwidge, P.R.; Kim, J. Surface Characteristics Influencing Bacterial Adhesion to Polymeric Substrates. *RSC Adv.* **2017**, *7*, 14254–14261. [CrossRef]
48. Zhang, L.; Chen, L. A Review on Biomedical Titanium Alloys: Recent Progress and Prospect. *Adv. Eng. Mater.* **2019**, *21*, 1801215. [CrossRef]
49. Ikegame, M.; Ejiri, S.; Okamura, H. Expression of Non-Collagenous Bone Matrix Proteins in Osteoblasts Stimulated by Mechanical Stretching in the Cranial Suture of Neonatal Mice. *J. Histochem. Cytochem.* **2019**, *67*, 107–116. [CrossRef]
50. Ansari, S.; Ito, K.; Hofmann, S. Alkaline Phosphatase Activity of Serum Affects Osteogenic Differentiation Cultures. *ACS Omega* **2022**, *7*, 12724–12733. [CrossRef]
51. Kligman, S.; Ren, Z.; Chung, C.-H.; Perillo, M.A.; Chang, Y.-C.; Koo, H.; Zheng, Z.; Li, C. The Impact of Dental Implant Surface Modifications on Osseointegration and Biofilm Formation. *J. Clin. Med.* **2021**, *10*, 1641. [CrossRef]
52. Li, P.; Yin, R.; Cheng, J.; Lin, J. Bacterial Biofilm Formation on Biomaterials and Approaches to Its Treatment and Prevention. *Int. J. Mol. Sci.* **2023**, *24*, 11680. [CrossRef]
53. Nandakumar, V.; Chittaranjan, S.; Kurian, V.M.; Doble, M. Characteristics of Bacterial Biofilm Associated with Implant Material in Clinical Practice. *Polym. J.* **2013**, *45*, 137–152. [CrossRef]
54. Sun, T.; Zhai, X.; Xu, Z.; Li, J.; Chen, X. Multifunctional Titanium Surface Coating for Oral Implants with Strong Resistance to Bacteria and Mineralization Ability. *Colloid Interface Sci. Commun.* **2023**, *52*, 100688. [CrossRef]
55. Persson, G.R.; Renvert, S. Cluster of Bacteria Associated with Peri-Implantitis. *Clin. Implant. Dent. Relat. Res.* **2014**, *16*, 783–793. [CrossRef]
56. Cui, Z.; Wang, P.; Gao, W. Microbial Dysbiosis in Periodontitis and Peri-Implantitis: Pathogenesis, Immune Responses, and Therapeutic. *Front. Cell. Infect. Microbiol.* **2025**, *15*, 1517154. [CrossRef]

Journal of
Functional Biomaterials

MDPI

Article

In Vitro Biological Properties Assessment of 3D-Printed Hydroxyapatite–Polylactic Acid Scaffolds Intended for Bone Regeneration

Eddy Shan [1,*], Cristina Chamorro [2], Ana Ferrández-Montero [3], Rosa M. Martin-Rodriguez [3], Begoña Ferrari [3], Antonio Javier Sanchez-Herencia [3], Leire Virto [2], María José Marín [2], Elena Figuero [1,2] and Mariano Sanz [1,2,*]

1 Section of Periodontology, Faculty of Odontology, Complutense University of Madrid, 28040 Madrid, Spain; elfiguer@ucm.es
2 Etiology and Therapy of Periodontal and Peri-implant Diseases (ETEP) Research Group, Complutense University of Madrid, 28040 Madrid, Spain; mjmarin@ucm.es (M.J.M.)
3 Institute of Ceramics and Glass (ICV), Spanish National Research Council (CSIC), 28049 Madrid, Spain; bferrari@icv.csic.es (B.F.)
* Correspondence: eddyshan@ucm.es (E.S.); marsan@ucm.es (M.S.)

Abstract: This study evaluated the biological performance in vitro of two 3D-printed hydroxyapatite (HA) and polylactic acid (PLA) composite scaffolds with two different infill densities (50% [HA-PLA50] and 70% [HA-PLA70]). Comparative analysis using MG-63 cell cultures evaluated the following: (1) integrity after exposure to various sterilization methods; (2) cell viability; (3) morphological characteristics; (4) cell proliferation; (5) cytotoxicity; (6) gene expression; and (7) protein synthesis. Ultraviolet radiation was the preferred sterilization method. Both scaffolds maintained adequate cell viability and proliferation over 7 days without significant differences in cytotoxicity. Notably, HA-PLA50 scaffolds demonstrated superior osteogenic potential, showing a significantly higher expression of collagen type I (COL1A1) and an increased synthesis of interleukins 6 and 8 (IL-6, IL-8) compared to HA-PLA70 scaffolds. While both scaffold types supported robust cell growth, the HA-PLA50 formulation exhibited enhanced bioactivity, suggesting a potential advantage for bone tissue engineering applications. These findings provide important insights for optimizing 3D-printed bone graft substitutes.

Keywords: tissue engineering; hydroxyapatite-polylactic acid; in vitro; scaffold; additive manufacturing; bone regeneration

1. Introduction

Dimensional changes in the alveolar ridge after tooth loss can lead to significant bone remodeling during the first year of healing, often preventing the placement of dental implants in a prosthetically driven position [1]. These clinical situations are currently managed with bone augmentation interventions, either simultaneous or staged with implant placement. These interventions follow tissue engineering principles, utilizing biomaterials and signaling molecules and/or cells while adhering to wound healing bases, such as primary wound closure, angiogenesis, space maintenance, and stability of the blood clot [2].

The most common bone augmentation approach has been guided bone regeneration, based on the use of barrier membranes combined with bone grafts or bone replacement grafts (autologous, xenogeneic, or allogeneic grafts) to achieve bone regeneration by promoting cell homing while maintaining the needed space for bone augmentation [3,4]. However, when using autografts, either blocks or particulate, bone availability is limited by anatomical constraints, and its harvesting is usually associated with increased morbidity and faster resorption [5]. Similarly, conventional xenografts and allogeneic grafts usually present difficulties in clinical handling and poorer bioactivity and dimensional stability [6]. These limitations, together with the current trends in avoiding human or animal-derived biomaterials, have prompted the research community to develop synthetically derived bone replacement grafts, which would provide the adequate handling and biological and physic-mechanical properties for the bone regeneration of the alveolar ridges [7].

Different metals, polymers, ceramics, and composite materials have been evaluated as biomaterials for tissue engineering applications aimed at bone regeneration [8,9]. Among these, polymers such as polylactic acid (PLA) have been evaluated, since they are biocompatible, biodegradable, and widely used as biomaterials in medical implant devices, as they are medical grade biomaterials (approved by the US Food and Drug Administration for medical use) [10]. However, PLA is limited by low bioactivity [7,11], accelerated resorption rates, and the release of acidic concentrations during its degradation [12]. Combining PLA with bioceramics like hydroxyapatite (HA) enhances its osteogenic bioactivity and counteracts the acidic environment produced during the in vitro degradation of PLA [13,14].

In recent years, the use of natural biomaterials as particulate or block forms, is being replaced by tissue engineering approaches using synthetic biomaterials where the scaffolds mimic the extracellular matrix of native bone, providing the framework for cell proliferation, cell differentiation and vascularization, thus promoting bone regeneration [15]. Scaffold properties, such as porosity, elasticity, and stability, have been shown to influence cell differentiation [11,16]. In fact, increased porosity has been associated with increased vascularity and a better supply of oxygen and nutrients, manifested by increased alkaline phosphatase activity, calcium mineralization, and faster bone formation [17].

These new biomaterials-based scaffolds and their fabrication methods are also being investigated to determine their mechanical properties [7,18] in search of customized solutions in bone regeneration. With new developments in imaging techniques and 3D printing, computer-based, additively manufactured scaffolds have been designed and fabricated using electrospinning and fused filament fabrication techniques [19–22].

Thus, in this research, we aim to characterize the in vitro biological properties of customizable HA-PLA scaffolds intended for bone regeneration fabricated with two different infill densities (50% vs. 70%).

2. Materials and Methods

2.1. Study Design

This in vitro investigation employs seven assays to examine the biological properties of HA-PLA scaffolds at two infill densities (50% and 70%). Experimental MG-63 cell lines from human osteosarcoma are positive controls, while the cells growing on the well surface act as negative controls. Ethics approval was not required for this study as it did not involve human or animal subjects.

For each assay, three to six biological replicates were carried out (except for the sterilization method assay, which used two replicates). All scaffolds were seeded simultaneously at the initial time point, and at each study interval (24 h, 48 h, 72 h, and 7 days), they were assessed and discarded.

2.2. Fabrication and Design of Scaffolds

Commercial HA-containing filaments (FOss HA, COLFEED4Print S.L., Madrid, Spain) with a diameter of 1.75 mm and nominal particle content of 20 vol.% HA were used to print customizable 3D scaffolds. The filaments consist of highly dispersed HA particles within a thermoplastic matrix of PLA, fabricated using a colloidal method [23]. Cylindrical scaffolds with a diameter of 9 mm and a height of 2.5 mm were printed using the FFF (Fused Filament Fabrication) technique on an NX PRO Dual printer (Tumaker, Gipuzkoa, Spain) with a nozzle diameter of 0.4 mm. The cylinders were designed with a 3D freeware (Tinkercad, Autodesk®, San Francisco, CA, USA), while slicing was performed with Ultimaker Cura v5.3.1 software (Ultimaker, Utrecht, The Netherlands), selecting linear infill patterns. The printing process involved setting the bed and nozzle temperatures to 40 and 165 °C, respectively, with a filament feed rate of 20 mm/s. A standardized scaffold morphology was employed for both the 50% and 70% infill scaffolds, labeled HA-PLA50 and HA-PLA70 (Figure 1). The different infill densities tested are translated into differences in porosity, total material volume, and printing times. Specifically, a higher infill density reduces scaffold porosity and increases structural integrity and mechanical strength, albeit with longer printing times and greater material consumption, while low infill density results in higher porosity, which may enhance nutrient diffusion and cell infiltration at the expense of mechanical properties.

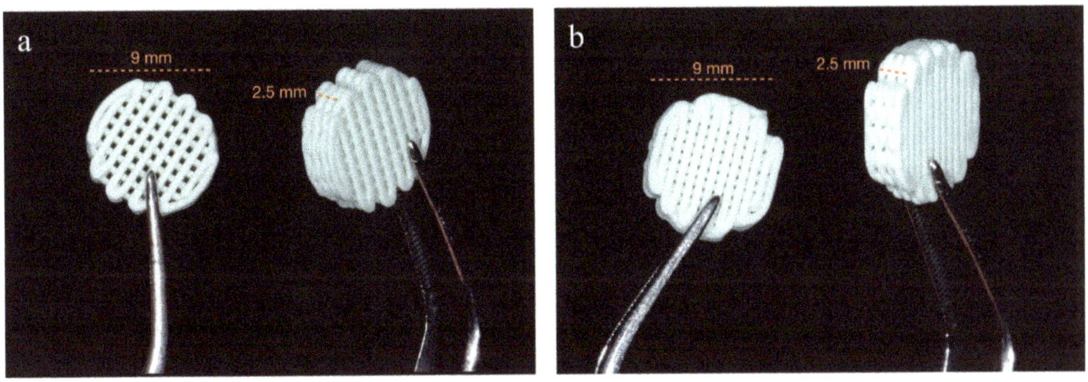

Figure 1. Scaffold designs, both with an analogous cylindrical geometry (9 mm diameter × 2.5 mm height). (**a**) HA-PLA50 scaffold design with a lower infill density (50%), exhibiting higher porosity due to wider spacing between deposited filaments. (**b**) HA-PLA70 scaffold design with a higher infill density (70%), which translates into a lower porosity and a denser internal architecture.

Particle size ranges between 1 and 2 μm (Figure A1). The high dispersion of HA and the absence of agglomerates are expected to facilitate the osteoconductive effect until the complete biodegradation of the scaffold [24].

2.3. Cell Culture

Experimental MG-63 cell lines derived from human osteosarcoma (ATCC, CRL-1427) were utilized to evaluate the biological properties of the scaffolds. The cells were cultured in 75 cm^2 flasks using Eagle's Minimum Essential Medium (MEM) with Earle's Balanced Salt Solution (Sigma-Aldrich, Saint Louis, MO, USA), supplemented with 10% fetal bovine serum (FBS) (Biowest, Nuaillé, France) and 1% penicillin–streptomycin (Gibco, Grand Island, NY, USA). The standard conditions for MG-63 cell incubation were set at 37 °C in a humidified atmosphere containing 5% CO_2. Subcultures were performed once 80% cell confluence was reached. Cryopreservation of the cells was conducted using liquid nitrogen (−196 °C).

2.4. Experimental Assays

2.4.1. Sterilization Method

Three methods of sterilization were compared: ultraviolet (UV) radiation (30 min per side), 70% ethanol (1 min + 3 PBS rinses) [25], and autoclave (121 °C, 1.10 atm, 20 min). For each method, two biological replicates of PLA-HA70 scaffolds were assessed in terms of scaffold integrity and success in attaining sterilization. Non-sterile scaffolds were used as a positive control, and culture plates with only media were used as negative controls.

The success of sterilization was evaluated though microbial culture. First, aliquots of each scaffold were obtained by submerging the scaffolds in 1 mL of PBS and vortexing for 1 min. Following this, a 1:10 (D1) dilution in PBS was prepared, and 100 µL of both the undiluted (D0) and D1 aliquots were cultured under aerobic conditions for 48 h and anaerobic conditions (80% N_2, 10% H_2, 10% CO_2 at 37 °C for 72 h) using Blood Agar Base No. 2 with 5% horse blood and Hemine/menadione culture media (Oxoid no. 2, Oxoid, Basingstoke, UK). The presence or absence of bacterial or fungal growth was determined by visual inspection under magnifying loupes. The integrity of the scaffolds after the sterilization process was evaluated using thermogravimetric analysis (TGA, model TAQ500, Minneapolis, MN, USA). This experiment was conducted under dynamic conditions at a rate of 10 °C/min from 20 °C to 500 °C in an argon atmosphere (40 mL/min).

2.4.2. Cell Proliferation

A water-soluble tetrazolium salt (WST-1) assay (Cell Proliferation Reagent, Roche, Basel, Switzerland) was conducted to evaluate cell proliferation at 24, 48, and 72 h, as well as at 7 days. Six biological replicates were utilized for each type of scaffold and study interval.

First, an initial density of 2×10^4 cells/well was seeded onto the scaffolds and incubated in a humid atmosphere with 5% CO_2 at 37 °C. After each study interval, the scaffolds were transferred into new wells containing 500 µL of culture media. The assay was conducted by adding 50 µL of WST-1 reagent and incubating under the same conditions for 4 h. The plates were gently vibrated for 1 min, and the supernatants were transferred to a new multi-well plate, which was measured using a microplate spectrophotometer (Multiskan SkyHigh, 2.00.35, Thermo Scientific, Waltham, MA, USA) at 440 nm (A_{440}) and 650 nm (A_{650}) as the reference wavelength following the manufacturer's protocol. The cell proliferation was calculated as (A_{440}–A_{650})—A_{blank}, where A_{blank} represents the absorbance obtained from wells with scaffolds and culture media but without seeded cells. The average value of all biological replicates was recorded as the representative datum.

2.4.3. Scanning Electron Microscopy (SEM)

A morphological analysis of the scaffold surfaces was conducted using SEM. An initial cell density of 2×10^4 cells/well was seeded onto each scaffold in 48-multiwell plates and incubated in a humid atmosphere at 5% CO_2 and 37 °C for 24 h and 7 days. Three biological replicates for each type of scaffold and study interval were utilized.

The cells were fixed with 2.5% glutaraldehyde plus 4% paraformaldehyde in PBS (pH 7.4). After each experimental timepoint, the culture medium was removed, and the scaffolds were completely submerged in the fixing solution for 2 h at room temperature. The fixing solution was then removed, and the scaffolds were kept in PBS overnight. The samples were then dehydrated using graded distilled water solutions (30—50—70—80—90—95% dH_2O).

After dehydration, the specimens were dried using a critical point drying method following the manufacturer's instructions (Leica EM CPD300, Leica Microsystems, Mannheim, Germany). The samples were subsequently sputtered with gold, and the scaffold topography was evaluated through SEM at a voltage of 7 kV (SEM-FEG Hitachi S-4800, Tokyo, Japan).

2.4.4. Cell Viability—Confocal Laser Scanning Microscopy (CLSM)

Cell viability was assessed using CLSM (Leica Microsystems, Milton Keynes, United Kingdom) and the LIVE/DEAD© viability/cytotoxicity kit for mammalian cells (Invitrogen, Molecular Probes, 2005, Carlsbad, CA, USA). Three biological replicates were conducted for each type of scaffold and time point.

An initial density of 2×10^4 cells/well was seeded onto the scaffolds using 48-well culture plates and incubated in a humid atmosphere at 5% CO_2 and 37 °C for 24 h and 7 days. The scaffolds were washed with PBS to reduce esterase activity. Next, 20 µL of a 2 mM EthD-1 solution was added to 10 mL of PBS, yielding a final concentration of 4 µM EthD-1). Then, 5 µL of calcein AM was added to the solution and thoroughly mixed via pipetting. This produced a working solution containing 2 µM calcein AM and 4 µM EthD-1. A total of 150 µL of solution was added directly to each well, followed by a 30 min incubation at room temperature before visualization under CLSM following the manufacturer's recommendations (Calcein AM: FITC filter, excitation 494 nm, emission 517 nm; EthD-1: RFP filter, excitation 528 nm, emission 617 nm).

A negative control, made up of scaffolds containing dead cells induced by incubation in 70% methanol for 15 min before CLSM analysis, was utilized to ensure the proper functioning of the resulting solution.

2.4.5. Cytotoxicity

A direct cytotoxicity assay was performed, in which MG-63 cells were placed onto the scaffolds and cultured for 24 h, 48 h, 72 h, and 7 days. The assay used an initial cell density of 6×10^4 cells/well, and a WST-1 assay (Cell Proliferation Reagent, Roche) was used to assess the cytotoxicity, following the protocol previously described (see cell proliferation assay).

2.4.6. Gene Expression

To assess osteogenic-related gene expression, a reverse transcription-quantitative polymerase chain reaction (RT-qPCR) was performed. MG-63 cells were seeded onto the scaffolds at an initial density of $2–3 \times 10^4$ cells/well and incubated for 24 h, 48 h, 72 h, and 7 days. Cells seeded directly onto the well were used as a control.

The total RNA was extracted using the RNeasy® Mini Kit (Qiagen, Germantown, MD, USA) according to the standard protocol. The RNA quality and quantity were assessed through spectrophotometry (NanoDrop Thermo Scientific) to ensure a ratio of $260/230 \geq 1.7$. It was determined that the RNA obtained by pooling the genetic material from eight scaffolds would constitute one biological replicate. A total of 800 ng of total RNA was used for reverse transcription with the SuperScript™ IV VILO™ Master Mix kit (Invitrogen) and the ezDNase kit (Invitrogen) for gDNA elimination, following the manufacturer's protocol. The qPCR analysis was conducted using KAPA SYBR® FAST (KAPA BioSystems, Wilmington, MA, USA). Relative gene expression was analyzed using the $2^{-\Delta\Delta Ct}$ method and normalized to the reference genes TATA-box binding protein (TBP) and Glyceraldehyde 3-phosphate dehydrogenase (GAPDH) [26]. The primers used for PCR amplification are shown in Appendix A (Table A1).

2.4.7. Protein Synthesis

The levels of matrix metalloproteinase 1 (MMP-1), receptor activator of nuclear factor kappa-B ligand (RANKL), macrophage colony-stimulating factor (M-CSF), interleukin

(IL)-6, and IL-8 were measured using a multiplex immunoassay (Luminex200, Luminex Corporation, Austin, TX, USA).

MG-63 cells were seeded at an initial density of 2×10^4 cells/well in 48 multi-well plates, which were incubated for 12 h to allow the cells to adhere to the scaffold surface. The scaffolds were then transferred to new plates and incubated for 24 h, 48 h, 72 h, and 7 days. Three biological replicates were conducted for each.

At each study interval, 500 µL of the supernatants were collected and stored at −20 °C until they were collectively analyzed. For quantifying the various proteins, three different kits were employed: MILLIPLEX® Human Cytokine/Chemokine/Growth Factor Panel A (HCYTA-60K-03) for IL-6, IL-8, and M-CSF; MILLIPLEX® MAP Human MMP Magnetic Bead Panel 2 (HMMP2MAG-55K-01) for MMP-1; and MILLIPLEX® MAP Human RANKL Magnetic Bead—Single Plex (HRNKLMAG-51K-01) for RANKL.

Prior to the analysis, the samples were centrifuged to eliminate cell remnants, and the undiluted supernatants were analyzed according to the manufacturer's protocol. The obtained data was assessed using xPonent software (version 4.2, Luminex Corporation, Austin, TX, USA).

2.5. Data Analysis

2.5.1. Outcome Variables

Quantitative data were collected for the cell proliferation assay (measured in absorbance units (UAs)), cytotoxicity (measured in UAs), gene expression assay (measured as relative gene expression using the $2^{-\Delta\Delta Ct}$ method), and protein synthesis assay (measured in pg/mL).

2.5.2. Statistical Analysis

Normality tests were carried out using the Shapiro–Wilk test and confirmed through distribution data of skewness, kurtosis, and boxplot diagrams. Data were presented as means and standard deviation (SD).

Comparisons within study intervals and between scaffolds and the control group were made using a one-way ANOVA test, followed by either Bonferroni's posthoc test or Dunnet's multiple comparisons test (based on the homogeneity of variances).

Statistical significance was established at $p < 0.05$. The analysis was conducted using IBM SPSS Statistics (version 29.0.1.1, IBM Corporation, New York, NY, USA).

3. Results

In all assays, a minimum sample size of three scaffolds for each infill density was tested. No cell culture contamination was observed during any of the assays.

3.1. Sterilization Method

HA-PLA50 scaffolds showed no bacterial or fungal growth in any of the 3 sterilization methods tested. However, in HA-PLA70 scaffolds, positive fungal growth was detected for the D0 cultures under aerobic conditions, but only in scaffolds sterilized with 70% ethanol.

The TGA test showed lower thermal degradation for scaffolds sterilized with UV radiation and 70% ethanol (Figure 2). However, scaffolds sterilized by autoclave showed a temperature degradation decrease of around 10–20 °C.

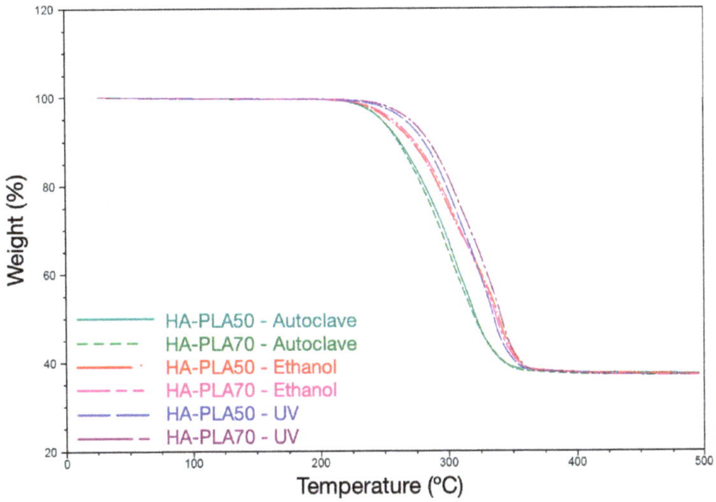

Figure 2. Thermogravimetry curves of HA-PLA50 and HA-PLA70 scaffolds after exposure to different sterilization methods. A higher rate of thermal degradation (represented as temperature-dependent weight loss) was detected for the scaffolds sterilized through autoclave, showing a faster weight loss with increasing temperatures.

3.2. Cell Proliferation

A similar cell proliferation rate was observed for both types of scaffolds, with a mean absorbance at 24 h of 0.33 AU (SD = 0.09) for HA-PLA70 and 0.36 AU (SD = 0.04) for HA-PLA50 ($p > 0.05$). At 7 days, the mean absorbance was 3.01 AU (SD = 0.28) for HA-PLA50 and 3.04 AU (SD = 0.39) for HA-PLA70 ($p > 0.05$) (Table 1) (Figure 3).

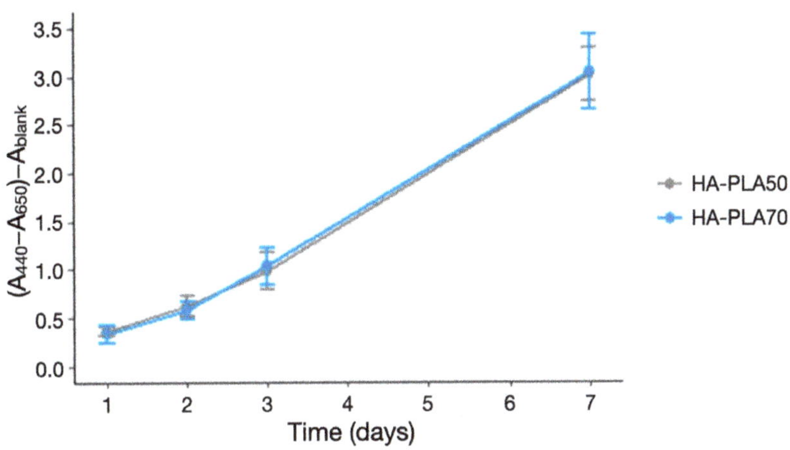

Figure 3. Cell proliferation of HA-PLA50 and HA-PLA70 scaffolds.

Table 1. Cell proliferation of HA-PLA50 and HA-PLA70 scaffolds measured by spectrophotometry using a WST-1 tetrazolium assay, measured at 24 h, 48 h, 72 h, and 7 days. No statistical differences between scaffolds were found for any of the study intervals. One-way ANOVA test and Dunnet's multiple comparisons test.

Infill	N	Absorbance (Mean (SD))				*p* Value (Between Time Points)		
		24 h	48 h	72 h	1 W	24 h vs. 48 h	24 h vs. 72 h	24 h vs. 1 W
HA-PLA50	6	0.36 (0.04)	0.62 (0.10)	0.96 (0.19)	3.01 (0.28)	0.014	0.006	<0.001
HA-PLA70	6	0.33 (0.09)	0.57 (0.09)	1.01 (0.19)	3.04 (0.39)	0.020	0.001	<0.001
p value (between scaffolds)		1.000	1.000	1.000	1.000			

3.3. Scanning Electron Microscopy

No differences in cell growth were observed between the scaffolds. The morphological features of the HA-PLA50 and HA-PLA70 scaffolds are shown in Figure 4. At 24 h, the scaffold surface exhibited a rough texture due to the HA particle topography, lacking micropores, and there was cell proliferation along it. The cells displayed a fibroblast-like morphology with cytoplasmic processes leading to interconnections between cells and connections to the scaffold. By day 7, the cell growth resulted in the formation of a dense cell layer over the scaffold surface.

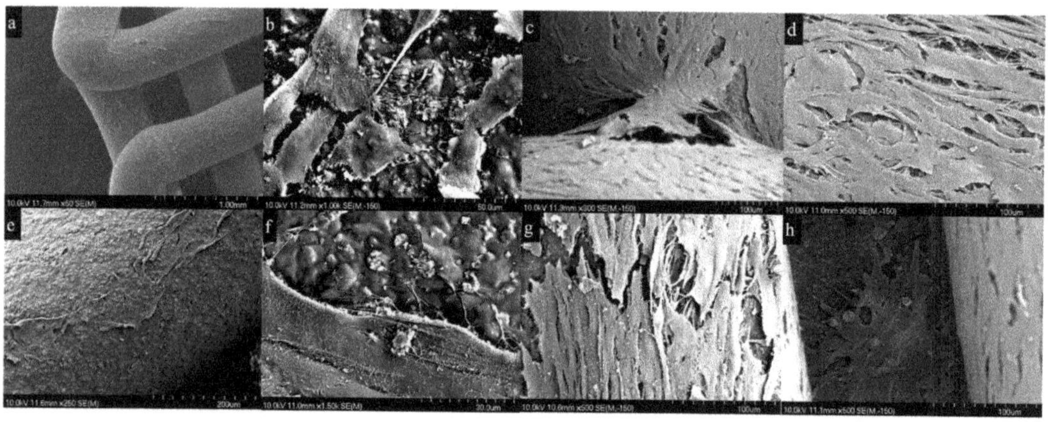

Figure 4. Scanning electron microscopy of HA-PLA50 (**a–d**) and HA-PLA70 (**e–h**) scaffolds seeded with MG-63 cells. (**a**) HA-PLA50 scaffolds at 24 h post-seeding at ×50 magnification. (**b**) HA-PLA50 scaffolds at 24 h post-seeding at ×1000 magnification. (**c**) HA-PLA50 scaffolds at 7 days post-seeding at ×300 magnification. (**d**) HA-PLA50 scaffolds at 7 days post-seeding at ×500 magnification. (**e**) HA-PLA70 scaffolds at 24 h post-seeding at ×250 magnification. (**f**) HA-PLA70 scaffolds at 24 h post-seeding at ×1500 magnification. (**g**,**h**) HA-PLA70 scaffolds at 7 days post-seeding at ×500 magnification. MG-63 cells exhibited a homogeneous distribution across the scaffold surface, characterized by a spindle-shaped morphology and cytoplasmic processes connecting the cells to one another and to the scaffold surface.

3.4. Cell Viability–CLSM

At 24 h, MG-63 cells were uniformly scattered along the surfaces of the HA-PLA50 and HA-PLA70 scaffolds, showing a low rate of dead cells. At 7 days, both scaffolds exhibited increased cell growth with a low death rate (Figure 5).

Figure 5. Confocal scanning microscopy of MG-63 cells stained with LIVE/DEAD dye, seeded onto HA-PLA50 and HA-PLA70 scaffolds after 24 h and 7 days of incubation. Both HA-PLA50 and HA-PLA70 scaffolds demonstrated high cell viability over the 7-day period. HA-PLA50 scaffolds: (**a**) 24 h; (**b**) 7 days. HA-PLA70 scaffolds: (**c**) 24 h; (**d**) 7 days. Control HA-PLA70 scaffolds: (**e**) 24 h; (**f**) 7 days. Live cells are shown in green and dead cells in red, with a magnification of 10×.

3.5. Cytotoxicity

In the direct cytotoxicity assay, no statistically significant changes in cytotoxicity were observed when comparing baseline and final time points for any of the scaffold's porosities, nor when comparing between the two groups ($p > 0.05$) (Table 2).

Table 2. Direct cytotoxicity assay using a WST-1 tetrazolium assay, measured through spectrophotometry at 24 h, 48 h, 72 h, and 7 days. One-way ANOVA test and Dunnet's multiple comparisons test.

Infill	N	Absorbance (Mean (SD))				*p* Value (Between Time Points)		
		24 h	48 h	72 h	1 W	24 h vs. 48 h	24 h vs. 72 h	24 h vs. 1 W
HA-PLA50	6	1.94 (0.28)	1.94 (0.20)	2.41 (0.51)	2.21 (0.11)	1.000	0.716	0.666
HA-PLA70	6	2.56 (0.19)	1.66 (0.50)	2.40 (0.37)	2.12 (0.17)	0.080	1.000	0.177
p value (between scaffolds)		0.071	0.974	1.000	0.999			

3.6. Gene Expression

Both types of scaffolds supported the growth of MG-63 cells, prompting the expression of OCN, OPN, ALPL, and COL1A1 at the mRNA level (Figure 6).

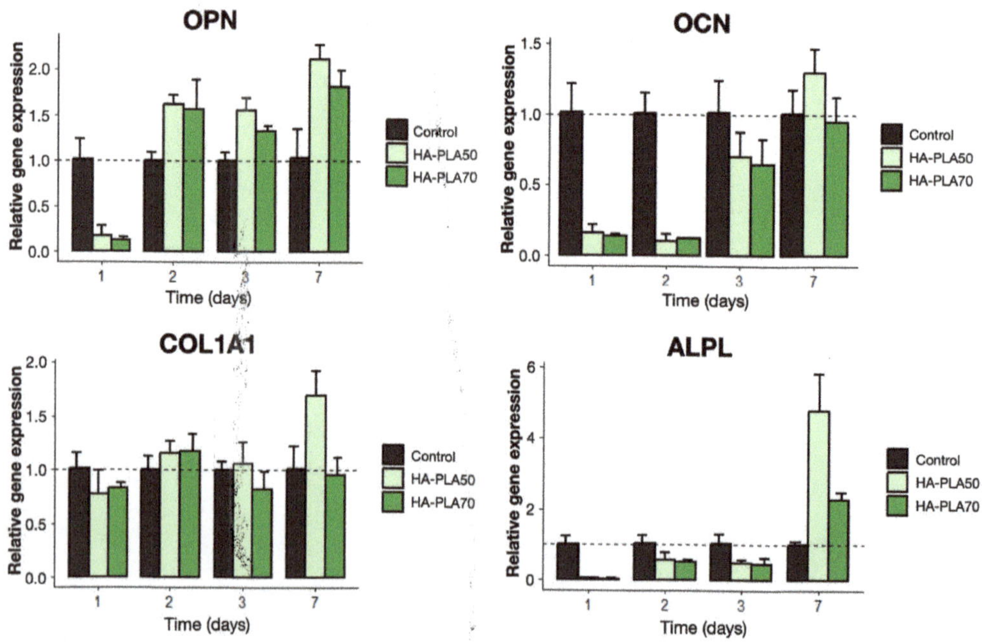

Figure 6. Relative gene expression of genes OPN, OCN, COL1A1, and ALPL, using GADPH and TBP as housekeeping genes, measured through RT-qPCR at 24 h, 48 h, 72 h, and 7 days. MG-63 cells growing on the well surface served as a positive control. OPN: osteopontin; OCN: osteocalcin; COL1A1: collagen type 1; ALPL: alkaline phosphatase.

For HA-PLA50 scaffolds, COL1A1, OCN, and OPN exhibited a significant increase in gene expression when comparing baseline and final time points ($p < 0.05$). In the case of HA-PLA70 scaffolds, there was a significant increase in gene expression over time for ALPL and OPN ($p < 0.05$). When comparing both types of scaffolds, statistically significant differences at the final time point could only be observed for the COL1A1 gene, favoring the HA-PLA50 group (1.70 [SD = 0.22] vs. 0.96 [SD = 0.16], at the 7-day interval) ($p < 0.05$) (Table A2).

3.7. Protein Synthesis

Out of five analytes, only MMP-1, IL-6, IL-8, and M-CSF produced detectable concentrations in the immunoassay analysis. RANKL concentrations remained below detection levels after the maximum incubation times (Figure 7).

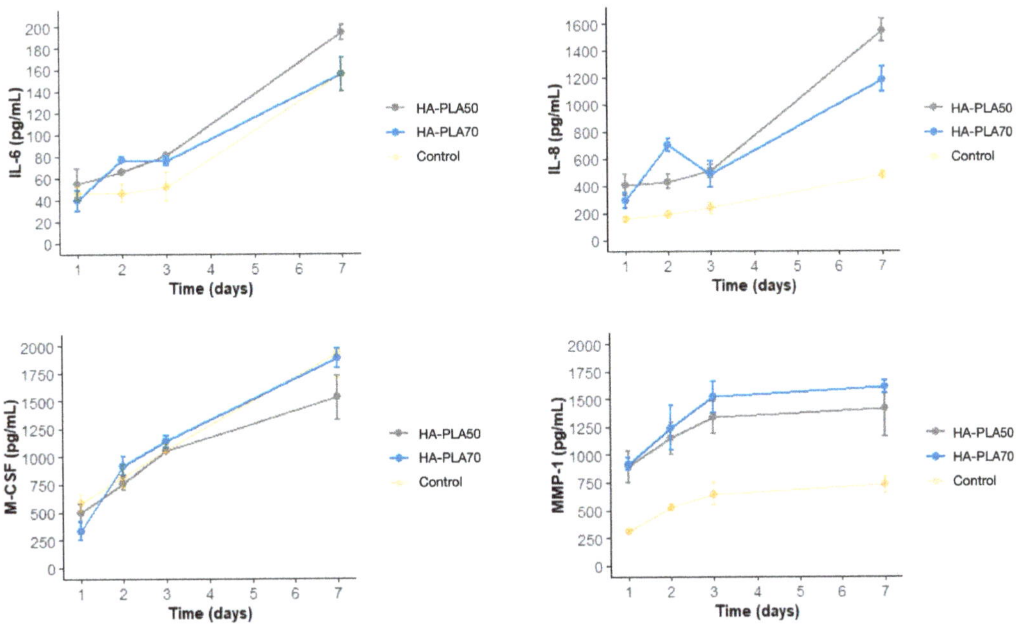

Figure 7. Protein synthesis in MG-63 cells for IL-6, IL-8, M-CSF, and MMP-1 at 24 h, 48 h, 72 h, and 7 days of incubation. This was measured using multiple immunoassays (Luminex® 200) on supernatants. The concentrations of RANKL were also evaluated; however, the levels obtained at each study interval did not reach detectable limits. MG-63 cells growing on the well surface served as a positive control. IL-6: interleukin 6; IL-8: interleukin 8; M-CSF: macrophage colony-stimulating factor; MMP-1: metalloproteinase 1.

All detectable analytes showed a significant increase ($p < 0.05$) in both types of scaffolds, except for MMP-1 in HA-PLA50 scaffolds, which revealed no statistically significant differences when comparing the baseline and final evaluations (Table A3).

When both scaffolds were compared, both IL-6 and IL-8 showed statistically significant differences, with higher concentrations for the HA-PLA50 scaffold (IL-6: 194.04 pg/mL [SD = 6.72] vs. 120.65 pg/mL [SD = 5.58] at the 7-day interval; IL-8: 1546.95 pg/mL [SD = 81.09] vs. 1187.54 pg/mL [SD = 90.89] at the 7-day interval) ($p < 0.05$). However, no differences between either type of scaffold or the control group were found for IL-6 ($p > 0.05$).

4. Discussion

This in vitro study evaluated the biological properties of two HA-PLA scaffolds with varying porosities, concentrating on their potential applications in tissue engineering for bone regeneration.

The evaluation of the most appropriate sterilization method identified ultraviolet radiation as the preferred technique for achieving successful sterilization while maintaining the integrity of the scaffolds. The use of MG-63 cells demonstrated successful cell proliferation on the surface of both scaffolds, confirming similar cell viability and proliferation on both types of scaffolds. Additionally, there were no differences in cytotoxicity between groups. Compared to HA-PLA70, the lower infill density (HA-PLA50) enhanced the expression of COL1A1. The protein concentrations of IL-8 at the final evaluation were also significantly higher in HA-PLA50 scaffolds compared to HA-PLA70 scaffolds. These results confirm adequate cell viability and proliferation of both tested HA-PLA scaffolds, although the one with lower infill density promoted a higher expression of osteogenic genes and IL-8, a significant chemokine and protein-binding cytokine.

Regarding the sterilization method, ethanol, UV radiation, and autoclave were selected for widespread adoption, especially for in vitro purposes since these scaffolds are designed for their use as point-of-care customized devices in craniofacial regeneration. The scaffolds were successfully sterilized without structural compromise using a 30 min exposure time per side of UV radiation. In fact, UV radiation is especially suited for biodegradable biomaterials like HA-PLA, as it minimizes risks of structural degradation or mechanical properties alterations associated with harsher sterilization techniques like autoclaving. However, there is still a lack of consensus on the optimal sterilization method for biodegradable scaffolds [27].

To analyze the biological performance of the scaffolds, we used MG-63 osteoblast-like cells to test cell viability and behavior when deposited on the scaffold surfaces. Both SEM and CSLM analyses confirmed adequate cell growth and adhesion over a 7-day incubation period, indicating that MG-63 cells adhered to both scaffold surfaces and successfully proliferated. Cells growing on both types of scaffolds exhibited a similar morphology, with cytoplasmic processes or filopodia interconnecting the cells and the material surface. This observation aligns with studies reporting similar experiments where MG-63 cells reached complete confluence over PLA scaffolds after 14 days [28–31]. Consistent with our approach, multiple investigations have evaluated the effect of scaffold structure and surface on cell proliferation and differentiation. For instance, Pamula et al. observed that for poly(L-lactide-co-glycolide), scaffolds with 83% porosity showed faster initial cell colonization during the first week of incubation in scaffolds featuring larger pore diameters (400 to 600 μm); however, this difference disappeared at the 2-week assessment [32]. Our results similarly highlight the importance of structural parameters, while extending these observations to HA-PLA composites with controlled infill densities.

The cytotoxicity assessment revealed no statistically significant differences between HAP-PLA50 and HA-PLA70 scaffolds at any timepoint nor across incubation periods within each scaffold. Similar results were reported by Gregor et al. using MG-63 cells, where no differences in cytotoxicity were observed among PLA scaffolds with varying porosities during a 5-day incubation period [30]. Therefore, both HA-PLA50 and HA-PLA70 scaffolds can be considered non-toxic to MG-63 cells.

The effects of infill density, porosity, and scaffold design on scaffold performance and cell behavior have been thoroughly examined in biomaterials research. This study observed variations in osteogenic gene expression and protein synthesis associated with lower infill densities (HA-PLA50 scaffolds). Chocholata et al. (2019) noted that during scaffold design, the pore size should be sufficiently large to allow for cell migration while remaining small enough to promote cell binding to the scaffold surface [11]. Huang et al. reported that PLA-nanoHA scaffolds with similar porosities (80%) showed enhanced cell adhesion and proliferation [13]. In this study, the expression of osteogenesis-related genes by MG-63 cells cultured on both types of scaffolds indicated sufficient biocompatibility and osteogenic capacity. HA-PLA50 scaffolds demonstrated a higher relative gene expression of COL1A1 compared to the higher porosity scaffolds. These findings are consistent with Gregor et al.'s results, which noted an increased collagen type 1 formation after 7 days in PLA scaffolds [30]. Therefore, these results suggest that infill density may influence cell adhesion, proliferation, and differentiation towards an osteogenic lineage.

In addition to gene expression, the study also examined protein synthesis, focusing on the secretion of different osteogenic markers. The protein concentrations of IL-8 were significantly higher in the HA-PLA50 scaffolds compared to the HA-PLA70 scaffolds. Interestingly, while elevated IL-8 could be associated with an increased immune response, this cytokine's high protein-binding activity could also reflect the scaffolds' potential to promote the recruitment of cells involved in the wound healing process and tissue regeneration, such as endothelial and immune-competent cells. Overall, the lower infill density of HA-PLA50 likely facilitated nutrient diffusion and cell–matrix interactions, explaining its superior protein/gene expression profile compared to denser HA-PLA70.

We recognize the limitations of this study, particularly the reduced sample size for each assay, dictated by limited biomaterial availability. The maximum incubation time for the experiments was set at 7 days to reduce the risk of microbial contamination; however, this may limit the insights gained from a longer degradation period. While SEM and CLSM demonstrated adequate cell–scaffold integration qualitatively, a quantitative morphometric analysis was not performed. Similarly, the assessment of osteogenic potential was conducted using molecular markers rather than histological staining (ALP/ARS), which could have provided additional data on mineralization. Additionally, the in vitro nature of this investigation does not allow for proper translation to more complex living models, emphasizing the need for in vivo and clinical studies to confirm these findings. Several 3D-printed devices have already been tested in rabbits [29], rat models [13,33], and ovine models [34], and recent case reports have shown successful outcomes in human regeneration of bone lesions in the jaw [35]. Nevertheless, despite these encouraging results, the evidence remains limited, and regulatory considerations continue to restrict their use in clinical settings [7].

However, the results of this investigation are relevant for future research using in vivo models, as 3D-printed synthetic composite bone replacement grafts provide highly suitable biological and mechanical properties for clinical applications. More specifically, HA-PLA50 and HA-PLA70 scaffolds have demonstrated biocompatibility and bioactivity when studied under strictly controlled experimental conditions. While these in vitro results offer initial evidence of their potential for tissue engineering applications, future research should evaluate the performance of these materials in vivo, where the complex interactions between the scaffold, host tissue, and immune response can be assessed more comprehensively.

5. Conclusions

This study demonstrates that both HA-PLA50 and HA-PLA70 scaffolds exhibit adequate biocompatibility, non-toxicity, and osteogenic potential in vitro, supporting their suitability for applications in bone regeneration.

Our key findings are as follows: (1) ultraviolet radiation is the optimal sterilization method, preserving scaffold integrity while ensuring cell viability; (2) both scaffolds supported significant MG-63 cell proliferation over 7 days, with no differences in cytotoxicity observed; (3) HA-PLA50 scaffolds outperformed HA-PLA70 in COL1A1 gene expression and synthesis of IL-6 and IL-8.

The HA-PLA50 and HA-PLA70 scaffolds have demonstrated biocompatibility and bioactivity when studied under strictly controlled experimental conditions. While these in vitro results offer initial evidence of their potential for tissue engineering applications, future research should evaluate the performance of these materials in vivo, where the complex interactions between the scaffold, host tissue, and immune response can be assessed more comprehensively.

Author Contributions: Conceptualization, M.S., C.C. and A.F.-M.; methodology, M.S., C.C., E.S. and A.F.-M.; validation, C.C., A.F.-M., R.M.M.-R. and E.S.; formal analysis, C.C., E.S., L.V. and M.J.M.; investigation, E.S., C.C., A.F.-M., R.M.M.-R. and L.V.; resources, A.F.-M., R.M.M.-R., B.F., A.J.S.-H. and M.S.; data curation, E.S. and E.F.; writing—original draft preparation, E.S.; writing—review and editing, C.C., A.F.-M., B.F., A.J.S.-H., E.F. and M.S.; visualization, E.S., C.C., A.F.-M., R.M.M.-R., B.F., A.J.S.-H., L.V., M.J.M., E.F. and M.S.; supervision, C.C., E.F. and M.S.; project administration, A.F.-M.,

B.F., A.J.S.-H. and M.S.; funding acquisition, M.S., E.F., A.F.-M., B.F. and A.J.S.-H. All authors have read and agreed to the published version of the manuscript.

Funding: This research was supported by the Spanish Centre for the Development of Industrial Technology (CDTI) through the grant SNEO-20211395 to COLFEED4Print S.L. which contracted the Complutense University of Madrid. A. Ferrandez-Montero is granted by the Comunidad de Madrid for "Atracción de Talento" project 2022-T1/IND-23973 and PID2022-137274NB-C31 (3DPOSTPERFORM) funded by MICIU/AEI/10.13039/501100011033 and by ERDF/EU.

Institutional Review Board Statement: Not applicable.

Informed Consent Statement: Not applicable.

Data Availability Statement: The original contributions presented in the study are included in the article, further inquiries can be directed to the corresponding authors.

Acknowledgments: The authors acknowledge the relevant help in software management and data acquisition from all laboratory technicians and researchers from the Laboratory of Microbiology at the Complutense University of Madrid, Spain and at the Centre for Biological Research Margarita Salas, CSIC, Madrid, Spain, involved in this project.

Conflicts of Interest: The authors declare the following financial interests/personal relationships which may be considered as potential competing interests: A. Ferrandez-Montero, A.J. Sanchez-Herencia, and B. Ferrari are inventors of the patent #PCT ES2019/070348, property of the Spanish National Research Council (CSIC), which is licensed to the startup COLFEED4Print S.L., provider of the filaments. A. Ferrandez-Montero, A.J. Sanchez-Herencia, and B. Ferrari are among the cofounders (10% share each) of the startup COLFEED4Print S.L. Other authors declare no conflicts of interest related to this study.

Abbreviations

The following abbreviations are used in this manuscript:

CLSM	Confocal scanning laser microscopy
COL1A1	Collagen type 1
GAPDH	Glyceraldehyde 3-phosphate dehydrogenase
HA	Hydroxyapatite
IL	Interleukin
M-CSF	Macrophage colony-stimulating factor
MMP	Matrix metalloproteinase
PLA	Polylactic acid
RANKL	Receptor activator of nuclear factor kappa-B ligand
RT-qPCR	Reverse transcription-quantitative polymerase chain reaction
SEM	Scanning electron microscopy
TBP	TATA-box binding protein
TGA	Thermogravimetry
UV	Ultraviolet
WST	Water-soluble tetrazolium

Appendix A

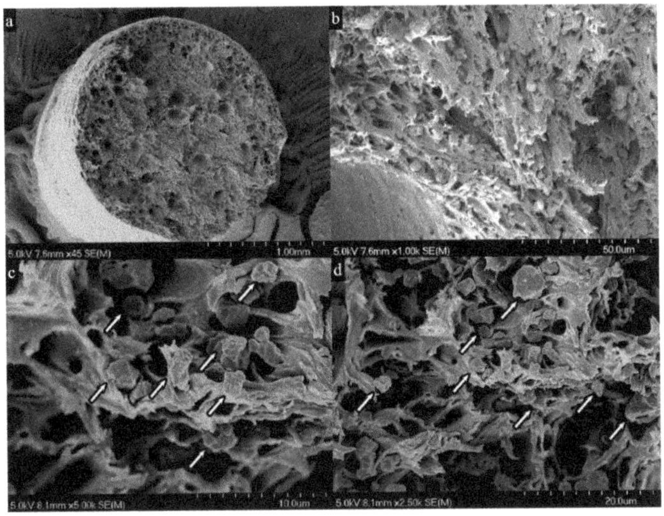

Figure A1. Surface of the fractured cross-section of the filament with the presence of closed pores of about 100 µm: (**a**) the filament is composed by the thermoplastic matrix, which is stretched due to strain during fracture; (**b**) individual particles of hydroxyapatite attached to the stretched walls of the polymer; (**c**,**d**) the arrows indicate inorganic particles highly dispersed into the PLA matrix without the presence of agglomerates as a consequence of the colloidal manufacturing approach.

Table A1. Primer sequences used to amplify each target and reference gene by RT-qPCR.

Gene	Primer	Sequence
TBP	Forward	5′-TGTATCCACAGTGAATCTTGGTTG-3′
	Reverse	5′-GGTTCGTGGCTCTCTTATCCTC-3′
GAPDH	Forward	5′-GTCTCCTCTGACTTCAACAGCG-3′
	Reverse	5′-ACCACCCTGTTGCTGTAGCCAA-3′
OPN	Forward	5′-CGAGGTGATATAGTGTGGTTTATGG-3′
	Reverse	5′-GCACCATTCAACTCCTCGCTTTC-3′
OCN	Forward	5′-CGCTACCTGTATCAATGGCTGG-3′
	Reverse	5′-CTCCTGAAAGCCGATGTGGTCA-3′
COL1A1	Forward	5′-GATTCCCTGGACCTAAAGGTGC-3′
	Reverse	5′-AGCCTCTCCATCTTTGCCAGCA-3′
ALPL	Forward	5′-GCTGTAAGGACATCGCCTACCA-3′
	Reverse	5′-CCTGGCTTTCTCGTCACTCTCA-3′

Table A2. Relative gene expression of OPN, OCN, COL1A1, and ALPL, using GADPH and TBP as housekeeping genes, measured through RT-qPCR at 24 h, 48 h, 72 h, and 7 days. One-way ANOVA and Bonferroni tests (for COL1A1 and OPN) and Dunnett's multiple comparisons test (for ALPL and OCN) were conducted.

Gene	Infill	N	Relative Gene Expression (Mean (SD))				*p* Value (Between Time Points)		
			24 h	48 h	72 h	1W	24 h vs. 48 h	24 h vs. 72 h	24 h vs. 1 W
ALPL	HA-PLA50	3	0.05 (0.01)	0.57 (0.21)	0.49 (0.08)	4.77 (1.04)	0.317	0.069	0.107
	HA-PLA70	3	0.04 (0.01)	0.50 (0.06)	0.47 (0.17)	2.28 (0.19)	0.033	0.294	0.017
	CONTROL	3	1.01 (0.22)	1.02 (0.25)	1.02 (0.26)	1.00 (0.08)	1.000	1.000	1.000
	p value (between scaffolds)	HA-PLA50 vs. HA-PLA70	0.034	1.000	1.000	0.326			
		HA-PLA50 vs. Control	0.324	0.681	0.414	0.161			
		HA-PLA70 vs. Control	0.320	0.409	0.438	0.032			
COL1A1	HA-PLA50	3	0.77 (0.21)	1.16 (0.11)	1.06 (0.21)	1.70 (0.22)	0.528	1.000	<0.001
	HA-PLA70	3	0.83 (0.05)	1.17 (0.16)	0.82 (0.16)	0.96 (0.16)	1.000	1.000	1.000
	CONTROL	3	1.01 (0.15)	1.00 (0.12)	1.00 (0.07)	1.02 (0.21)	1.000	1.000	1.000
	p value (between scaffolds)	HA-PLA50 vs. HA-PLA70	1.000	1.000	1.000	<0.001			
		HA-PLA50 vs. Control	1.000	1.000	1.000	0.002			
		HA-PLA70 vs. Control	1.000	1.000	1.000	1.000			
OCN	HA-PLA50	3	0.15 (0.06)	0.10 (0.05)	0.70 (0.17)	1.29 (0.17)	0.998	0.185	0.030
	HA-PLA70	3	0.13 (0.02)	0.12 (0.00)	0.64 (0.18)	0.95 (0.17)	0.985	0.255	0.098
	CONTROL	3	1.01 (0.20)	1.00 (0.14)	1.01 (0.22)	1.00 (0.16)	1.000	1.000	1.000
	p value (between scaffolds)	HA-PLA50 vs. HA-PLA70	1.000	1.000	1.000	0.632			
		HA-PLA50 vs. Control	0.100	0.040	0.843	0.841			
		HA-PLA70 vs. Control	0.113	0.062	0.732	1.000			
OPN	HA-PLA50	3	0.16 (0.11)	1.62 (0,09)	1.55 (0.14)	2.11 (0.16)	<0.001	<0.001	<0.001
	HA-PLA70	3	0.12 (0.03)	1.57 (0.32)	1.33 (0.05)	1.81 (0.19)	<0.001	<0.001	<0.001
	CONTROL	3	1.01 (0.22)	1.00 (0.09)	1.00 (0.09)	1.03 (0.32)	1.000	1.000	1.000
	p value (between scaffolds)	HA-PLA50 vs. HA-PLA70	1.000	1.000	1.000	1.000			
		HA-PLA50 vs. Control	<0.001	0.024	0.182	<0.001			
		HA-PLA70 vs. Control	<0.001	0.057	1.000	0.002			

Table A3. Protein synthesis of MG-63 cells for IL-6, IL-8, M-CSF, and MMP-1 was measured at 24 h, 48 h, 72 h, and 7 days of incubation using multiple immunoassays (Luminex®). For IL-8, a one-way ANOVA test and Bonferroni test were applied, while Dunnett's multiple comparisons test was utilized for IL-6 and M-CSF MMP-1.

Protein Synthesis	Infill	N	Concentration pg/mL (Mean (SD))				p Value (Between Time Points)		
			24 h	48 h	72 h	1 W	24 h vs. 48 h	24 h vs. 72 h	24 h vs. 1 W
IL-6	HA-PLA50	3	55.25 (13.32)	66.09 (1.09)	80.68 (1.14)	194.04 (6.72)	0.954	0.461	0.007
	HA-PLA70	3	39.33 (9.37)	77.07 (2.47)	75.66 (4.28)	120.65 (5.58)	0.117	0.107	0.009
	CONTROL	3	45.40 (6.05)	46.18 (8.13)	52.22 (13.42)	155.57 (15.62)	1.000	1.000	0.027
	p value (between scaffolds)	HA-PLA50 vs. HA-PLA70	0.916	0.077	0.813	0.003			
		HA-PLA50 vs. Control	0.992	0.314	0.395	0.300			
		HA-PLA70 vs. Control	1.000	0.128	0.533	0.360			
IL-8	HA-PLA50	3	408.04 (75.85)	434.42 (48.83)	510.72 (48.75)	1546.95 (81.09)	1.000	1.000	<0.001
	HA-PLA70	3	293.65 (56.01)	701.98 (50.19)	484.16 (95.73)	1187.54 (90.89)	<0.001	0.051	<0.001
	CONTROL	3	155.10 (20.08)	188.73 (30.11)	234.43 (41.37)	479.21 (29.63)	1.000	1.000	<0.001
	p value (between scaffolds)	HA-PLA50 vs. HA-PLA70	1.000	<0.001	1.000	<0.001			
		HA-PLA50 vs. Control	0.002	0.003	<0.001	<0.001			
		HA-PLA70 vs. Control	0.655	<0.001	0.002	<0.001			
M-CSF	HA-PLA50	3	492.79 (84.50)	755.61 (59.26)	1055.86 (6.84)	3054.39 (396.82)	0.183	0.050	0.046
	HA-PLA70	3	333.43 (80.04)	918.58 (84.30)	1137.25 (55.17)	3756.68 (175.99)	0.016	0.004	0.001
	CONTROL	3	582.21 (74.77)	821.08 (100.85)	1065.47 (31.64)	3870.45 (477.06)	0.371	0.031	0.043
	p value (between scaffolds)	HA-PLA50 vs. HA-PLA70	0.673	0.539	0.640	0.542			
		HA-PLA50 vs. Control	0.981	0.999	1.000	0.710			
		HA-PLA70 vs. Control	0.229	0.989	0.827	1.000			
MMP-1	HA-PLA50	3	893.41 (140.99)	1146.76 (143.42)	1333.12 (145.00)	2823.84 (513.31)	0.749	0.257	0.133
	HA-PLA70	3	919.62 (56.51)	1242.56 (200.89)	1519.94 (140.59)	3212.64 (114.59)	0.592	0.089	0.001
	CONTROL	3	307.15 (17.50)	525.09 (30.18)	646.04 (101.22)	1451.68 (138.29)	0.016	0.178	0.031
	p value (between scaffolds)	HA-PLA50 vs. HA-PLA70	1.000	1.000	0.944	0.976			
		HA-PLA50 vs. Control	0.121	0.105	0.052	0.262			
		HA-PLA70 vs. Control	0.012	0.159	0.021	0.002			

References

1. Schropp, L.; Wenzel, A.; Kostopoulos, L.; Karring, T. Bone healing and soft tissue contour changes following single-tooth extraction: A clinical and radiographic 12-month prospective study. *Int. J. Periodontics Restor. Dent.* **2003**, *23*, 313–323.
2. Wang, H.L.; Boyapati, L. "PASS" principles for predictable bone regeneration. *Implant. Dent.* **2006**, *15*, 8–17. [CrossRef] [PubMed]

3. Sanz-Sanchez, I.; Ortiz-Vigon, A.; Sanz-Martin, I.; Figuero, E.; Sanz, M. Effectiveness of Lateral Bone Augmentation on the Alveolar Crest Dimension: A Systematic Review and Meta-analysis. *J. Dent. Res.* **2015**, *94*, 128S–142S. [CrossRef] [PubMed]
4. Urban, I.A.; Montero, E.; Amerio, E.; Palombo, D.; Monje, A. Techniques on vertical ridge augmentation: Indications and effectiveness. *Periodontol 2000* **2023**, *93*, 153–182. [CrossRef] [PubMed]
5. Rogers, G.F.; Greene, A.K. Autogenous bone graft: Basic science and clinical implications. *J. Craniofac. Surg.* **2012**, *23*, 323–327. [CrossRef]
6. Mir-Mari, J.; Wui, H.; Jung, R.E.; Hammerle, C.H.; Benic, G.I. Influence of blinded wound closure on the volume stability of different GBR materials: An in vitro cone-beam computed tomographic examination. *Clin. Oral Implant. Res.* **2016**, *27*, 258–265. [CrossRef]
7. Ivanovski, S.; Breik, O.; Carluccio, D.; Alayan, J.; Staples, R.; Vaquette, C. 3D printing for bone regeneration: Challenges and opportunities for achieving predictability. *Periodontol 2000* **2023**, *93*, 358–384. [CrossRef]
8. Wang, C.; Huang, W.; Zhou, Y.; He, L.; He, Z.; Chen, Z.; He, X.; Tian, S.; Liao, J.; Lu, B.; et al. 3D printing of bone tissue engineering scaffolds. *Bioact. Mater.* **2020**, *5*, 82–91. [CrossRef]
9. Obregon, F.; Vaquette, C.; Ivanovski, S.; Hutmacher, D.W.; Bertassoni, L.E. Three-Dimensional Bioprinting for Regenerative Dentistry and Craniofacial Tissue Engineering. *J. Dent. Res.* **2015**, *94*, 143S–152S. [CrossRef]
10. Persson, M.; Lorite, G.S.; Kokkonen, H.E.; Cho, S.W.; Lehenkari, P.P.; Skrifvars, M.; Tuukkanen, J. Effect of bioactive extruded PLA/HA composite films on focal adhesion formation of preosteoblastic cells. *Colloids Surf. B Biointerfaces* **2014**, *121*, 409–416. [CrossRef]
11. Chocholata, P.; Kulda, V.; Babuska, V. Fabrication of Scaffolds for Bone-Tissue Regeneration. *Materials* **2019**, *12*, 568. [CrossRef]
12. Cao, S.S.; Li, S.Y.; Geng, Y.M.; Kapat, K.; Liu, S.B.; Perera, F.H.; Li, Q.; Terheyden, H.; Wu, G.; Che, Y.J.; et al. Prefabricated 3D-Printed Tissue-Engineered Bone for Mandibular Reconstruction: A Preclinical Translational Study in Primate. *ACS Biomater. Sci. Eng.* **2021**, *7*, 5727–5738. [CrossRef]
13. Huang, J.; Xiong, J.; Liu, J.; Zhu, W.; Chen, J.; Duan, L.; Zhang, J.; Wang, D. Evaluation of the novel three-dimensional porous poly (L-lactic acid)/nano-hydroxyapatite composite scaffold. *Biomed. Mater. Eng.* **2015**, *26* (Suppl. S1), S197–S205. [CrossRef]
14. Ferrández-Montero, A.; Ortega-Columbrans, P.; Eguiluz, A.; Sanchez-Herencia, A.J.; Detsch, R.; Boccaccini, A.R.; Ferrari, B. Biocompatible colloidal feedstock for material extrusion processing of bioceramic-based scaffolds. *Polym. Compos.* **2024**, *45*, 7237–7255. [CrossRef]
15. Kinoshita, Y.; Maeda, H. Recent developments of functional scaffolds for craniomaxillofacial bone tissue engineering applications. *Sci. J.* **2013**, *2013*, 863157. [CrossRef]
16. Bartnikowski, M.; Klein, T.J.; Melchels, F.P.; Woodruff, M.A. Effects of scaffold architecture on mechanical characteristics and osteoblast response to static and perfusion bioreactor cultures. *Biotechnol. Bioeng.* **2014**, *111*, 1440–1451. [CrossRef]
17. Abbasi, N.; Ivanovski, S.; Gulati, K.; Love, R.M.; Hamlet, S. Role of offset and gradient architectures of 3-D melt electrowritten scaffold on differentiation and mineralization of osteoblasts. *Biomater. Res.* **2020**, *24*, 2. [CrossRef]
18. Bartnikowski, M.; Vaquette, C.; Ivanovski, S. Workflow for highly porous resorbable custom 3D printed scaffolds using medical grade polymer for large volume alveolar bone regeneration. *Clin. Oral Implant. Res.* **2020**, *31*, 431–441. [CrossRef]
19. Yu, N.; Nguyen, T.; Cho, Y.D.; Kavanagh, N.M.; Ghassib, I.; Giannobile, W.V. Personalized scaffolding technologies for alveolar bone regenerative medicine. *Orthod. Craniofac. Res.* **2019**, *22* (Suppl. S1), 69–75. [CrossRef]
20. Giannitelli, S.M.; Accoto, D.; Trombetta, M.; Rainer, A. Current trends in the design of scaffolds for computer-aided tissue engineering. *Acta Biomater.* **2014**, *10*, 580–594. [CrossRef]
21. Abbasi, N.; Abdal-Hay, A.; Hamlet, S.; Graham, E.; Ivanovski, S. Effects of Gradient and Offset Architectures on the Mechanical and Biological Properties of 3-D Melt Electrowritten (MEW) Scaffolds. *ACS Biomater. Sci. Eng.* **2019**, *5*, 3448–3461. [CrossRef]
22. Smith, J.A.; Mele, E. Electrospinning and Additive Manufacturing: Adding Three-Dimensionality to Electrospun Scaffolds for Tissue Engineering. *Front. Bioeng. Biotechnol.* **2021**, *9*, 674738. [CrossRef]
23. Ferrandez-Montero, A.; Lieblich, M.; Benavente, R.; González-Carrasco, J.L.; Ferrari, B. New approach to improve polymer-Mg interface in biodegradable PLA/Mg composites through particle surface modification. *Surf. Coat. Technol.* **2020**, *383*, 125285. [CrossRef]
24. Oberbek, P.; Bolek, T.; Chlanda, A.; Hirano, S.; Kusnieruk, S.; Rogowska-Tylman, J.; Nechyporenko, G.; Zinchenko, V.; Swieszkowski, W.; Puzyn, T. Characterization and influence of hydroxyapatite nanopowders on living cells. *Beilstein J. Nanotechnol.* **2018**, *9*, 3079–3094. [CrossRef]
25. Ribeiro, M.M.; Neumann, V.A.; Padoveze, M.C.; Graziano, K.U. Efficacy and effectiveness of alcohol in the disinfection of semi-critical materials: A systematic review. *Rev. Lat. Am. Enferm.* **2015**, *23*, 741–752. [CrossRef]
26. Jin, P.; Zhao, Y.; Ngalame, Y.; Panelli, M.C.; Nagorsen, D.; Monsurró, V.; Smith, K.; Hu, N.; Su, H.; Taylor, P.R.; et al. Selection and validation of endogenous reference genes using a high throughput approach. *BMC Genom.* **2004**, *5*, 55. [CrossRef]
27. Dai, Z.; Ronholm, J.; Tian, Y.; Sethi, B.; Cao, X. Sterilization techniques for biodegradable scaffolds in tissue engineering applications. *J. Tissue Eng.* **2016**, *7*, 2041731416648810. [CrossRef]

28. Rodovalho, A.J.R.L.; Barbosa, W.T.; Vieira, J.L.; Oliva, C.A.d.; Gonçalves, A.P.B.; Cardoso, P.d.S.M.; Modolon, H.B.; Montedo, O.R.K.; Arcaro, S.; Hodel, K.V.S.; et al. Influence of size and crystallinity of nanohydroxyapatite (nHA) particles on the properties of Polylactic Acid/nHA nanocomposite scaffolds produced by 3D printing. *J. Mater. Res. Technol.* **2024**, *30*, 3101–3111. [CrossRef]
29. Chen, X.; Gao, C.; Jiang, J.; Wu, Y.; Zhu, P.; Chen, G. 3D printed porous PLA/nHA composite scaffolds with enhanced osteogenesis and osteoconductivity in vivo for bone regeneration. *Biomed. Mater.* **2019**, *14*, 065003. [CrossRef]
30. Gregor, A.; Filova, E.; Novak, M.; Kronek, J.; Chlup, H.; Buzgo, M.; Blahnova, V.; Lukasova, V.; Bartos, M.; Necas, A.; et al. Designing of PLA scaffolds for bone tissue replacement fabricated by ordinary commercial 3D printer. *J. Biol. Eng.* **2017**, *11*, 31. [CrossRef]
31. Karanth, D.; Puleo, D.; Dawson, D.; Holliday, L.S.; Sharab, L. Characterization of 3D printed biodegradable piezoelectric scaffolds for bone regeneration. *Clin. Exp. Dent. Res.* **2023**, *9*, 398–408. [CrossRef]
32. Pamula, E.; Filova, E.; Bacakova, L.; Lisa, V.; Adamczyk, D. Resorbable polymeric scaffolds for bone tissue engineering: The influence of their microstructure on the growth of human osteoblast-like MG 63 cells. *J. Biomed. Mater. Res. A* **2009**, *89*, 432–443. [CrossRef]
33. Bahraminasab, M.; Arab, S.; Ghaffari, S. Osteoblastic cell response to Al_2O_3-Ti composites as bone implant materials. *Bioimpacts* **2022**, *12*, 247–259. [CrossRef]
34. Vaquette, C.; Mitchell, J.; Fernandez-Medina, T.; Kumar, S.; Ivanovski, S. Resorbable additively manufactured scaffold imparts dimensional stability to extraskeletally regenerated bone. *Biomaterials* **2021**, *269*, 120671. [CrossRef]
35. Ivanovski, S.; Staples, R.; Arora, H.; Vaquette, C.; Alayan, J. Alveolar bone regeneration using a 3D-printed patient-specific resorbable scaffold for dental implant placement: A case report. *Clin. Oral Implant. Res.* **2024**, *35*, 1655–1668. [CrossRef]

Journal of
Functional Biomaterials

Article

Morphometric, Biomechanical and Macromolecular Performances of β-TCP Macro/Micro-Porous Lattice Scaffolds Fabricated via Lithography-Based Ceramic Manufacturing for Jawbone Engineering

Carlo Mangano [1,†], Nicole Riberti [2,†], Giulia Orilisi [3], Simona Tecco [4], Michele Furlani [5,*], Christian Giommi [6], Paolo Mengucci [7], Elisabetta Giorgini [6] and Alessandra Giuliani [3,*]

1 Department of Dental Sciences, University Vita Salute San Raffaele, 20132 Milan, Italy; camangan@gmail.com
2 Department of Clinical and Molecular Sciences, Marche Polytechnic University, 60126 Ancona, Italy; n.riberti@staff.univpm.it
3 Department of Clinical Specialistic and Dental Sciences, Marche Polytechnic University, 60126 Ancona, Italy; g.orilisi@univpm.it
4 I.R.C.C.S. San Raffaele Hospital, University Vita Salute San Raffaele, 20132 Milan, Italy; tecco.simona@hsr.it
5 Department of Biomedical Sciences and Public Health, Marche Polytechnic University, 60126 Ancona, Italy
6 Department of Life and Environmental Science, Marche Polytechnic University, 60131 Ancona, Italy; c.giommi@pm.univpm.it (C.G.); e.giorgini@staff.univpm.it (E.G.)
7 Department Science and Engineering of Materials, Environment and Urban Planning, Marche Polytechnic University, 60131 Ancona, Italy; p.mengucci@staff.univpm.it
* Correspondence: m.furlani@staff.univpm.it (M.F.); a.giuliani@staff.univpm.it (A.G.)
† These authors contributed equally to this work.

Abstract: Effective bone tissue regeneration remains pivotal in implant dentistry, particularly for edentulous patients with compromised alveolar bone due to atrophy and sinus pneumatization. Biomaterials are essential for promoting regenerative processes by supporting cellular recruitment, vascularization, and osteogenesis. This study presents the development and characterization of a novel lithography-printed ceramic β-TCP scaffold, with a macro/micro-porous lattice, engineered to optimize osteoconduction and mechanical stability. Morphological, structural, and biomechanical assessments confirmed a reproducible microarchitecture with suitable porosity and load-bearing capacity. The scaffold was also employed for maxillary sinus augmentation, with postoperative evaluation using micro computed tomography, synchrotron imaging, histology, and Fourier Transform Infrared Imaging analysis, demonstrating active bone regeneration, scaffold resorption, and formation of mineralized tissue. Advanced imaging supported by deep learning tools revealed a well-organized osteocyte network and high-quality bone, underscoring the scaffold's biocompatibility and osteoconductive efficacy. These findings support the application of these 3D-printed β-TCP scaffolds in regenerative dental medicine, facilitating tissue regeneration in complex jawbone deficiencies.

Keywords: β-TCP; lithography-based ceramic 3D printing; bone regeneration; scaffold; jawbone engineering; maxillary sinus augmentation; microarchitecture; osteocyte lacunae network

1. Introduction

Rehabilitation of edentulous patients with dental implants can be challenging due to insufficient alveolar bone volume resulting from alveolar atrophy and maxillary sinus pneumatization. In such anatomical conditions, achieving primary stability is often difficult owing to the lack of adequate cortical bone. Biomaterials can augment the body's innate capacity for bone regeneration, as newly forming bone requires a scaffold to facilitate and support the regenerative process.

Various biomaterials have been investigated for use in bone regeneration procedures, including demineralized freeze-dried bone allografts, calcium carbonate, bioactive glass, polymers such as polylactic acid (PLA) and polyglycolic acid (PGA), bovine-derived xenografts and peptides, calcium sulfate, bovine deproteinized bone, and hydroxyapatite. Although autologous bone remains the gold standard due to its superior osteogenic potential, constraints including donor site morbidity, limited availability and the need for additional surgical procedures pose significant disadvantages. Consequently, synthetic biomaterials have been developed to overcome these limitations [1]. In particular, concerns regarding disease transmission associated with allografts and xenografts have driven research toward synthetic bone substitutes that aim to replicate the physical and chemical properties of native bone tissue, with the goal of achieving osteoconduction, osteoinduction, and osteointegration [2,3].

An ideal biomaterial should possess specific biological and clinical characteristics. Biologically, it should facilitate mesenchymal cell recruitment via host-derived growth factors and exert bioactive effects to promote ossification. Additionally, it should provide a three-dimensional scaffold supporting vascular ingrowth and osteoprogenitor cell migration and be bioresorbable over time. Clinically, the material should be user-friendly, cost-effective, and radiographically evaluable throughout the healing process. To facilitate monitoring of resorption and substitution, the biomaterial should be radiopaque.

Bone engineering within a scaffold involves cellular recruitment, infiltration from surrounding bone tissue, and vascularization [4,5]. High porosity has been shown to enhance osteogenesis, as confirmed by numerous studies. Bioceramics that mimic natural bone can combine favorable mechanical properties with an interconnected porous architecture, making them suitable as delivery vehicles for cells. Structurally, pores larger than 100–150 μm are necessary to ensure adequate vascularization and tissue ingrowth [6]. Moreover, an optimal biomaterial should degrade gradually, being fully replaced by vital bone tissue, with the resorption rate matched to the rate of new bone formation; excessively rapid degradation can impair regenerative outcomes [7,8].

Rapid Prototyping (RP) is an additive manufacturing technology that constructs three-dimensional structures through layer-by-layer deposition of materials. One significant advantage of RP techniques is their capacity to produce patient-specific scaffolds based on cone-beam computed tomography (CBCT) data, which is particularly beneficial for complex defect geometries [9]. Lithography-based ceramic manufacturing (LCM) is an additive manufacturing process that uses a photopolymerization technique to create ceramic parts. It involves building up layers of a UV-curable ceramic suspension (slurry) via photolithography, followed by debinding and sintering to form a dense ceramic object. This method allows for the fabrication of complex geometries that are difficult to achieve with traditional ceramic manufacturing processes. LCM offers an excellent approach for fabricating bone substitute scaffolds that replicate biomimetic architectures with optimal interconnected macro-porosities [10]. The microenvironment can be further optimized through control of sintering parameters. This LCM process opens new avenues for the shaping of ceramic scaffolds: it not only utilizes the same ceramic material as traditional production methods and is capable of producing components composed of 100% ceramic material despite the presence of photopolymers during the process, but also ensures that both precision and density are maintained while exhibiting mechanical properties comparable to conventionally fabricated parts. By employing this approach, it is possible to significantly reduce the time-to-market, thereby supporting the goals of personalized medicine. This technique has already been tested in a case report involving alveolar ridge augmentation with Hydroxyapatite (HA) scaffolds [11] and in a clinical trial for the rehabilitation of the atrophic maxilla [12], although the characterization of the β-TCP ceramic block grafts used in these studies was not performed.

In this context, recent interest has been directed towards β-tricalcium phosphate (β-TCP) printed scaffolds: their biodegradable nature enables them to serve as temporary bone substitutes, providing mechanical support and facilitating biological activities during tissue regeneration [8,13]. Various recent approaches for fabricating porous β-TCP ceramic scaffolds with micro-porous struts have been proposed [14]; however, some methods are significantly influenced by the mechanical properties of porous β-TCP scaffolds with non-uniform pore sizes. Additive manufacturing techniques, such as 3D printing, have been suggested as promising strategies to produce highly porous β-TCP scaffolds with optimized pore size, porosity, and mechanical characteristics. A patient-specific β-TCP scaffold for alveolar ridge augmentation has been explored in a case report; however, no information has been reported on how the scaffold was produced [15]. Indeed, a rigorous characterization of the LCM-printed β-TCP scaffolds is currently missing in the literature.

Extensive research indicates that the three-dimensional optimization of macro- and micro-structures enhances the bioactivity of bone graft substitutes and is critical for successful bone tissue engineering. While the CAD drawings define the scaffold macro-porosity, the printing parameters was proved to totally determine the microporosity of the scaffold [16]. Macro-porosities with pore sizes between 300 and 700 μm are essential for neovascularization and new bone formation without compromising mechanical integrity. An interconnected pore network is vital for cellular infiltration, while microporosity and surface micro-roughness positively influence osteoblast adhesion, proliferation, and differentiation [17].

This multidisciplinary experimental study aimed to verify the morphometric, biomechanical and macromolecular performances of a new β-TCP scaffold for jawbone engineering, fabricated with an innovative macro/micro-porous lattice via lithography-based ceramic manufacturing. The performance of the scaffolds was evaluated in comparison to data from previous studies and native bone. Mechanical and biological assessments offered valuable insights for optimizing scaffold design to enhance bone regeneration efforts.

2. Materials and Methods

2.1. Design and Production

Nine (n = 9) cubic TCP scaffolds were fabricated using a TCP slurry (LithaBone™ TCP 300, Lithoz, Vienna, Austria), as previously described [18]. Briefly, the scaffolds were produced via additive manufacturing using the CeraFab 7500 system (Lithoz, Vienna, Austria), which selectively solidifies the slurry by photopolymerization. Each 25 μm layer of a photoactive polymer was exposed to blue LED light at a resolution of 50 μm in the x/y-plane. The green body of the scaffold was built in a layer-by-layer manner [19]. Following fabrication, the green body was carefully detached from the build platform using a razor blade, and then cleaned with LithaSol 20™ (Lithoz, Vienna, Austria) and pressurized air to remove residual powder.

The polymeric binder, responsible for binding the ceramic particles, was decomposed through a thermal treatment regime, followed by sintering to densify the ceramic structure. The sintering process involved a dwell time exceeding >48 h at temperatures of >1000 °C, allowing for controlled microporosity via partial sintering. Variations in green body shrinkage due to different maximum sintering temperatures were compensated by dimensional adjustments in all three axes, ensuring that the final macro- and microarchitecture remained consistent post-sintering. The sintered scaffolds were packaged for integration into the surgical workflow and utilized as bone substitute implants with gamma sterilization.

The scaffold geometry, with indication of the nominal overall volume (V), actual TCP volume (V_0), and porosity percentage (%), is reported in Figure 1.

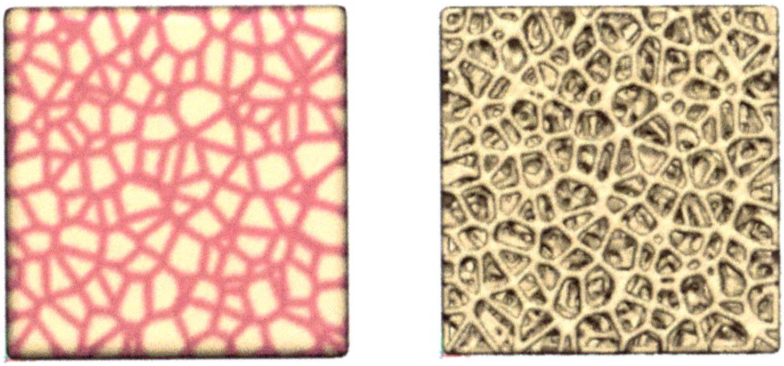

$V = 999.3969\ \text{mm}^3$ $V_0 = 407.3190\ \text{mm}^3$

$$Porosity = 1 - \frac{V_0}{V} = 0.5924 \rightarrow around\ 59\%$$

Figure 1. Nominal scaffold geometry: indication of the nominal overall volume (V), actual TCP volume (V_0), and porosity percentage (%).

2.2. Clinical Case Description

A 60-year-old male patient required implant-supported prosthetic rehabilitation of the left posterior maxillary region. Cone-beam computed tomography (CBCT) revealed significant alveolar bone atrophy, necessitating maxillary sinus floor elevation to facilitate implant placement. During the initial consultation, comprehensive clinical and occlusal examinations were conducted, along with periapical radiographs and CBCT imaging.

A routinary procedure of maxillary sinus augmentation was performed in the atrophic posterior maxilla utilizing a three-dimensionally printed β-tricalcium phosphate (β-TCP) graft (KLS Martin SE & Co. KG, Mühlheim an der Donau, Germany). The sinus augmentation procedure adhered to the classical lateral window technique as described by Tatum OH (1986) [20].

The surgical approach involved a horizontal crestal incision combined with two vertical releasing incisions extending beyond the mucogingival junction, followed by reflection of a full-thickness mucoperiosteal flap to expose the lateral wall of the maxillary sinus. An osteotomy of approximately 1.0 ± 0.1 cm was delineated and isolated using piezo surgical instruments. The bony window was carefully removed and preserved in a sterile saline solution. The Schneiderian membrane was then meticulously elevated and protected with dedicated elevators to prevent perforation.

A block of 3D-printed β-TCP, fabricated via the same additive manufacturing process described previously, was inserted into the sinus cavity to serve as the graft material. Following graft placement, the bony window was repositioned to restore the lateral sinus wall, and the surgical site was closed with sutures to ensure complete flap adaptation. Postoperative antibiotic therapy was administered.

After a healing period of six months, a secondary surgical procedure was performed for re-entry. During this classical routinary procedure, a 10 mm bone core is normally harvested (biopsy) using a trephine bur (diameter 2.5 mm, length 15 mm) under copious saline irrigation to insert a dental implant (Megagen Implant Co., Gyeongbuk, Republic of Korea).

So, in this case, a dental implant was placed precisely at the site of the biopsy, guided by a cone-beam computed tomography (CBCT) template for guided bone regeneration. Three months following implant placement, definitive prosthetic rehabilitation was completed with zirconia-ceramic fixed prostheses.

The patient gave his informed consent to use the harvested bone for observational analyses (Macromolecular Fourier Transform InfraRed Imaging; histology; X-ray micro-computed tomography, Phase-contrast X-ray micro-computed tomography; Convolutional Neural Network for Osteocyte Lacunae Analysis; Bone Mineralization Analysis).

In addition, the preliminary in vitro results obtained by mechanical tests of scaffolds and by X-ray exams of scaffolds (mechanical tests, SEM analysis, X-ray diffraction, X-ray micro-computed tomography) were also retrospectively analyzed.

The protocol was approved by the Ethical Committee "Comitato Etico Territoriale Lombardia 1", Milan, Italy (CET 462-2024, 9 December 2024).

2.3. Mechanical Tests

Compressive tests were conducted on all the scaffolds in a continuous mode until fracture, utilizing a Material Testing Stage (MTS1) integrated within the Bruker Sky-Scan 1174 experimental chamber, replacing the standard sample holder. Throughout the testing process, the scaffolds were subjected to load application controlled by specialized software; the corresponding load–deformation curve was displayed in real time on the screen, with the deformation measured accurately by a high-precision displacement sensor. The MST1 stage has the following characteristics: maximum force: 440 N; displacement sensor accuracy: ±0.01 mm; load measurement accuracy = ±4 N (±1% of the full range); maximum object diameter: 20 mm; maximum object height (in compression mode): 23 mm.

All β-TCP cubic scaffolds were submitted to compressive tests: $n = 4$ samples were loaded in the direction of lithography-based manufacturing, while the remaining $n = 5$ samples were loaded in a random direction during this test.

2.4. Structural and Morphological Characterization

2.4.1. SEM Analysis

Scanning electron microscope (SEM) observations were performed by Tescan Vega 3 (Tescan Company, Brno, Czech Republic) in order to observe the β-TCP scaffold surface and inner structure, before and after fracture due to compression loading.

2.4.2. X-Ray Diffraction

Structural characterization of β-TCP raw powder and scaffolds was carried out by X-ray diffraction (XRD) using a Bruker D8 Advance diffractometer (Billerica, MA, USA) operating at V = 40 kV and I = 40 mA, with Cu-Kα radiation, in the angular range 2θ = 5–80°. Pattern analysis was performed by DIFFRAC. EVA (Version 7, provided by Bruker, Billerica, MA, USA)) software with the PDF2 database of the International Centre for Diffraction Data (ICDD). Rietveld refinement of XRD patterns was carried out by the ReX Powder Diffraction software, version 0.9.4 [21] after calibration of the instrumental broadening by reference Al_2O_3 powder. The crystallographic structure of the $Ca_3(PO_4)_2$ (Whitlockite) phase with nominal lattice parameters a = 1.0429 nm and c = 3.7380 nm (ICDD file n. 9-169) used in Rietveld refinement was obtained from the Crystallography Open Database (COD, http://www.crystallography.net/cod/, accessed on 22 April 2025).

2.4.3. X-Ray Micro-Computed Tomography (XμCT)

The nine scaffolds were investigated by X-ray micro-computed tomography (XμCT), using a desktop device Bruker SkyScan 1174 μ-CT system (SkyScan-Bruker, Antwerp, Belgium). The projection settings used were the following: acceleration voltage of 50 kV;

beam current of 800 μA; aluminum filter (thickness of 1 mm); pixel size equal to 11.5 μm; rotation of 180° in steps of 0.2°; exposure time per projection equal to 10 s. On average, the scanning process took about 3 h. To convert the projections into transversal slices, NRecon software (version 1.6.10.2, Bruker, Billerica, MA, USA) was used, employing the following correction settings: ring artifact reduction (3); smoothing (5); beam hardening (50%).

The 3D microstructural analysis was carried out using the BoneJ plugin within FIJI [22] to evaluate the following morphological parameters: the BV/TV specific volume (Vol.%), measuring the overall mineralized (β-TCP) volumetric density, i.e., the ratio between the mineralized volume and the total volume of the scaffold; the mean trabecular thickness (Tb.Th; μm) and the mean trabecular spacing (Tb.Sp; μm), measuring the mean distance between two trabeculae. The 3D orientation and arrangement indices (namely, the Connectivity Density (Conn.D; px^{-3}), the Anisotropy Degree (DA), and the Fractal Dimension (FD)) were also calculated. The Conn.D parameter is designed to estimate the number of connected structures, i.e., trabeculae in a network. This connectivity measurement is related to a topological number χ known as Euler number. Mathematically, connectivity is defined $= 1 - (\chi + \Delta\chi)$, where χ describes the shape or structure of a topological space and the term $\Delta\chi$ corrects for the change in the topology of an object, when it is cut to pieces; the input image must be 3D and binary; the resulting parameter is the Connectivity density (Conn.D): number of elements per unit volume, showing higher values for better-connected trabeculae and lower values for poorly connected ones. DA was used to quantify the directionality of the trabeculae; it evaluates whether the trabeculae have a certain orientation, or if they are randomly aligned. DA = 0.0 means that the structure is completely isotropic; thus, the trabeculae have no preferential directionality; DA = 1.0 means that there is a prevailing orientation in the structure of the scaffold. The fractal dimension (FD) is a statistical measure that quantifies the complexity of the trabecular pattern. It is determined from a series of images using the box-counting method, where a grid of decreasing size is overlaid on the image, and the number of boxes that contain at least one foreground voxel is tallied. As the box size becomes smaller and the grid becomes more detailed, the number of boxes covering the structure increases in a fractal manner. In three dimensions, the FD value ranges from 2, indicating a planar (2D) distribution, to 3, representing a fully three-dimensional structure. The box-counting algorithm was configured with the following initial parameters [23]: box initial size [px]: 48; smallest box size [px]: 6; box scale factor: 1.2; grid translation: 0.

Three of the previous nine scaffolds were retested by XμCT after the compressive load test, using the same experimental parameters. The same morphometric parameters previously described were also calculated on these samples, with the exclusion of the Tb.Sp, unreliable after the sample fracture.

The same XμCT (SkyScan-Bruker, Antwerp, Belgium) device was used to investigate the patient's biopsy using the following experimental settings: acceleration voltage of 50 kV; beam current of 800 μA; aluminum filter (thickness of 1 mm); pixel size equal to 6.5 μm; rotation of 180° in steps of 0.4°; exposure time per projection equal to 11 s. The 3D volume reconstruction was achieved employing the following correction settings: ring artifact reduction (3); smoothing (3); beam hardening (30%). In this biopsy, the same morphometric parameters investigated in the as-produced scaffolds were analyzed. In addition, the percentage ratios between bone/biomaterial in the process of ossification/residual biomaterial was quantified.

2.4.4. Phase-Contrast X-Ray Micro-Computed Tomography (PhC-XμCT)

Phase-contrast X-ray micro-computed tomography (PhC-XμCT) was carried out at the SYRMEP beamline of the ELETTRA Synchrotron Radiation facility (Trieste, Italy). An analysis was performed on the patient's biopsy already investigated by desktop XμCT to obtain morphometric information on the osteocyte lacunae network and on the newly formed bone mineralization distribution. The method used for PhC-XμCT measurements differs from the XμCT analysis used to assess scaffold morphometric parameters. In standard XμCT, the imaging relies on differences in absorption contrast, whereas in this specific synchrotron phase-contrast setup, the imaging is based on phase shifts of the X-ray wave as it interacts with the sample. For X-rays, the refractive index n of the material is a complex number equal to $n = 1 - \delta + i\beta$. In this equation, δ is the decrement of the real part of the complex refractive index n, while the imaginary part β describes the material absorption. For peak energies used to investigate biological tissues, δ is about three orders of magnitude greater than β, thus resulting in higher sensitivity of the phase-contrast approach respect to absorption contrast.

The experiment was set with the following parameters: the imaging system captured a total of 1800 projections over 360° sample rotation, with exposure time per projection of 450 ms; the white beam was filtered with 1 mm thick silicon (Si) plate leading to a peak energy of 19.3 keV; sample-to-detector distance was set to 100 mm in order to enable phase contrast imaging. The raw images were collected by using an Orca Flash 4.0 CMOS camera (2048 × 2048 pixels, physical pixel size 6.5 μm × 6.5 μm) coupled with a 17 μm GGG scintillator screen. Owing to the optical magnification system used in combination with the detector, the effective pixel size selected for imaging was 0.9 μm × 0.9 μm, yielding a field of view of 1.8 mm × 1.8 mm.

After the acquisition, the patient's biopsy was reconstructed using the STP software (SYRMEP Tomo Project, v.112022a) [24]. The parameters selected for the slice reconstruction were Rivers ring removal (wd = 3) for ring removal correction, Paganin TIE-Hom algorithm for the phase retrieval step [25], and the Filtered Back Projection algorithm, in combination with the Shepp-Logan filter, for the ultimate reconstruction step. In detail, phase retrieval was performed by applying Paganin's algorithm with a fixed δ/β ratio of 100. Finally, the reconstructed stack comprising 2048 images was converted in 8-bit images.

2.4.5. Convolutional Neural Network for Osteocyte Lacunae Analysis

The image processing was conducted by means of the Dragonfly software [26] (Dragonfly 2022.2, Comet Technologies Canada Inc., Montreal, QC, Canada), exploiting a deep learning tool which applies artificial intelligence to the segmentation process. Moreover, the Scalar Generator tool was used to extract quantitative information on eight specific regions of interest (ROIs—box dimension: 534 × 534 × 534 μm^3) virtually extracted from the patient biopsy. In particular, two sub-volumes were extracted from both the apical and coronal portions, and four sub-volumes were extracted from the central portion.

The network employed on these portions of bone had already been previously trained and applied [27]. In order to detect osteocyte lacunae and discriminate them from the bone matrix and the background, a Pre-Trained U-Net was chosen (one of the AI tools of the Dragonfly software), and hyperparameters (Table S1) were set for the training phase. The use of this neural network model enabled the precise detection of the osteocyte lacunae in an automatic way and short time [28], mapping the full biopsy.

Class count refers to the number of components that the network must discriminate; the depth level is the number of layers that capture contextual information in the two phases of encoding and decoding; the initial filter count represents the filter count at the first convolutional layer; the patch size is the number of sub-sections in which the dataset will be cut; the stride ratio specifies the overlap between adjacent patches; the batch size determines the number of patches in a batch; the epoch's number indicates the number of iterations; the loss function measures the error between the neural network's prediction and reality; the optimization algorithm is used to update the parameters of the model, to minimize the prediction errors; the metric function is used to judge the model's performance; early stopping is used to stop iterations once the estimation error drops below a certain threshold; and the learning rate is a parameter which controls the step with which we arrive at the final network optimization result. Generally, this last parameter reduces as one becomes closer to optimal performance; we chose to have it adapted during training and not to keep it fixed. The percentage division of the dataset is 90% for training and 10% for testing.

Network performance was evaluated by applying the training to each ROI and visually verifying that it accurately segmented each osteocyte lacuna. The region of interest (ROI) of the lacunae selected and the Scalar Generator tool was applied; it considered each lacuna individually; afterwords, morphometric parameters were measured in each osteocyte lacuna.

According to the literature [29,30], only lacunae within the volumetric range of (80–980) μm^3 can be considered functional; thus, only lacunae in this volumetric range were considered.

The following shape complexity features (SCFs) were evaluated in each osteocyte lacuna to quantify the degree of detail or complexity that its shape possesses: the mean lacuna volume (LacV—μm^3); the mean lacunar volume/surface ratio (LacV/S—μm) [31]; the 3D aspect ratio (LacAR), which returns the ratio between the lacuna shortest and longest axes; the Sphericity (LacS), which quantifies the extent to which the lacuna resembles a sphere by calculating the ratio of the surface area of an equal-volume sphere to the actual surface area of the lacuna; and the specific number of lacunae (LacNr—mm^{-3}), which refers to the number of lacunae within a 1 mm^3 volume of bone matrix.

2.4.6. Bone Mineralization Analysis

Nine specific regions of interest (ROIs—box dimension: $600 \times 600 \times 600$ $pixel^3$), i.e., three for each axial (coronal, central, apical) portions, were virtually extracted from the patient biopsy.

Starting from the reconstructed complex refractive index n that is linearly related to the mass density (mg/cm^3), the apparent bone mineralization distribution (BMD^r) was calculated in each ROI of the biopsy, following the Roschger approach [32]. The BMD^r parameters were calculated within the mineralized domain; since the reconstructed complex refractive index might be biased due to the constant ratio δ/β and the use of a white beam, absolute values of bone mass density (calcium concentrations − Ca weight %) could not be retrieved; hereafter, the superscript r denotes relative values for all the parameters. Indeed, the Paganin phase retrieval algorithm [25] is based on the assumption of a monochromatic beam and works as an approximation in the present analysis. However, since ROIs were similar in terms of size and composition, relative differences in bone mineralization distribution between them were appreciated. For this reason, the analysis was performed with reference to the distribution of pixel frequencies in the 0–255 range of 8-bits.

Three parameters were extracted for each detected peak using the PeakFit software (v. 4.12, Systat Software, San Jose, CA, USA): the mean relative mass density (BMD^r—mean), the full width at half maxima of the distribution (BMD^r—fwhm), and the peak area percentage (BMD^r—area %).

2.5. Morpho-Chemical Characterization of the Patient's Biopsy

The morpho-chemical characterization of the patient's biopsy was performed combining histological analysis with Fourier Transform Infrared Imaging (FTIRI) spectroscopy.

The analyses were carried out at the Advanced Research Instrumentation Laboratory, Department of Life and Environmental Sciences, Università Politecnica delle Marche (Ancona, Italia). Thin adjacent sections (~5 μm thickness) were cut by the FFPE sample and deposited alternatively onto glass slides for the histological evaluation and CaF_2 optical windows (38 mm × 26 mm × 1 mm size) for the vibrational analysis. All sections were left to air dry for 20 min.

2.5.1. Histology

Histological sections were stained overnight at room temperature using a solution of Alizarin red S (Sigma-Aldrick, Milan, Italy) and Alcian blue (Sigma-Aldrick, Milan, Italy) in a 1:100 ratio according to a modified protocol described by Walker and Kimmel [33]. The two staining solutions were prepared as follows. Considering Alizarin red S, 20 mg of powder was dissolved in 2 mL of 0.5% KOH. Regarding Alcian blue, a stock solution was produced dissolving 20 mg of powder in 5 mL of 70% ethanol. Then, 50 μL of the Alcian blue stock solution was diluted in a solution of 70% ethanol containing 0.0122 g of $MgCl_2$ at a final volume of 1 mL. After staining overnight, the sections were then washed in distilled water for 1 min and dehydrated in the ascendant alcohol series (70%, 80%, 95% and 100%; 1 min per step), and cleared in Xylene for 10 min; then, a coverslip was mounted by using the Eukitt reagent (Bio Optica, Milan, Italy). The sample was left to dry overnight and then visualized under an optical microscope, the Zeiss Axio Imager A.2 (Zeiss, Oberkochen, Germany). A microscopic overview of the tissue was obtained at 50× magnification, while the details of the sample were acquired with higher magnification at 200× magnification.

2.5.2. Fourier Transform InfraRed Imaging (FTIRI) Analysis

The FTIR analysis was carried out by using an INVENIO-R interferometer, coupled with a Hyperion 3000 Vis-IR microscope (Bruker Optics, Ettlingen, Germany) and equipped with a Focal Plane Array (FPA) detector operating at liquid nitrogen temperature (Bruker Optics, Ettlingen, Germany). The FPA detector is well suitable for the analysis of non-homogeneous biological samples, such as tissues and cells, since it allows for the collection of IR images which provide information at the morpho-chemical level, correlating, in each point of the sample, the morphological features with the macromolecular composition [34–36].

The microphotograph of each section was acquired by using a television camera, and the regions of interest (ROIs) were selected based also on information from histological evidence. IR images were acquired in transmission mode in the 4000–900 cm^{-1} range, with a spectral resolution of 4 cm^{-1}; every IR image was 164 × 164 $μm^2$ size and it was composed by 4096 pixel/spectra, with a spatial resolution of 2.56 × 2.56 $μm^2$; each spectrum was the result of 256 scans. The background spectrum of a clean area of the CaF_2 optical window was always acquired before each IR image acquisition. Raw IR images obtained in this way were then subjected to the Atmospheric Compensation and Vector Normalization routines in the whole spectral range of acquisition, respectively, for discarding the contribution of atmospheric carbon dioxide and water vapor and avoiding artifacts deriving from thickness differences.

IR images were then integrated under specific spectral intervals to generate false color images, which allow to identify, within the mapped areas, the regions characterized by a different biochemical composition: an arbitrary color scale was used, with blue color indicating areas with the lowest absorbance values, while pink/violet those with the highest ones.

The IR spectra representative of these different areas were extracted, cut in the 1800–900 cm^{-1} range, vector-normalized, and two-points baseline linear fitted. The height of specific peaks, representative of the inorganic and organic matrices, was used to calculate definite spectral parameters.

The acquisition of IR images and all the spectral treatments were performed by the software OPUS 7.1 (Bruker Optics, Ettlingen, Germany).

2.6. Statistical Analysis

XμCT, PhC-XμCT, and FTIRI data were submitted to statistical analysis. Descriptive statistics were analyzed using the software package Prism 10.4.1 (GraphPad Software, San Diego, CA, USA). Data distributions were checked for normality by the D'Agostino & Pearson, Anderson–Darling, Kolmogorov–Smirnov and Shapiro–Wilk tests, and homogeneity of variances was assessed by the F-test. Multivariate analyses were performed by Spearman's rank matrices. The correlations were classified as "very weak" ($\rho = 0 \div 0.20$), "weak" ($\rho = 0.20 \div 0.40$), "moderate" ($\rho = 0.40 \div 0.60$), "strong" ($\rho = 0.60 \div 0.80$) and "very strong" ($\rho = 0.80 \div 1.0$). "Strong" and "very strong" correlations with $p < 0.05$ were considered statistically significant. A p-value of <0.05 was accepted as statistically significant for all tests. * $p < 0.05$; ** $p < 0.01$; *** $p < 0.001$; **** $p < 0.0001$.

3. Results

The proposed study was conducted in two phases: the first phase involved the analysis of nine cubic scaffolds as fabricated; the second phase entailed the examination of a patient's biopsy specimen obtained from a sinus lift procedure performed using the same biomaterial.

In the initial phase, the workflow involved several steps: first, the as-fabricated β-TCP scaffold was characterized using scanning electron microscopy (SEM) and X-ray diffraction (XRD) to assess its surface morphology and crystalline structure. Subsequently, micro-computed tomography (XμCT) scans were acquired to perform morphometric analysis of the scaffold's microarchitecture. A compressive load test was then performed until failure to evaluate mechanical performance. Following mechanical testing, XμCT imaging was repeated on the fractured specimen to analyze the failure modes and structural integrity. Finally, the fractured sample was subjected to SEM analysis to investigate fracture surfaces and microstructural features in detail.

In the second phase, the biopsy obtained from the patient was initially subjected to XμCT imaging to acquire morphometric data regarding the mineralized tissue, focusing on the microarchitecture of the trabecular structure. Subsequently, the same sample was imaged using phase-contrast imaging based on synchrotron radiation (PhC-XμCT) to analyze the network of osteocyte lacunae within the newly formed bone in contact with the residual β-TCP graft biomaterial. Finally, the biopsy was sectioned into slices suitable for histological analysis and FTIRI to investigate tissue organization at the macromolecular level.

3.1. Crystalline Structure of the β-TCP Scaffold

X-ray diffraction (XRD) patterns of the raw powder used for scaffold production and the scaffold top surface are reported in Figure S1a, along with the Rietveld refinement results, as shown in Figure S1b,c. Patterns are shown in the reduced angular range $2\theta = 35–70°$, where the most intense peaks of β-TCP are found. The experimental lattice parameters of the β-TCP scaffold and powder estimated from the Rietveld refinement are reported in Figure S1b and Figure S1c, respectively. Such lattice parameters are in close agreement with the nominal values reported in the PDF2-ICDD file for the β-TCP phase.

3.2. Morphometric Analysis of the β-TCP Scaffold

3.2.1. Surface Morphology

SEM images of scaffolds surfaces are reported in Figure 2 in both fractured (Figure 2a–c,e) and not fractured (Figure 2d) conditions. The observation of the fracture surfaces showed that the fracture is primarily brittle, more frequent along the branches of the trabecular structure rather than at the intersections of the nodes. Regarding the unbroken surface, the printing patterns are clearly visible, with a spacing between the layers that is close to the nominal value of 25 μm (Figure 2d). On the fractured surface, instead, an increase in microporosity is observed (Figure 2e), with pores measuring between 1 and 5 μm in diameter, which was also confirmed by EDS analysis, which reported weight percentages of Ca and P not normalized that are slightly lower in the fractured areas than in not fractured (Figure S2a,b). The amounts of Ca and P (in at.%) obtained by EDS analysis performed on the complete set of scaffolds, within experimental uncertainties the Ca/P atomic ratio, are unvaried, suggesting unchanged stoichiometry of β-TCP in all the analyzed samples. It is worth noting that the values of the Ca/P atomic ratio reported in Figure S2a for a representative sample, within the experimental error, agree with the nominal value for stoichiometric β-TCP (Ca/P = 1.5), as obtained from the chemical formula $Ca_3(PO_4)_2$.

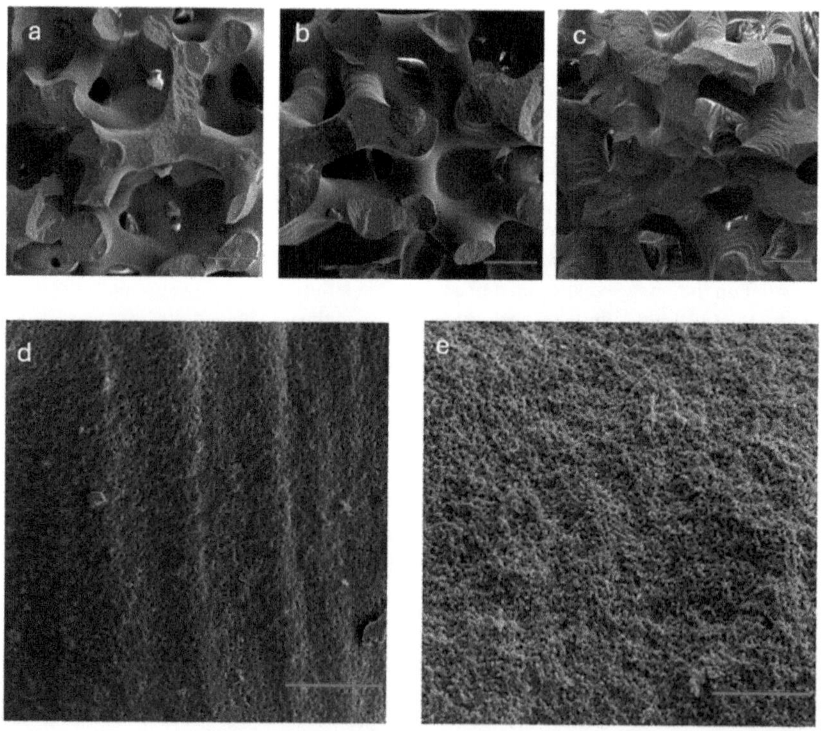

Figure 2. SEM analysis of β-TCP scaffolds (before and after fracture). (**a–c**) The observation of the fracture surfaces of three representative samples indicates a predominantly brittle fracture, more pronounced along the branches rather than at the nodes of the trabecular structure. Red bar: 500 μm; (**d**) non-fractured surface: the printing patterns are evident with layer spacing consistent with the nominal value of 25 μm. Red bar: 50 μm; (**e**) fractured surface: an increased microporosity (1–5 μm, in diameter) was observed. Red bar: 50 μm.

3.2.2. Microarchitecture

All nine scaffolds exhibited a microarchitecture, as shown in Figure 3a, that closely resembled the original nominal design. The internal microstructure is illustrated in Supplementary Video S1, which showcases a representative sample from the group. The morphometric parameters of the scaffolds, obtained via XμCT prior to compressive loading, are summarized in Table 1. These parameters were calculated using the BoneJ plugin—an ImageJ (v. 1.53q) extension specialized for bone image analysis [37].

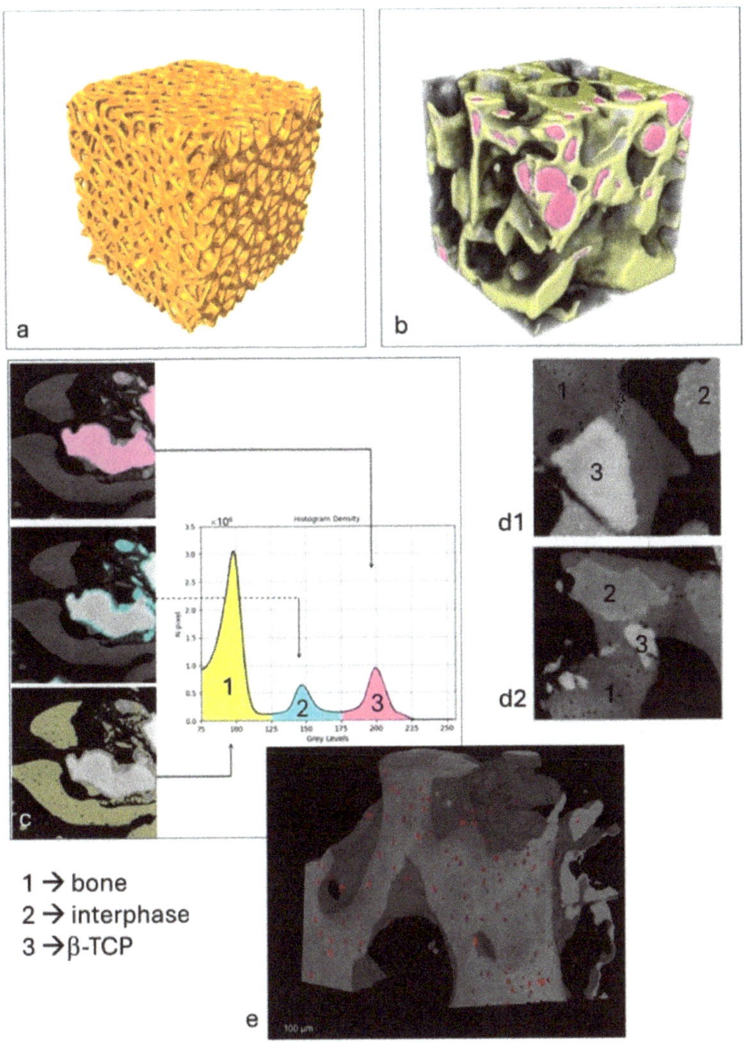

Figure 3. XμCT and PhC-XμCT imaging analysis: (**a**) scaffold microarchitecture in a representative sample; (**b**) 3D XμCT reconstruction of a sub-volume in the patient's biopsy: yellow: newly formed bone; pink: residual β-TCP; (**c**) histogram of densities in a representative area of the patient's biopsy: peak 1 = newly formed bone; peak 2 = interphase; peak 3 = residual β-TCP. (**d1**,**d2**) Other representative areas of the patient's biopsy as imaged by PhC-XμCT with evidence of the biomaterial integration and resorption; (**e**) representative sub-volume of the patient's biopsy: osteocyte lacunae segmented by the neural net are represented in red.

Pre-loading assessments revealed that the scaffolds were homogeneous, with reproducible morphometric metrics and shape complexity, as indicated by low standard deviations (SD). This suggests highly isotropic structures, with an anisotropy DA close to zero. The trabeculae exhibited an average thickness of approximately 400 μm and an average inter-trabecular distance of about 550 μm, with a volume fraction near 50%. Additionally, parameters such as fractal dimension and connectivity demonstrated high stability, further confirming the reproducibility of the manufacturing process.

Furthermore, considering the significant influence of pore dimensions and distribution on the bioactive properties of the scaffold, an analysis was performed to determine both the spatial pore distribution and the pore size distribution. The results, depicted through color-coding in Figure 4a and the histogram illustrating the pore size distribution in Figure 4b, collectively demonstrate a broad range of pore sizes, with dimensions exceeding 100 μm and an average value of approximately 500 μm, in accordance with the literature recommendations for achieving osteoconductivity.

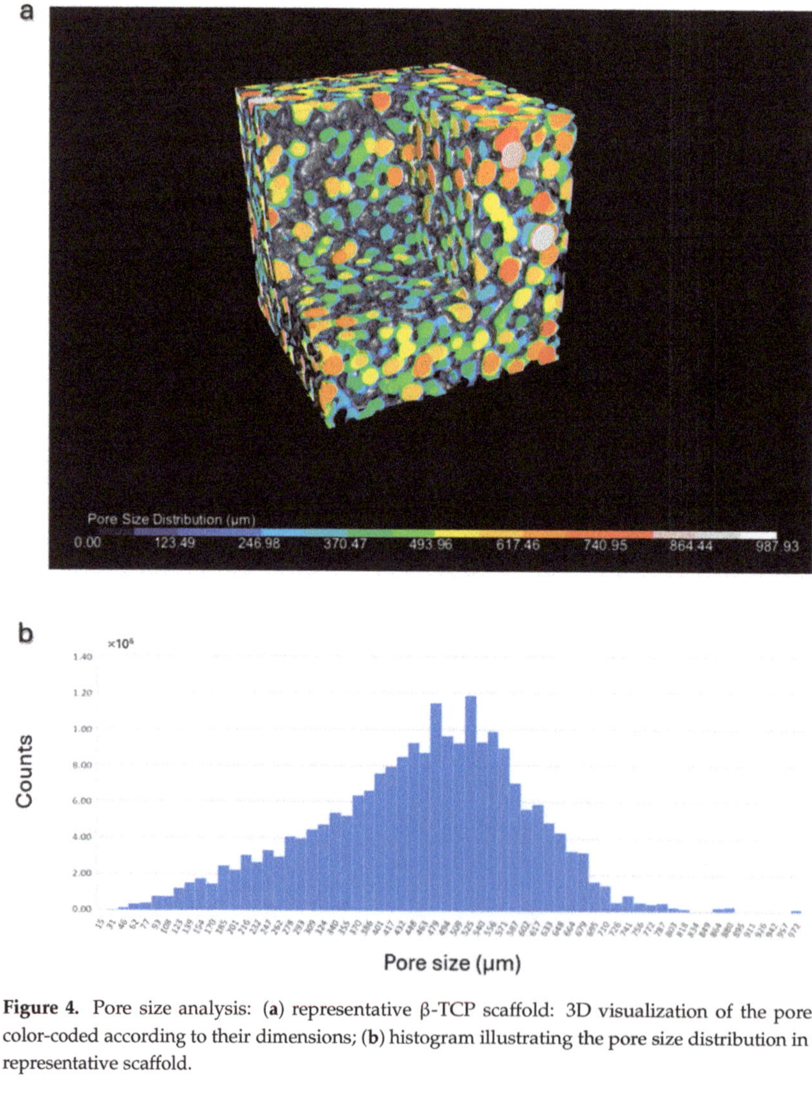

Figure 4. Pore size analysis: (**a**) representative β-TCP scaffold: 3D visualization of the pores, color-coded according to their dimensions; (**b**) histogram illustrating the pore size distribution in a representative scaffold.

Table 1. β-TCP scaffold (as produced) microarchitectural parameters—descriptive statistics.

β-TCP Scaffold (as Produced)	Mean	Std.Dev.	95% CI of Mean
BV/TV (%)	48	4	45 to 51
Porosity (%)	52	4	49 to 55
Tb.Th Mean (μm)	387	16	375 to 399
Tb.Th Std Dev (μm)	97	10	89 to 104
Tb.Th Max (μm)	702	60	656 to 748
Tb.Sp Mean (μm)	546	93	475 to 618
Tb.Sp Std Dev (μm)	330	137	225 to 435
Tb.Sp Max (μm)	2377	624	1897 to 2856
DA	0.195	0.015	0.183 to 0.207
Fractal Dimension	2.52	0.08	2.46 to 2.59
Conn. D (mm^{-3})	7	2	6 to 9

Post-compressive loading XμCT analyses are detailed in Table S2, which reports parameters including volume percentage, trabecular thickness, fractal dimension, and connectivity after fracture. Notably, the structure exhibited increased anisotropy post-loading, as evidenced by an elevated DA, likely attributable to the applied mechanical stress.

3.3. Compressive Load Tests

The mechanical response of scaffolds to compressive tests is shown in Figure 5. Samples tested in the build direction almost show a similar behavior, except for sample #1, which shows oscillations of the curve before fracture (Figure 5a). Such oscillations are typical of scaffolds built layer by layer by additive manufacturing techniques. On compression, the failure of one or two struts of the layer causes the collapse of the entire layer that occurs in correspondence with a minimum of the curve. The collapse of the layer produces a compacted structure responsible for the curve shape after this event. This mechanism is repeated layer by layer, causing the oscillating shape of the curve. As the collapse of the layers occurs from the outside to the inside during compression, the general shape of the first part of the curve is not perturbed. However, the slope of this part of the curve up to about 170 N load may differ from sample to sample, due to small differences in the positioning of the sample on the compression stage.

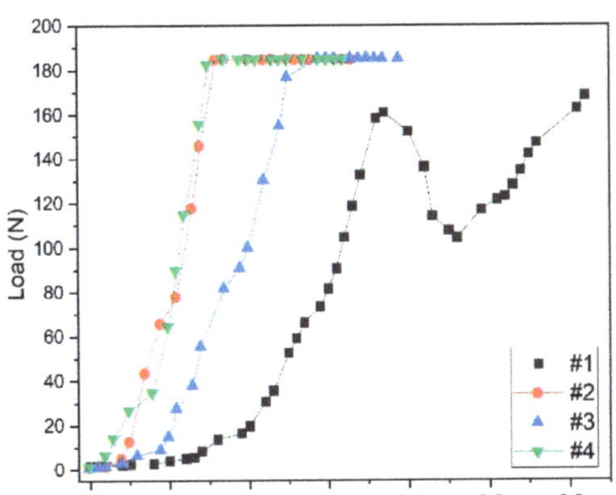

(a)

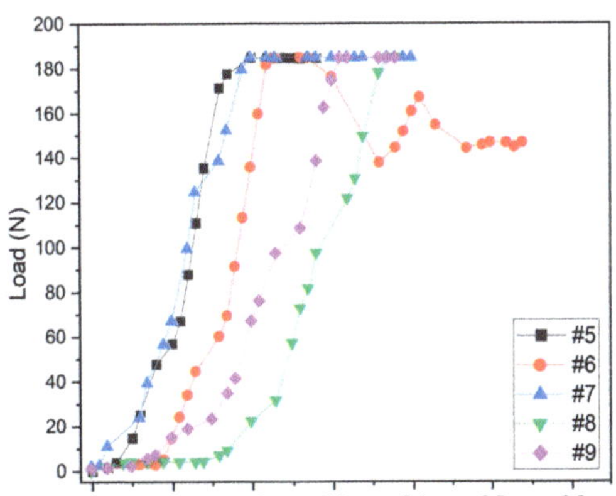

(b)

Figure 5. Compressive load tests: (**a**) mechanical behavior of samples tested in the build direction, (**b**) mechanical behavior of samples tested in the random direction.

The mechanical behavior of samples tested in random direction is shown in Figure 5b. Also, in this case, the general shape of the curves is similar, except for sample #6, which was probably tested in the build direction because of the oscillations of the curve. The maximum applied load for all samples is about 180 N, and the maximum displacement before fracture is about 0.4 mm. It is worth noting that the two oscillating curves of samples #1 (Figure 5a) and #6 (Figure 5b) reach the maximum value of about 0.6 mm displacement before fracture.

3.4. Spearman Correlative Analysis

The Spearman correlation analysis (Supplementary Figure S3a), examining the relationships between scaffold morphometric parameters and mechanical loading indices across the nine investigated scaffolds, identified the following significant correlations: (a) trabecular thickness (Tb.Th) exhibits a strong positive correlation with Young's modulus ($\rho = +0.70$, $p = 0.043$); (b) maximum trabecular thickness (Tb.Th.Max) shows a strong positive correlation with maximum load capacity ($\rho = +0.78$, $p = 0.017$); (c) the anisotropy degree (DA) is strongly positively correlated with specific energy absorption ($\rho = +0.68$, $p = 0.050$).

3.5. The Patient's Biopsy

3.5.1. Histological Analysis

The analysis of the histological images obtained by Alcian blue and Alizarin red S staining is described below. A grey-colored region lacking staining and attributable to the β-TCP scaffold was observed (white asterisk; Figure 6A), while the bone tissue was well evident in red, owing to the binding with hydroxyapatite (white arrow heads, Figure 6B–F). Some small light blue colored areas were also found, probably associated with collagen deposition (yellow hashes, Figure 6A–F). Interestingly, it was possible to detect some regions light red stained (yellow asterisks, Figure 6C,D,F) suggesting the beginning of the mineralization process and hence the formation of new bone.

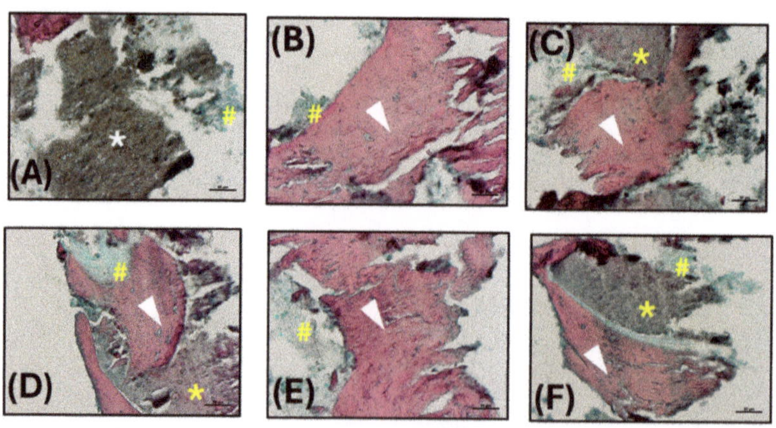

Figure 6. Histological analysis: (**A–F**) Histological images obtained by Alcian blue and Alizarin red S staining (200× magnification; scale bars 50 μm). Hydroxyapatite deposition is red stained (white arrow heads), while the collagen component is blue (yellow hashes). The β-TCP appears not stained, in grey (white asterisk), or lightly stained in red in the region of new bone formation (yellow asterisks).

3.5.2. Microarchitecture of the Trabecular Structure

The bioptic specimen analyzed via bench-top XμCT (Figure 3b) exhibited a microarchitecture characterized by the following parameters: the specific volume percentage (Vol.%)—representing the combined volume of newly formed bone and residual mineralized β-TCP scaffold—was approximately 55.4%. The mean trabecular thickness (Tb.Th) was 195 ± 74 μm, with a maximum value of 428 μm, while the mean trabecular spacing (Tb.Sp) was 194 ± 88 μm, reaching up to 379 μm. Additionally, the connectivity density (Conn.D; mm^{-3}) was measured at 104, the degree of anisotropy (DA) was 0.327, and the fractal dimension (FD) was calculated to be 2.44.

Furthermore, the examined biopsy sample, analyzed by PhC-XμCT with synchrotron radiation, shows three distinct absorption peaks, each corresponding to phases with different densities (Figure 3c,d). The first peak is attributed to mature bone tissue and corresponds to the phase with the lowest density (Phase 1, in Figure 3c,d). The second peak represents an interphase, which will probably disappear for regenerative periods much longer than six months after grafting (Phase 2, in Figure 3c,d). The third and densest peak indicates residual β-tricalcium phosphate (β-TCP) (Phase 3, in Figure 3c,d). The volumetric percentage of the newly formed bone was measured to be approximately 28.4 (±14.2) vol%, and the interphase was around 15.4 (±11.4) vol%, while the residual biomaterial was found to be approximately 11.6 (±10.4) vol%. The large standard deviations for the three parameters suggest a non-homogeneity of regeneration within the different biopsy areas. This confirms the fact that the regenerative process is not yet complete after 6 months of biomaterial grafting. The Spearman correlation analysis investigating the relationships among bone, interphase, and biomaterial indices across the nine regions analyzed in the biopsy (Supplementary Figure S3b) revealed the following significant correlations: (a) the bone peak volume percentage (bone-vol%) shows a strong negative correlation with the interphase peak volume percentage (interphase-vol%) ($\rho = -0.73$, $p = 0.031$); (b) the interphase peak mean density (interphase-mean) exhibits a strong negative correlation with the interphase peak volume percentage (interphase-vol%) ($\rho = -0.80$, $p = 0.013$).

3.5.3. Osteocyte Lacunae Network in the Patient's Biopsy

Figure 3e shows, in a representative sub volume, the distribution of osteocyte lacunae; most of the lacunae have a size ranging from 400 to 500 μm^3, variously distributed among the different areas of the sample, in the bone phase portion.

The lacunar SCFs that were evaluated by the synchrotron analysis are reported in Table 2, introducing mean, standard deviation and 95% confidence interval as descriptive parameters.

Table 2. Osteocyte lacunae network shape complexity features (SCF) in the patient's biopsy—descriptive statistics.

Patient's Biopsy—SCFs	Mean	Std.Dev.	95% CI of Mean
LacVol (μm^3)	426	42	391 to 461
LacV/S (μm)	1.24	0.05	1.20 to 1.28
LacAR	0.334	0.019	0.318 to 0.350
LacS	0.825	0.013	0.815 to 0.836
LacNr (mm^{-3}) × 10^4	1.82	0.32	1.55 to 2.09

Spearman's correlation analysis examining the relationships between the shape complexity characteristics of osteocyte lacunae (Supplementary Figure S3c) in the nine ROIs of the examined biopsy identified a significant positive correlation between LacV and LacV/S

(ρ = +0.90, p = 0.005), confirming that larger lacunae correspond to more spherical shapes, i.e., presumably younger lacunae.

3.5.4. Macromolecular Analysis

For better comprehension of the hyperspectral imaging analysis that follows, the IR spectra of pure β-TCP and bone hydroxyapatite (HA) were collected (Supplementary Figure S4). As expected, β-TCP shows different spectral features respect to a mature bone. In fact, in the IR spectrum of pure β-TCP, only the bands associated with the vibrations of phosphate groups are present (~1117 cm^{-1}, ~1018 cm^{-1}, ~970 cm^{-1}, and ~944 cm^{-1}) [38]. Conversely, the IR spectrum of a bone sample is characterized by the Amide I (~1656 cm^{-1}), II (~1551 cm^{-1}), and III (~1282 cm^{-1}, ~1239 cm^{-1}, and ~1203 cm^{-1}) bands of the organic matrix (proteins, and collagen) and by the peaks of phosphates (~1100 cm^{-1}, and ~1032 cm^{-1}, attributable, respectively, to poor and well-crystallized stoichiometric hydroxyapatite, and ~960 cm^{-1}) and carbonates (~1451 cm^{-1}, and ~1415 cm^{-1}) of the inorganic component [39].

The hyperspectral imaging analysis that follows was performed on specific regions of interest (hereinafter named ROI-#) identified on the sections cut from patient's biopsy, as well as on the base of the histological evidence. In ROI-1, the presence of two different tissue areas stands out clearly (Figure 7A). Both regions are mainly composed of hydroxyapatite, as suggested by the spectral profiles in Figure 7B. This is also confirmed by false color images calculated on the bands at ~1412 cm^{-1}, attributable to inorganic carbonates (Figure 7C), ~1100 cm^{-1} (Figure 7D), and ~1032 cm^{-1} (Figure 7E); the latter two were assigned to phosphate groups, respectively, in poor and well-crystallized stoichiometric HA. The orange/red color in Figure 7C suggests a homogeneous distribution of carbonate groups and confirms the presence of bone tissue in the whole analyzed area. Nevertheless, the different colorations in Figure 7D,E uphold a distinct degree of crystallization: the lower left part of Figure 7D (red/pink colored), corresponding to the red asterisk region, is characterized by the presence of poor HA, while the upper right part of Figure 7E (red/pink colored), corresponding to the blue asterisk area, is composed of well-crystallized stoichiometric HA. Finally, to ascertain the quality of the bone tissue, the following spectral parameters were analyzed [39,40]: (i) the Mineral Maturity index (MM), calculated as ratio between the intensities of the peaks at ~1032 cm^{-1} and ~1100 cm^{-1}, which is diagnostic for the level of bone crystallinity; (ii) the Carbonates-to-Phosphates ratio (C/P), obtained by ratioing the intensities of the peaks at ~1415 cm^{-1} and ~1032 cm^{-1}, which indicates the level of substitution of carbonates in the apatite lattice; (iii) the Mineral-to-Matrix ratio (AI/P), calculated as ratio between the intensities of the peaks at ~1656 cm^{-1} and ~1032 cm^{-1}, which signifies the level of mineralization of the bone and can be related to its mechanical performance. In the region identified by the blue asterisk, the MM was higher ($p < 0.05$), while the C/P and AI/P were lower (respectively, $p < 0.05$ and $p < 0.01$), indicating, in this area, a higher level of bone crystallinity and a lower amount of carbonate groups in the lattice apatite (Figure 7F).

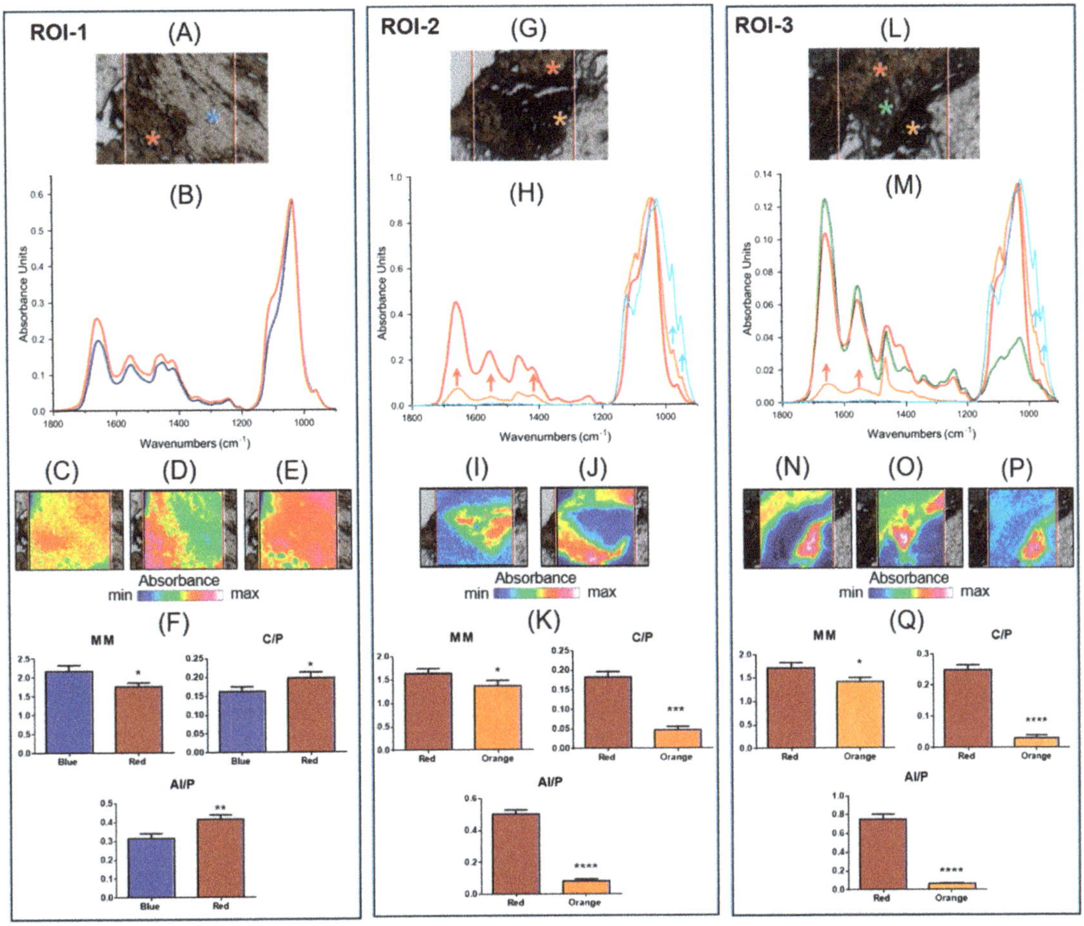

Figure 7. FTIR hyperspectral Imaging analysis: ROI-1: (**A**) Microphotograph of ROI-1 (red box): red and blue asterisks indicate the presence of areas with different macromolecular composition. (**B**) Absorbance IR spectra representative of the two areas (red and blue lines). False color images obtained integrating IR images under the bands at (**C**) ~1412 cm^{-1}. (**D**) ~1100 cm^{-1}, and (**E**) ~1032 cm^{-1}. (**F**) Spectral parameters: Mineral Maturity Index (MM), Carbonates-to-Phosphates ratio (C/P), and Mineral-to-Matrix ratio (AI/P). ROI-2: (**G**) Microphotograph of ROI-2 (red box): red and orange asterisks indicate the presence of two regions with different macromolecular composition. (**H**) Absorbance IR spectra representative of the two areas (red and orange lines) and of pure β-TCP (light blue line). False color images obtained integrating under the bands at (**I**) ~970 cm^{-1}, and (**J**) ~1032 cm^{-1}. (**K**) Spectral parameters: Mineral Maturity Index (MM), Carbonates-to-Phosphates ratio (C/P), and Mineral-to-Matrix ratio (AI/P). ROI-3: (**L**) Microphotograph of ROI-3 (red box): red, green, and orange asterisks indicate the presence of areas with different macromolecular composition. (**M**) Absorbance IR spectra representative of the three areas (red, green, and orange lines) and of pure β-TCP (light blue line). False color images obtained integrating under the bands at (**N**) ~1032 cm^{-1}, (**O**) ~1656 cm^{-1}, and (**P**) ~970 cm^{-1}. (**Q**) Spectral parameters: Mineral Maturity Index (MM), Carbonates-to-Phosphates ratio (C/P), and Mineral-to-Matrix ratio (AI/P). False color images were built by using an arbitrary color scale, blue color indicating areas with the lowest absorbance values (min absorbance), while pink/violet those with the highest ones (max absorbance). As regards the statistical analysis, data are shown as mean ± SD; asterisks over histograms indicate statistically significant differences (* $p < 0.05$; ** $p < 0.01$; *** $p < 0.001$, and **** $p < 0.0001$).

ROI-2 is also characterized by two different areas (red and orange asterisks in Figure 7G). In Figure 7H, the IR spectra representative of the two regions are shown together with that of pure β-TCP (light blue line). The analysis of the spectral profiles suggests, in the red asterisk region, the presence of bone tissue with poorly crystallized HA. Conversely, the IR spectrum of the orange asterisk region is characterized not only by the bands typical of a bone (such as the bands at ~1656 cm^{-1} and ~1551 cm^{-1}, named AI and AII, and associated with the organic matrix, and ~1412 cm^{-1} attributable to the inorganic carbonates in the apatite lattice; red arrows), but also by the peaks at ~970 cm^{-1}, and ~943 cm^{-1}, related to β-TCP (light blue arrows). This was also confirmed by false color images obtained integrating IR images, respectively, under the bands at ~970 cm^{-1} (Figure 7I) and ~1032 cm^{-1} (Figure 7J). All these findings allow us to hypothesize, in the orange asterisk region, the beginning of the mineralization process induced by β-TCP, as also suggested by the analysis of the spectral parameters (Figure 7K); in fact, in the orange asterisk region, the MM, C/P and AI/P were lower (respectively, $p < 0.05$, $p < 0.001$, and $p < 0.0001$) with respect to those calculated in the red asterisk one, confirming, in this area, a new immature bone with low crystallinity and a low amount of carbonates and organic matrix.

The third investigated region, named ROI-3, appears more complex due to the presence of various biochemical components (Figure 7L). The analysis of the spectral profiles (Figure 7M) indicates, in the red asterisk region, the presence of bone tissue, with poorly crystallized HA; conversely, the orange asterisk region shows both the peaks associated with the organic matrix (such as AI and AII; red arrows), and those related to β-TCP at ~970 cm^{-1} and ~943 cm^{-1} (light blue arrows), indicating the presence of new bone formation. Interestingly, an additional region at the interface of those already described was evidenced (green asterisk) characterized as extracellular matrix, as suggested by the high intensities of Amide I, II and III bands, and by the low intensity of a convoluted band attributable to phosphates. The same results were evidenced by false color images integrated under the bands at ~1032 cm^{-1} (Figure 7N), ~1656 cm^{-1} (Figure 7O), and ~970 cm^{-1} (Figure 7P). As expected, the spectral parameters MM, C/P, and AI/P showed lower values in the orange asterisk region, confirming, in this area, the presence of new bone formation with scarce crystallinity and a low amount of carbonates and organic components (Figure 7Q).

4. Discussion

The body of literature on bone substitute biomaterials (BSBs) proposed for jawbone regeneration is vast, reflecting an ongoing pursuit to identify ideal scaffolds capable of effectively substituting autologous bone. This need is particularly pronounced in sinus lift elevation procedures, which are performed when an insufficient bone volume is available for the placement of dental implants. Outcomes associated with these BSBs exhibit considerable variability, influenced not only by individual patient characteristics but also by specific initial surgical conditions, the chemical composition of the biomaterial, and its resorption kinetics and profile (as outlined in Table S3). While some biomaterials, endowed with favorable biocompatibility and biomechanical properties, promote early healing compatible with immediate implant loading, they often exhibit minimal or even negligible resorption over time, potentially leading to several complications, including infections or granuloma formation, if the residual biomaterial is not fully integrated or if unassimilated fragments are present [41,42]. Moreover, impaired implant osseointegration may occur if the biomaterial is not fully absorbed or integrated into the bone, resulting in inadequate anchorage of the dental implant and an increased risk of implant failure [43]. Conversely, other materials favor rapid resorption to facilitate native bone regeneration but may lack adequate mechanical stability during the critical early phases of healing, thus precluding immediate functional loading [44]. The ongoing challenge, therefore, remains the development of a biomaterial that simultaneously balances biological integration, mechanical stability, and controlled resorption, addressing the multifaceted demands of jawbone regeneration.

This study highlights the potential of lithography-based manufactured β-tricalcium phosphate (β-TCP) scaffolds as viable biomaterials for sinus lift elevation, particularly in complex cases involving severe maxillary atrophy. The integration of advanced additive manufacturing techniques, detailed structural characterization, and in vivo outcomes provides a comprehensive understanding of scaffold performance and biological tissue response.

From a material perspective, crystalline analysis through X-ray diffraction (XRD) confirmed the exclusive presence of the β-TCP phase, with no significant deviations in stoichiometry. Scanning electron microscopy observations revealed the presence of a uniform microporous lattice, with pores of average diameters of a few micrometers. Moreover, after compressive loading, it revealed the presence of brittle fracture, primarily along trabecular branches but almost excluding the trabecular nodes. The microarchitectural analysis demonstrated that the additive manufacturing process reliably reproduced the intended porosity and trabecular features, without significant deviations between one sample and another, with average macro/micro porosity conducive to vascularization and biological infiltration. In follow-up studies, the Brunauer–Emmett–Teller (BET) theory [45] will be employed for surface area analysis. This will enable a more precise characterization of the scaffold's surface properties, thereby enhancing our understanding of its interactions with the biological environment.

The mechanical performance of biomaterials used for sinus augmentation is critically important to the success of this surgical procedure. Numerous studies have evaluated the effectiveness of various grafting materials, revealing significant differences in outcomes related to bone regeneration and postoperative complications. While the maximum compressive load supported by these scaffolds is slightly lower than the values measured on lithography-based manufactured biphasic calcium phosphate scaffolds and comparable to that of those that are traditionally sintered, it has also been shown that the complete reabsorption of β-TCP, combined with a more suitable microarchitecture of the regenerated tissue, makes this scaffold extremely promising for maxillary bone engineering [10,46].

Moreover, post-mechanical compressive testing indicated the preservation of macrostructural features, although increased anisotropy suggested some structural reorganization under load, which is consistent with the correlation between microstructural parameters and mechanical resilience observed in Spearman's analysis. By Spearman's analysis, Young's modulus was observed to be strongly positively correlated to trabecular thickness; this occurs because thicker trabeculae provide greater resistance to compression and applied mechanical forces, enhancing the scaffold's ability to withstand loads without deforming. This greatly mimics the natural trabecular microarchitecture of bone that is designed to efficiently distribute forces: as trabecular thickness increases, the bone becomes less susceptible to microfractures and other deformations that would reduce its strength. This results in greater strength and, consequently, a higher Young's modulus. Moreover, the anisotropy degree of the scaffold was found to strongly positively correlate with specific energy absorption; indeed, anisotropic scaffolds can be designed to better withstand specific mechanical loads, such as compression, by aligning the structure in ways that enhance energy absorption. This helps to optimize the scaffold's performance, particularly in dynamic or high-stress environments, like in peri-implant bone, by enabling it to absorb and dissipate energy more effectively.

Biologically, the in vivo evaluation corroborated the biocompatibility and osteoconductive of the β-TCP scaffold, as evidenced by histological analysis and combined XμCT and synchrotron imaging. The detection of nascent mineralized bone tissue, indicated by histological staining, suggests the initiation of osteogenesis within the scaffold, while XμCT and PhC-XμCT analyses revealed the ongoing resorption of residual β-TCP concurrent with new bone formation. Volumetric quantification demonstrated a complex regenerative landscape, with substantial heterogeneity in both mineralization stages and tissue integration, reflecting the typical temporal dynamics of bone healing and scaffold resorption while maintaining overall functional microarchitecture. The osteocyte lacunae analysis via synchrotron phase-contrast XμCT and AI segmentation revealed that the osteocyte network exhibits morphological features consistent with ongoing bone maturation, with lacunae sizes and shapes suggesting variable stages of osteoblast-derived new bone. The positive correlation between lacunar volume and sphericity underscores the relationship between osteocyte morphology and bone vitality, offering potential biomarkers for tissue regenerative quality.

Furthermore, morpho-chemical characterization via Fourier Transform Infrared Imaging (FTIRI) depicted a landscape of mineralization maturity, with spectral parameters indicating heterogeneous levels of bone matrix organization and mineral crystallinity. Regions with higher mineral maturity indices and well-consolidated crystalline domains highlight the progressive ossification and maturation trajectory following grafting, aligning with the objective of achieving functional loading through dental implants.

Overall, this integrated approach, combining scaffold design and biomechanical characterization, in vivo biological response, and advanced imaging, underscores the potential of lithography-based manufactured β-TCP scaffolds in maxillofacial regenerative procedures, particularly sinus lift augmentation. Future research should focus on optimizing scaffold production to furtherly standardize microarchitectural parameters and, consequently, mechanical performance, extending in vivo observation periods to monitor long-term remodeling.

In any case, the outcomes of this study contribute to the validation of the proposed scaffold. This constitutes a necessary, yet distinct and preliminary phase, which precedes the evaluation of clinical applicability—a process that, by definition, entails the establishment of specific inclusion and exclusion criteria for participant selection, the definition of intervention (test) and control groups, and the implementation of procedures such as randomization, blinding, and the identification of predefined endpoints.

5. Conclusions

This study rigorously investigated the synergistic advantages of combining LCM production processes with β-TCP material for bone grafts, focusing on their clinical applicability in maxillofacial regenerative procedures, specifically sinus lift augmentation. Through a multidisciplinary approach, this research quantified key properties of the resulting β-TCP scaffolds, revealing several critical findings:

Microarchitectural uniformity and tissue integration: The scaffolds demonstrated uniform porosity and reliably reproduced trabecular-like structures. This architecture is highly advantageous for promoting vascularization and facilitating tissue ingrowth, crucial aspects for successful bone regeneration.

Mechanical performance: While it exhibited slightly lower compressive strength compared to biphasic calcium phosphate alternatives, the scaffold's mechanical properties were deemed sufficient for clinical application. Furthermore, positive correlations were observed between trabecular thickness and Young's modulus, and between anisotropy and specific energy absorption. These correlations suggest optimized mechanical resilience under physiological loads, indicating the scaffold's capacity to withstand the forces encountered in the maxillofacial region.

Biocompatibility and osteoconductivity: This study confirmed the scaffold's biocompatibility and osteoconductivity, evidenced by active osteogenesis and concurrent scaffold resorption. This indicates the material's ability to integrate with host tissue and promote new bone formation while gradually degrading.

Regenerative dynamics and bone maturation: Heterogeneous mineralization and tissue integration patterns were observed, reflecting typical regenerative dynamics. The scaffold maintained its functional architecture throughout this process. Importantly, lacunar analysis indicated ongoing bone maturation, with lacunar morphology identified as a potential marker for assessing regenerative quality.

Future research should prioritize optimizing scaffold production to further standardize microarchitectural parameters and enhance mechanical performance. Additionally, extending in vivo observation periods will be crucial for monitoring long-term remodeling processes, and exploring functional outcomes in broader patient populations will provide a more comprehensive understanding of clinical efficacy.

Supplementary Materials: The following supporting information can be downloaded at: https://www.mdpi.com/article/10.3390/jfb16070237/s1, Table S1: Convolutional neural network. Used training parameters to maximize the performance; Figure S1: X-ray diffraction spectrum of the β-TCP powder and scaffold; Figure S2: EDS data analysis of β-TCP scaffolds; Video S1: 3D reconstruction of a representative β-TCP scaffold; Table S2: β-TCP scaffold (after compressive loading) microarchitectural parameters—descriptive statistics; Figure S3: Spearman's correlative matrixes. (a) β-TCP scaffold (as produced): microarchitectural and compressive loading parameters; (b) retrieved biopsy (after 6 months of grafting): microarchitectural and bone mineral density parameters; (c) retrieved biopsy (after 6 months of grafting): osteocyte lacunae network shape complexity features; Figure S4: Absorbance IR spectra of β-TCP (light blue line), and bone hydroxyapatite (red line). The spectra are displayed in the 1800–900 cm^{-1} spectral range; the position (in terms of wavenumbers, cm^{-1}) of the most significant peaks are indicated at the top; Table S3: Biomaterials performance in sinus lift augmentation—microarchitecture in clinical cases.

Author Contributions: Conceptualization, C.M. and A.G.; data curation, S.T. and P.M.; formal analysis, G.O., C.G., P.M., E.G. and A.G.; funding acquisition, C.M.; investigation, N.R., G.O., M.F. and C.G.; methodology, N.R., G.O., S.T., M.F., P.M. and E.G.; project administration, A.G.; resources, C.M., S.T. and A.G.; software, N.R. and M.F.; supervision, E.G. and A.G.; validation, G.O. and S.T.; visualization, N.R. and C.G.; writing—original draft, C.M., N.R., C.G., P.M., E.G. and A.G.; writing—review and editing, G.O., S.T. and M.F. All authors have read and agreed to the published version of the manuscript.

Funding: This research received no external funding.

Institutional Review Board Statement: This study was conducted in accordance with the Declaration of Helsinki and approved by the Ethical Committee "Comitato Etico Territoriale Lombardia 1", Milan, Italy (protocol code CET 462-2024 and date of approval 9 December 2024).

Informed Consent Statement: Informed consent was obtained from all subjects involved in this study.

Data Availability Statement: The original contributions presented in this study are included in this article and in the Supplementary Materials; further inquiries can be directed to the corresponding authors.

Acknowledgments: The authors acknowledge Frank Reinauer and Aksu Adem—KLS Martin Group, Mühlheim an der Donau, Baden-Württemberg, Deutschland—for providing all the β-TCP scaffolds studied in this investigation. The authors do not have any financial interest, either directly or indirectly, in the products or information mentioned or elaborated within this paper. This research did not receive any specific grant from funding agencies in the public, commercial, or not-for-profit-sectors. We acknowledge Elettra Sincrotrone Trieste for providing access to its synchrotron radiation facilities and Giuliana Tromba for assistance in using beamline SYRMEP. The 3D reconstructions, AI-based quantitative data and the Supplementary Video for this paper were generated using Dragonfly software, Version 2022.2, Comet Technologies Canada Inc., Montreal, Canada; software available at https://dragonfly.comet.tech/ (accessed on 22 April 2025).

Conflicts of Interest: The authors declare no conflicts of interest.

Abbreviations

The following abbreviations are used in this manuscript:

β-TCP	Beta Tricalcium Phosphate
LCM	Lithography-based ceramic manufacturing
RP	Rapid Prototyping
CBCT	Cone-beam computed tomography
CAD	Computer-aided design
XRD	X-ray diffraction
ICDD	International Centre for Diffraction Data
SEM	Scanning electron microscopy
XμCT	X-ray micro-computed tomography
MTS1	Material Testing Stage
PhC-XμCT	Phase-contrast X-ray micro-computed tomography
STP	Syrmep Tomo Project
TIE	Transport of Intensity Equation
SCF	Shape complexity features
ROI	Regions Of Interest
I	infraRed
FTIRI	Fourier Transform InfraRed Imaging
BV/TV	Ratio between mineralized volume and total scaffold volume
Tb.Th	Mean trabecular thickness
Tb.Sp	Mean trabecular spacing
Conn.D	Connectivity Density
DA	Anisotropy Degree
FD	Fractal Dimension
LacV	Mean Lacuna Volume
LacV/S	Mean Lacunar Volume/Surface ratio
LacAR	3D Lacunar Aspect Ratio
LacS	Lacunar Sphericity
LacNr	Lacunar density
BMD^r	Relative mass density

References

1. Iezzi, G.; Piattelli, A.; Giuliani, A.; Mangano, C.; Manzon, L.; Degidi, M.; Iaculli, F.; Scarano, A.; Filippone, A.; Perrotti, V. Molecular, Cellular and Pharmaceutical Aspects of Bone Grafting Materials and Membranes During Maxillary Sinus-Lift Procedures. Part 1: A General Overview. *Curr. Pharm. Biotechnol.* **2017**, *18*, 19–32. [CrossRef] [PubMed]
2. Zhao, R.; Yang, R.; Cooper, P.R.; Khurshid, Z.; Shavandi, A.; Ratnayake, J. Bone Grafts and Substitutes in Dentistry: A Review of Current Trends and Developments. *Molecules* **2021**, *26*, 3007. [CrossRef] [PubMed]
3. Chaudhari, S.; Khade, A.; Girase, V.; Dhatrak, P. A Systematic Review on Bone Grafts and Biomaterials Substitutes for Bone Regeneration. *J. Phys. Conf. Ser.* **2024**, *2837*, 012033. [CrossRef]
4. Todd, E.A.; Mirsky, N.A.; Silva, B.L.G.; Shinde, A.R.; Arakelians, A.R.L.; Nayak, V.V.; Marcantonio, R.A.C.; Gupta, N.; Witek, L.; Coelho, P.G. Functional Scaffolds for Bone Tissue Regeneration: A Comprehensive Review of Materials, Methods, and Future Directions. *J. Funct. Biomater.* **2024**, *15*, 280. [CrossRef]
5. Gatto, M.L.; Cerqueni, G.; Furlani, M.; Riberti, N.; Tognoli, E.; Denti, L.; Leonardi, F.; Giuliani, A.; Mattioli-Belmonte, M.; Mengucci, P. Influence of Trabecular Geometry on Scaffold Mechanical Behavior and MG-63 Cell Viability. *Materials* **2023**, *16*, 2342. [CrossRef]

6. Nikolova, M.P.; Chavali, M.S. Recent Advances in Biomaterials for 3D Scaffolds: A Review. *Bioact. Mater.* **2019**, *4*, 271–292. [CrossRef]
7. Farjaminejad, S.; Farjaminejad, R.; Hasani, M.; Garcia-Godoy, F.; Abdouss, M.; Marya, A.; Harsoputranto, A.; Jamilian, A. Advances and Challenges in Polymer-Based Scaffolds for Bone Tissue Engineering: A Path Towards Personalized Regenerative Medicine. *Polymers* **2024**, *16*, 3303. [CrossRef]
8. Wei, S.; Ma, J.-X.; Xu, L.; Gu, X.-S.; Ma, X.-L. Biodegradable Materials for Bone Defect Repair. *Mil. Med. Res.* **2020**, *7*, 54. [CrossRef]
9. Bahraminasab, M. Challenges on Optimization of 3D-Printed Bone Scaffolds. *Biomed. Eng. Online* **2020**, *19*, 69. [CrossRef]
10. Mangano, C.; Mangano, F.; Gobbi, L.; Admakin, O.; Iketani, S.; Giuliani, A. Comparative Study between Laser Light Stereo-Lithography 3D-Printed and Traditionally Sintered Biphasic Calcium Phosphate Scaffolds by an Integrated Morphological, Morphometric and Mechanical Analysis. *Int. J. Mol. Sci.* **2019**, *20*, 3118. [CrossRef]
11. Perez, A.; Lazzarotto, B.; Marger, L.; Durual, S. Alveolar Ridge Augmentation with 3D-Printed Synthetic Bone Blocks: A Clinical Case Series. *Clin. Case Rep.* **2023**, *11*, e7171. [CrossRef] [PubMed]
12. de Almeida Malzoni, C.M.; Gonçalves, V.; Possari, J.; Junior, E.M. The Use of 3D Ceramic Block Graft Compared with Autogenous Block Graft for Rehabilitation of the Atrophic Maxilla: A Randomized Controlled Clinical Trial. *Trials* **2022**, *23*, 903. [CrossRef] [PubMed]
13. Ghayor, C.; Chen, T.-H.; Bhattacharya, I.; Özcan, M.; Weber, F.E. Microporosities in 3D-Printed Tricalcium-Phosphate-Based Bone Substitutes Enhance Osteoconduction and Affect Osteoclastic Resorption. *Int. J. Mol. Sci.* **2020**, *21*, 9270. [CrossRef] [PubMed]
14. Mohammed, A.H.M.; Shariff, K.A.; Abu Bakar, M.H.; Mohamad, H. Recent Methods in Fabricating Porous β-Tricalcium Phosphate Scaffolds: A Mini Review. *Mater. Today Proc.* **2022**, *66*, 2702–2704. [CrossRef]
15. Schönegg, D.; Essig, H.; Al-Haj Husain, A.; Weber, F.E.; Valdec, S. Patient-Specific Beta-Tricalcium Phosphate Scaffold for Customized Alveolar Ridge Augmentation: A Case Report. *Int. J. Implant. Dent.* **2024**, *10*, 21. [CrossRef]
16. Montelongo, S.A.; Chiou, G.; Ong, J.L.; Bizios, R.; Guda, T. Development of Bioinks for 3D Printing Microporous, Sintered Calcium Phosphate Scaffolds. *J. Mater. Sci. Mater. Med.* **2021**, *32*, 94. [CrossRef]
17. Ivanovski, S.; Breik, O.; Carluccio, D.; Alayan, J.; Staples, R.; Vaquette, C. 3D Printing for Bone Regeneration: Challenges and Opportunities for Achieving Predictability. *Periodontol. 2000* **2023**, *93*, 358–384. [CrossRef]
18. Chen, T.-H.; Ghayor, C.; Siegenthaler, B.; Schuler, F.; Rüegg, J.; De Wild, M.; Weber, F.E. Lattice Microarchitecture for Bone Tissue Engineering from Calcium Phosphate Compared to Titanium. *Tissue Eng. Part. A* **2018**, *24*, 1554–1561. [CrossRef]
19. Schwentenwein, M.; Homa, J. Additive Manufacturing of Dense Alumina Ceramics. *Int. J. Appl. Ceram. Technol.* **2015**, *12*, 1–7. [CrossRef]
20. Tatum, H. Maxillary and Sinus Implant Reconstructions. *Dent. Clin. N. Am.* **1986**, *30*, 207–229. [CrossRef]
21. Bortolotti, M.; Lutterotti, L.; Lonardelli, I. *ReX*: A Computer Program for Structural Analysis Using Powder Diffraction Data. *J. Appl. Crystallogr.* **2009**, *42*, 538–539. [CrossRef]
22. Schindelin, J.; Arganda-Carreras, I.; Frise, E.; Kaynig, V.; Longair, M.; Pietzsch, T.; Preibisch, S.; Rueden, C.; Saalfeld, S.; Schmid, B.; et al. Fiji: An Open-Source Platform for Biological-Image Analysis. *Nat. Methods* **2012**, *9*, 676–682. [CrossRef] [PubMed]
23. Fazzalari, N.L.; Parkinson, I.H. Fractal Dimension and Architecture of Trabecular Bone. *J. Pathol.* **1996**, *178*, 100–105. [CrossRef]
24. Brun, F.; Massimi, L.; Fratini, M.; Dreossi, D.; Billé, F.; Accardo, A.; Pugliese, R.; Cedola, A. SYRMEP Tomo Project: A Graphical User Interface for Customizing CT Reconstruction Workflows. *Adv. Struct. Chem. Imaging* **2017**, *3*, 4. [CrossRef]
25. Paganin, D.; Mayo, S.C.; Gureyev, T.E.; Miller, P.R.; Wilkins, S.W. Simultaneous Phase and Amplitude Extraction from a Single Defocused Image of a Homogeneous Object. *J. Microsc.* **2002**, *206*, 33–40. [CrossRef]
26. Makovetsky, R.; Piche, N.; Marsh, M. Dragonfly as a Platform for Easy Image-Based Deep Learning Applications. *Microsc. Microanal.* **2018**, *24*, 532–533. [CrossRef]
27. Riberti, N. Artificial Intelligence in Dentistry: Innovative Tools and Advanced Analysis of Bone and Soft Tissues. Ph.D. Thesis, University G.d'Annunzio, Chieti-Pescara, Italy, 2025.
28. Silveira, A.; Greving, I.; Longo, E.; Scheel, M.; Weitkamp, T.; Fleck, C.; Shahar, R.; Zaslansky, P. Deep Learning to Overcome Zernike Phase-Contrast NanoCT Artifacts for Automated Micro-Nano Porosity Segmentation in Bone. *J. Synchrotron Radiat.* **2024**, *31*, 136–149. [CrossRef]
29. van Oers, R.F.M.; Wang, H.; Bacabac, R.G. Osteocyte Shape and Mechanical Loading. *Curr. Osteoporos. Rep.* **2015**, *13*, 61–66. [CrossRef]
30. Wu, V.; van Oers, R.F.M.; Schulten, E.A.J.M.; Helder, M.N.; Bacabac, R.G.; Klein-Nulend, J. Osteocyte Morphology and Orientation in Relation to Strain in the Jaw Bone. *Int. J. Oral. Sci.* **2018**, *10*, 2. [CrossRef]
31. Lindblad, J. Surface Area Estimation of Digitized 3D Objects Using Weighted Local Configurations. *Image Vis. Comput.* **2005**, *23*, 111–122. [CrossRef]
32. Roschger, P.; Paschalis, E.P.; Fratzl, P.; Klaushofer, K. Bone Mineralization Density Distribution in Health and Disease. *Bone* **2008**, *42*, 456–466. [CrossRef] [PubMed]

33. Walker, M.; Kimmel, C. A Two-Color Acid-Free Cartilage and Bone Stain for Zebrafish Larvae. *Biotech. Histochem.* **2007**, *82*, 23–28. [CrossRef] [PubMed]
34. Zadka, Ł.; Chrabaszcz, K.; Buzalewicz, I.; Wiercigroch, E.; Glatzel-Plucińska, N.; Szleszkowski, Ł.; Gomułkiewicz, A.; Piotrowska, A.; Kurnol, K.; Dzięgiel, P.; et al. Molecular profiling of the intestinal mucosa and immune cells of the colon by multi-parametric histological techniques. *Sci. Rep.* **2021**, *11*, 11309. [CrossRef] [PubMed]
35. Belloni, A.; Montanari, E.; Sagrati, A.; Lorenzi, T.; Balloni, A.; Busardò, F.P.; Notarstefano, V.; Fabri, M.; Giorgini, E. Novel Insights from Fourier-Transform InfraRed Imaging on the Morpho-Chemical Profile of Human Corpus Callosum. *Appl. Sci.* **2023**, *13*, 3954. [CrossRef]
36. Togni, L.; Furlani, M.; Belloni, A.; Riberti, N.; Giuliani, A.; Notarstefano, V.; Santoni, C.; Giorgini, E.; Rubini, C.; Santarelli, A.; et al. Biomolecular Alterations Temporally Anticipate Microarchitectural Modifications of Collagen in Oral Tongue Squamous Cell Carcinoma. *iScience* **2024**, *27*, 110303. [CrossRef]
37. Domander, R.; Felder, A.A.; Doube, M. BoneJ2—Refactoring Established Research Software. *Wellcome Open Res.* **2021**, *6*, 37. [CrossRef]
38. Xidaki, D.; Agrafioti, P.; Diomatari, D.; Kaminari, A.; Tsalavoutas-Psarras, E.; Alexiou, P.; Psycharis, V.; Tsilibary, E.; Silvestros, S.; Sagnou, M. Synthesis of Hydroxyapatite, β-Tricalcium Phosphate and Biphasic Calcium Phosphate Particles to Act as Local Delivery Carriers of Curcumin: Loading, Release and In Vitro Studies. *Materials* **2018**, *11*, 595. [CrossRef]
39. Paschalis, E.P.; Mendelsohn, R.; Boskey, A.L. Infrared Assessment of Bone Quality: A Review. *Clin. Orthop. Relat. Res.* **2011**, *469*, 2170–2178. [CrossRef]
40. Castorina, F.; Masi, U.; Giorgini, E.; Mori, L.; Tafuri, M.A.; Notarstefano, V. Evidence for Mild Diagenesis in Archaeological Human Bones from the Fewet Necropolis (SW Libya): New Insights and Implications from ATR–FTIR Spectroscopy. *Appl. Sci.* **2023**, *13*, 687. [CrossRef]
41. Liu, Q.; Liang, Q.; Chu, H.; Chen, P.; Jiang, L.; Liu, W.; Yang, Z.; Rong, M. The Treatment of Sinus Graft Infection after Sinus Floor Elevation: A Series of Four Case Reports. *J. Oral Implantol.* **2024**, *50*, 87–94. [CrossRef]
42. Park, W.-B.; Okany, K.P.; Park, W.; Han, J.-Y.; Lim, H.-C.; Kang, P. Atypical and Late-Developed Sinus Graft Complications Following Maxillary Sinus Augmentation: Successful Management with Guided Bone Regeneration. *Medicina* **2024**, *60*, 1246. [CrossRef] [PubMed]
43. Kochar, S.P.; Reche, A.; Paul, P. The Etiology and Management of Dental Implant Failure: A Review. *Cureus* **2022**, *14*, e30455. [CrossRef] [PubMed]
44. Łuczak, J.W.; Palusińska, M.; Matak, D.; Pietrzak, D.; Nakielski, P.; Lewicki, S.; Grodzik, M.; Szymański, Ł. The Future of Bone Repair: Emerging Technologies and Biomaterials in Bone Regeneration. *Int. J. Mol. Sci.* **2024**, *25*, 12766. [CrossRef] [PubMed]
45. Brunauer, S.; Emmett, P.H.; Teller, E. Adsorption of Gases in Multimolecular Layers. *J. Am. Chem. Soc.* **1938**, *60*, 309–319. [CrossRef]
46. Ji, C.; Qiu, M.; Ruan, H.; Li, C.; Cheng, L.; Wang, J.; Li, C.; Qi, J.; Cui, W.; Deng, L. Transcriptome Analysis Revealed the Symbiosis Niche of 3D Scaffolds to Accelerate Bone Defect Healing. *Adv. Sci.* **2022**, *9*, e2105194. [CrossRef]

Journal of
Functional Biomaterials

MDPI

Article

Investigating the Osteoregenerative Properties of *Juglans regia* L. Extract on Mesenchymal Stem Cells and Osteoblasts Through Evaluation of Bone Markers: A Pilot Study

Alina Hanga-Fărcaș [1], Gabriela Adriana Filip [2], Simona Valeria Clichici [2], Laura Grațiela Vicaș [3,*], Olga Şoritău [4,*], Otilia Andercou [4], Luminița Fritea [5] and Mariana Eugenia Mureșan [5]

1 Doctoral School of Biomedical Sciences, University of Oradea, 1 University Street, 410087 Oradea, Romania; alina.hanga@didactic.uoradea.ro
2 Department of Physiology, Iuliu Hatieganu University of Medicine and Pharmacy, 8 Victor Babes Street, 400347 Cluj-Napoca, Romania; gabriela.filip@umfcluj.ro (G.A.F.); sclichici@umfcluj.ro (S.V.C.)
3 Department of Pharmacy, Faculty of Medicine and Pharmacy, University of Oradea, 10, 1 December Square, 410073 Oradea, Romania
4 Radiobiology and Tumor Biology Laboratory, Oncological Institute "Prof. Dr. I. Chiricuță", 34–36 Republicii Str., 400015 Cluj-Napoca, Romania; otiliabarbos2006@yahoo.com
5 Department of Preclinical Discipline, Faculty of Medicine and Pharmacy, University of Oradea, 10, 1 December Square, 410073 Oradea, Romania; lfritea@uoradea.ro (L.F.); mmuresan@uoradea.ro (M.E.M.)
* Correspondence: lvicas@uoradea.ro (L.G.V.); olgasoritau@yahoo.com (O.S.)

Abstract: Bone tissue regeneration is a complex process that takes place at the level of osteoblasts derived from mesenchymal cells and occurs under the action of multiple signaling pathways and through the expression of osteoregenerative markers. The leaf extract of *Juglans regia* L. (JR) is rich in polyphenols with demonstrated osteoregeneration effects. In the present study, we investigated the extract's effects on three types of cells with various stages of differentiation: adult mesenchymal stem cells (MSCs), osteoblasts at low passage (O6) and osteoblasts at advanced passage (O10). To assess the efficacy of the walnut leaf extract, in vitro treatments were performed in comparison with ellagic acid (EA) and catechin (CAT). The osteoregenerative properties of the leaf extract were evaluated in terms of cell viability, bone mineralization (by staining with alizarin red) and the expression of osteogenesis markers such as osteocalcin (OC), osteopontin (OPN), dentin matrix acidic phosphoprotein 1 (DMP1) and collagen type 1A. Another compound implicated in oxidative stress response, but also a bone homeostasis regulator, nuclear factor erythroid 2-related factor 2 (NRF2), was studied by immunocytochemistry. Together with collagen amount, alkaline phosphatase (ALP) activity and NF-kB levels were measured in cell lysates and supernatants. The obtained results demonstrate that JR treatment induced osteogenic differentiation and bone mineralization, and it showed protective effects against oxidative stress.

Keywords: osteoblast; mesenchymal stem cells; bone regeneration; bone biomarkers

1. Introduction

Fractures, bone infections, primary bone tumors and bone metastasis, and osteoporosis are bone lesions with a great health impact and high socioeconomic expenses. Repairing bone injuries with new bone formation is a process involving the activity of osteoprogenitor cells and osteoclasts, accompanied by a dynamic inflammatory response [1].

Bone is composed of a specialized cell network made up of osteocytes, osteoblasts, osteoclasts, MSCs, hematopoietic cells (HSCs) and immune cells, which are in continuous communication and collaboration. At the center of this network are osteocytes, found inside the bone matrix and constantly communicating with the osteoblasts and osteoclasts on the surface, providing a well-proportioned remodeling process through signaling pathways. MSCs differentiate into osteoblasts under the influence of some molecules such as hormones, cytokines and various growth factors, representing potential for bone regeneration. Immune cells also play an active role; there is an equilibrium between proinflammatory and anti-inflammatory signals, affecting bone metabolism [2].

During the process of proliferation and differentiation of osteoblasts from MSCs, they secrete different proteins—growth factors, transcription factors and enzymes—each of them with distinct periods of expression during osteoblast maturation. Some of the most important expression markers are Runx2, ALP, collagen type I (Col-1), OC, OPN and transforming growth factor-β (TGF-β) [3].

The process of osteoblastogenesis in vitro from MSCs can be divided into three overlapping steps, with the expression of two transcription factors being mandatory: Runt-related transcription factor 2 (Runx2) and osterix (OSX). In the first phase, MSCs undergo proliferation and the expressions of collagen 1 and OPN are amplified. The second phase of matrix maturation occurs when proliferation begins to decrease. In this phase, collagen 1 is still produced and secreted in the extracellular space, and ALP activity is significantly amplified. OSX expression also increases, being essential in the ossification process, and Runx2 expression decreases because it inhibits mineralization. The last phase is mineralization, in which collagenase expression reaches its maximum, being necessary in the process of remodeling the newly formed bone matrix [4]. Runx2 is also necessary for non-collagenous protein expression such as OC and bone sialoprotein (BSP). Their expression increases with pre-osteoblast transformation into fully differentiated osteoblasts. BSP binds to collagen I, resulting in hydroxyapatite crystals' formation and initiating bone mineralization, while OC determines the binding of calcium and hydroxyapatite. Osteoblasts surround themselves with their own matrix, forming an osteoid. Later, osteoblasts differentiate into osteocytes or sometimes they can line cells or undergo apoptosis [5].

Osteopontin (OPN), a non-collagen phosphoprotein mainly secreted by osteoblasts, has a role in bone metabolism and hemostasis. OPN is involved in the proliferation, migration and adhesion of bone-related cells [6]. Produced by osteocytes, dentin matrix protein (DMP1) is an acidic phosphorylated extracellular matrix non-collagenous bone protein involved in the mineralization of the bone matrix by binding to Ca^{2+} molecules [7]. DMP1 is a significant mineralization regulator and thereby a regulator of responses mediated by osteocytes to mechanical load [8]. Another important player in regulating bone homeostasis is nuclear factor erythroid 2-related factor 2 (NRF2). NRF2 agonists have protective effects on osteoblasts, osteocytes and stem cells, while osteoclast differentiation is inhibited by resistance to oxidative stress [9]. The process of bone remodeling is continuous and takes place throughout life, representing an equilibrium between processes of bone resorption carried out by osteoclasts and processes of bone formation performed by osteoblasts. This balance can be disrupted by inflammation and oxidative stress, which facilitate osteoclastogenesis and inhibit osteoblastogenesis. In inflammation, cytokines have an important role, disrupting normal remodeling, separating the two processes of bone resorption and bone formation [10]. The transcription factor nuclear factor kappa B (NF-kB) is a regulator of cellular processes, inflammatory processes and the immune response activated by proinflammatory cytokines: tumor necrosis factor (TNF) and interleukin-17. It also represents a significant contribution in the differentiation processes of stem cells and PDLSCs (human periodontal ligament tissue-derived MSCs) [11–14].

The most common bone disease, osteoporosis, is a complex pathology whose main mechanism is the bone resorption/bone formation disequilibrium. This is why research is focused on finding effective remedies to treat this condition. The menopausal status is the most well-known cause of osteoporosis, where low levels of estrogen interrupt bone remodeling. Estrogen binds to estrogen receptors at the level of osteoblasts and favors the synthesis of osteoprotegerin (OPG), the "decoy receptor" of the receptor activator of the NF-κB ligand (RANKL) [15]. Low estrogen levels will alter the estrogen target genes' expression, which will induce an increase in the secretion of TNF and interleukins IL-1, IL-6 and IL-1β. Also, due to the low level of OPG, RANKL will bind to the receptor activator of NF-κB (RANK) on osteoclasts and will promote their maturation and bone resorption [16]. Osteoporosis in allopathic medicine is commonly treated with hormone therapies like estrogen or testosterone replacement, or with inhibitors of osteoclasts (bisphosphonates). Severe osteoporosis with high risk of fractures is treated with parathyroid hormone analogs. These treatments, although they have achieved certain effects in clinical treatment, besides being quite expensive, can be associated with some unwanted adverse effects, one of them being osteonecrosis of the jaw [17]. For this reason, numerous studies have focused on traditional medicine, aiming toward osteoporosis prevention and treatment without severe complications. Various phytochemical compounds extracted from several plants present huge potential in the bone regeneration process because of their pro-estrogenic activity, antioxidant capacity, anti-inflammatory property and modulation of pathways for bone signaling [18–20]. Many approaches to osteoporosis treatment and prevention with phyto-biocompounds are presented in the literature. The obtained effects are induced by certain chemically active ingredients from plants (secondary metabolites) like resveratrol, naringin and ginsenoside. Traditional Chinese medicine describes such herbs and their extracts with proven effects in in vivo and in vitro studies on osteoporosis [21].

Juglans regia L. (JR), or walnut, is known to possess anti-inflammatory and antioxidant activities. All the plant's parts are used in traditional medicine, not only for its nutritional properties, but also as a curative choice in a great diversity of diseases: fungal, viral and bacterial infections, oncological diseases, type 2 diabetes and pathologies associated with arterial hypertension [22]. Walnut leaves are employed in natural medicine especially for treating some pathological conditions such as skin ulcers, venous insufficiency, hyperhidrosis, skin irritation and hemorrhoidal disease [23]. Until now, walnut leaves' osteoregenerative effects have not been studied extensively. According to Pang X. et al., JR leaves exert a positive effect on bone metabolism, particularly on osteogenic gene expression in human bone marrow MSCs [24]. In our previous work, we have proved the relevant antioxidant effect and bone protective activity of JR leaf extract, as well as the accumulation of Ca^{2+} and phosphate (PO_4^{3-}) in the bone structure in a rat model [25].

The objective of the present study was to assess the potential of JR leaf extract in bone regeneration by assessing the effect on MSCs from bone marrow and primary osteoblasts in two stages of differentiation (pre-osteoblast and osteoblast) exposed to osteogenic medium (Scheme 1). In order to delineate the effects of EA and CAT, two biocompounds contained in the JR leaf extract in different proportions, JR's effect was compared with EA and CAT treatments.

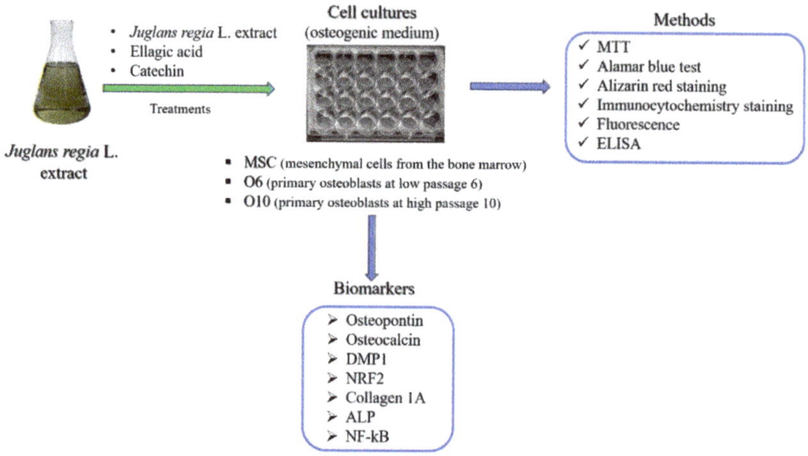

Scheme 1. The experimental design of this study.

2. Materials and Methods

2.1. Cell Cultures

Three primary cell lines were used: adult MSCs from bone marrow, primary osteoblasts at low passage (passage 6-called O6) and primary osteoblasts at high passage (O10).

Human osteoblasts were purchased from a human patient with the approval of the Ethics Council of the County Emergency Clinical Hospital of Oradea, Romania (no. 7700/02.03.2023). Bone fragments were harvested in aseptic conditions from a comminuted femoral neck fracture from a patient aged 77 years old. The bone fragments were immersed in transport medium (DMEM/F12 with 10% bovine serum) and were processed within 24 h of harvesting by mechanical fragmentation, obtaining bone pieces with sizes of 2 × 2 mm. An enzymatic digestion step with collagenase IV 1 mg mL^{-1} was applied for 2 h, followed by inactivation of enzyme activity with medium containing 10% FBS. Several washes of fragments were performed with PBS (phosphate-buffered saline) and complete medium. The cells detached during mechanical and enzymatic processing were collected, and supernatants were filtered with 70 nm Filcons meshes (Dako). The obtained cell suspension and bone fragments were centrifuged at 1200 rpm for 10 min, and the pellets were resuspended in complete standard medium consisting of DMEM high glucose/F12 without phenol red 15% fetal calf serum (FCS), 1% non-essential amino acids (NEAs), 1% antibiotic (penicillin–streptomycin) and 2 mM L^{-1} glutamine. All reagents were purchased from Sigma-Aldrich (St Louis, MO, USA). After 14 days, when fibroblastoid cells migrated around the bone fragments, the first passage was performed. The cells showed a proliferation rate that determined the passage to be performed at 3–5 days. In our study, cells at two different passages were used, passage 6 and passage 10, which under the influence of the osteogenic differentiation medium showed distinct morphologies. Osteogenic medium (OS medium) was composed of DMEM high glucose/F12 without phenol red 15% FCS, 1% NEA, 1% antibiotic (penicillin–streptomycin), 2 mM L^{-1} glutamine, ascorbic acid phosphate (50 μg mL^{-1}), dexamethasone (20 μM) and β-glycerophosphate (10 mM).

Adults MSCs derived from human bone marrow from the iliac crest, isolated and characterized in a previous study by our team [26], were obtained from an orthopedic patient during a total hip replacement, with the patient giving informed consent and the ethical legal laws being adhered to. The cells were thawed from a batch stored in liquid nitrogen and expanded using standard stem cell medium consisting of DMEM with 4.5 g glucose/F12-HAM (1/1) with 15% FBS (fetal bovine serum), 2 mM L^{-1} glutamine, 1% penicillin–streptomycin solution, 1% NEA solution, 55 μM β-mercaptoethanol and 1 mM sodium pyruvate. All reagents used in cell cultures were provided by Sigma-Adrich. Isolated MSCs were positive for surface pluripotent stem cell markers Nanog, Oct ¾, SOX2, SSEA-4, CD29 and CD105 [26]. The experimental protocol was approved by the Ethical Committee of "Iuliu Hatieganu" University (16 September November 2021).

Treatment experiments were performed with the walnut extract (JR), EA and CAT aqueous solutions, previously filtered with 0.22 nm filters, with the purpose of sterilizing the solutions. CAT was solubilized in double-distilled water to obtain a stock solution of 3.22 mg mL^{-1}. The CAT working concentration was 0.32 μg mL^{-1}, corresponding to the quantification of the contained CAT concentration in walnut extracts. The EA solution had a concentration of 1%. The 10% solution of walnut extract was obtained as presented in our previous study [25].

2.2. Establishing the Working Doses of Walnut Extract (JR), EA and CAT

The cytotoxicity levels of JR extract, EA and CAT were investigated by using the MTT assay. MTT is a tetrazolium salt (3-[4,5-dimethylthiazol-2-yl]-2,5-diphenyl tetrazolium bromide) used to highlight viable cells. MTT solution dissolved in saline buffers without phenol red is yellow. Mitochondrial dehydrogenases of viable cells cleave the tetrazolium ring, producing purple formazan crystals that are insoluble in aqueous solutions. The crystals are dissolved in dimethyl sulfoxide (DMSO) or in acidified isopropanol, resulting in a purple solution whose optical density can be measured spectrophotometrically. An increase or decrease in cell numbers also produces a concomitant change in the amount of formazan formed, indicating the degree of cytotoxicity caused by the test material. Adult MSCs from the bone marrow and osteoblastic cells in two stages of differentiation (pre-osteoblast—O6 and osteoblast—O10) were seeded on flat-bottom 96-well plates from Nunclon (Nunc) at a density of 10^4 cells in 200 μL standard medium/well. After 24 h, when the cells reached a subconfluence of 60–70%, serial dilutions of JR extract, EA and CAT were used. The determinations were performed in triplicate in two independent experiments. The cells' viability response to the three compounds was investigated after a longer period of time because in a previous experiment it was observed that after 24 h the effects of the treatments on cell viability and proliferation were minimal. After treatments, the plates were incubated for 72 h. The MTT test consisted of removing the medium from the surface of the cells and adding 100 μL of MTT solution/well (concentration of 1 mg mL^{-1}). The plates were then incubated for 1 h in the dark at 37 °C and 5% CO_2 to allow the MTT to be metabolized by mitochondrial reductases into formazan crystals. The MTT solution was removed and 150 μL DMSO/well was added, homogenizing the samples by using a shaker at 37 °C for 5 min. The resulting purple formazan was dissolved, and optical densities were determined at 570 nm using a BioTek Synergy 2 microplate reader (Winooski, VT, USA).

2.3. The alamarBlue Viability Test on Long-Term Cultures Treated with JR Extract, EA and CAT

The alamarBlue assay was employed for the viability determination of the three cell lines cultivated over a longer period (the analysis was performed at 3 and 6 days). The two treatments were performed at 24 h and 72 h, inducing osteogenic differentiation with OS medium. Cells were seeded on 96-well plates with a flat bottom at 10^4 cells in 200 μL OS differentiation medium/well. After 24 h, when the cells reached a subconfluence of 60–70%, the treatments with walnut extract, EA and CAT were administered at the same dilution x26.6 with the following concentrations: JR extract (0.5 mg mL^{-1}), EA (0.05 mg mL^{-1}) and CAT (0.15 μg mL^{-1}). At 72 h and 6 days, the medium was extracted from the wells and

100 μL of medium with 10% alamarBlue was added. After one hour of incubation at 37 °C and 5% CO_2 in the dark, the plates were analyzed with the Biotek Synergy 2 fluorescence plate reader using the 560/590 nm (excitation/emission) filter. Statistical analysis was performed with the two-way ANOVA program, comparing the values at 3 days with those at 6 days for each treatment and the control at each time interval with the values obtained in the treated cells.

2.4. Evaluation of Mineralization by Alizarin Red Staining

Alizarin red staining was performed for all three cell lines to observe the differences in behavior between the different stages of differentiation of osteoblastic cells: MSC, O6 and O10. They were seeded in plates with 24 wells with a cell density of 5×10^4 cells/well in 1 mL/well of OS differentiation medium. The cells were cultured for 14 days, and 4 treatments were performed with the same dose of JR, EA and CAT (dilution × 26.6). At the end of cultivation, the cells were fixed with 4% paraformaldehyde for 20 min and washed three times with PBS, then with double-distilled water. Exposure of 20 min to 2% alizarin red solution (pH 4.2) was applied with gentle agitation on an orbital shaker. The staining solution was discarded and extensive washes with double-distilled water were performed, with the last wash using PBS. Microscopic images were taken to visualize the calcium deposits under an inverted phase microscope, Zeiss Axiovert D1, using an AxioCAM MRC color camera. Quantification of mineralization was performed with a 10% cetylpyridinium chloride (CPC) solution, a detergent that solubilizes calcium deposits (violet coloration that can be read on a microplate reader at 562 nm). PBS was extracted from the wells, and 10% CPC solution was added, with 15 min incubation on an orbital shaker. Aliquots were extracted into a 96-well plate and readings were performed on a Biotek Synergy2 reader at a wavelength of 562 nm.

2.5. Evaluation of Osteogenic Differentiation by Immunocytochemical Staining

MSC, O6 and O10 cells were seeded on 16-well Nunc chamber slides and cultured in osteoinductive medium for 10 days with two treatments each of JR extract (0.5 mg mL^{-1}), EA (0.05 mg mL^{-1}) and CAT (0.15 μg mL^{-1}). Cells were fixed with 4% paraformaldehyde (20 min), then permeabilized with 0.02% Triton X100 solution for 15 min. Non-specific fixation was blocked with 10% bovine albumin for 15 min. Exposure to primary monoclonal antibodies was performed overnight at 4 °C. The monoclonal antibodies used in this experiment were anti-human mouse IgG1 (Invitrogen-Thermo Fisher Scientific, Waltham, MA, USA), anti-human mouse IgG2a osteocalcin (Santa Cruz Biotechnologies, Heidelberg, Germany), anti-human mouse monoclonal IgG1 collagen 1a (Santa Cruz Biotechnologies), anti-human DMP1 (Invitrogen) and Anti-NRF2 polyclonal rabbit antibody (ThermoFischer Scientific) (1:100 dilution).

Secondary antibodies labeled with FITC (Santa Cruz Biotechnogies, Heidelberg, Germany) were applied to the samples using the same dilution as the corresponding primary antibody for 45 min at room temperature in the dark. The dilution ratio of primary and secondary antibodies was 1:50, except for the Anti-NRF2 antibody where the dilution was 1:100. Nuclei staining was performed with a mounting medium containing DAPI. Fluorescence images were taken with a Nikon Elipse E600 fluorescence microscope using a color digital camera under the same exposure conditions at different wavelengths (346 nm, 488 nm).

The mean green fluorescence intensity (represented as arbitrary units) of the captured OPN and OC images were normalized to nuclei number using ImageJ (v1.52P, NIH).

2.6. Determination of Collagen Levels from Supernatants and Cell Lysates

For this study, MSC, O6 and O10 cells were seeded in 6-well plates at a cell density of 2×10^5 cells/well in 2 mL of osteogenic medium. Three treatments with JR extract (0.5 mg mL^{-1}), EA (0.05 mg mL^{-1}) and CAT (0.15 µg mL^{-1}) were performed. Supernatants were collected on days 0, 3, 7 and 10. Finally, on day 10, cell lysis was performed using CelLytic™ MT Cell Lysis Reagent (Sigma Aldrich Chemicals GmbH, St Louis, MO, USA). All collected samples were kept in a freezer at −80 °C until analysis. The Collagen Assay Kit (Sigma Aldrich), a simple and sensitive method in fluorescence, was used for collagen dosage. In the first step of this procedure, the collagen in the samples was enzymatically digested into peptides under incubation for 60 min at 37 °C. The obtained N-terminal glycine-containing peptides subsequently reacted with the dye reagent to form a fluorescent complex for 10 min at 37 °C. The fluorescence intensity of the samples, directly proportional to the collagen concentration, was measured at 375/465 nm using a microplate reader under fluorescence at $\lambda ex = 375/\lambda em = 465$ nm, Biotek Synergy2. Extrapolation of the obtained results was carried out from the standard curve using GraphPad Prism 5 software, obtaining collagen concentrations in µg mL^{-1}. Comparison between control values at time 0 (T0) and the obtained values at different points of time (3, 7 and 10 days) was performed with one-way analysis of variance followed by the "Bonferroni post test".

2.7. Assessment of ALP Activity

The effects of JR, EA and CAT were evaluated in terms of ALP activity in culture medium supernatants and from cell lysates of MSC, O6 and O10 cells. The cultivation conditions and treatments were the same as those described for collagen dosing, with cultivation of cells in osteogenic medium and three treatments with JR extract (0.5 mg mL^{-1}), EA (0.05 mg mL^{-1}) and CAT (0.15 µg mL^{-1}). Supernatants were collected on days 0, 7 and 10 and cell lysates on day 10.

The activity of ALP was assessed by using the Alkaline Phosphatase Detection Kit, Fluorescence (Sigma Aldrich), whose principle is based on the hydrolyzation of ALP from samples of p-nitrophenyl phosphate into a yellow-colored product with absorbance at 405 nm. The rate of reaction is directly proportional to the enzyme activity. The samples were thawed at room temperature. In a plate with 96 wells, 20 µL of each sample was added in duplicate, using as a negative control medium DMEM/F-12 without phenol red and as a positive control different concentrations of the control enzyme. The 96-well plate was incubated at 65 °C for 10 min to minimize background enzyme activity, followed by sudden cooling on ice for 2 min. The substrate solution at a concentration of 10 mM (disodium salt of 4-methylumbelliferyl phosphate) was added to the wells containing 20 µL of dilution buffer and 160 µL of fluorescent assay buffer. After a short agitation, fluorescence intensity readings were taken on the Biotek Synergy 2 microplate fluorescence reader. The fluorometer was set to 360 nm excitation and 440 nm emission. The results were processed statistically using the analysis software GraphPad Prism 5 and one-way analysis of variance followed by the "Bonferroni post test".

2.8. Evaluation of NFkB p105 Levels from Supernatants and Cell Lysates

The human nuclear factor NF-kappa-B p105 subunit levels were measured by using a sandwich-ELISA method (Elabscience® Human NF-κBp105 ELISA Kit-Houston, Texas, USA) with medium supernatants and cell lysates from MSC, O6 and O10 cells cultivated for 10 days in osteogenic medium with exposure to JR extract (0.5 mg mL^{-1}), EA (0.05 mg mL^{-1}) and CAT (0.15 µg mL^{-1}). Supernatants were collected on days 0, 7 and 10 and cell lysates on day 10. The samples, standards and negative controls were added to the pre-coated plate with a specific antibody for human NF-κBp105 and incubated for 90 min at 37 °C. A biotinylated detection antibody was added to each well with an incubation of 1 h at 37 °C. The optical density was measured spectrophotometrically at a wavelength of 450 nm using the Biotek Synergy 2 microplate reader (Winooski, VT, USA). The concentrations in μg mL^{-1} of human NF-κBp105 protein were extrapolated from the standard curve.

2.9. Statistical Analysis

For statistical analysis, we employed the software GraphPad Prism version 5. The data in the graphs are expressed as mean values with their standard deviations. For multiple comparisons among groups, one-way ANOVA or two-way ANOVA was used, followed by the "Bonferroni post test", with statistical significance set at $p \leq 0.05$. Using the one-way ANOVA test for normally distributed variables followed by post tests to assess multiple comparisons represents a conservative method that controls the overall error rate by dividing the level of significance by the number of comparisons being made. The Bonferroni post test was used when a large number of pairwise comparisons were analyzed.

3. Results

3.1. Establishment of Osteoblast Culture

Applying the mechanical and enzymatic processing protocol, the appearance of the first migrated cells to the periphery of the bone fragments was observed in the primary culture after 4–7 days. The first passage was performed after 14 days, when confluence was around 60–70%. After two to three passages, the proliferation rate was increased and a more homogenous fibroblastoid-like shape resembled pre-osteoblasts at passage 6. By applying an osteogenic cultivation medium containing ascorbic acid, β-glycerophosphate and dexamethasone, the cells' morphology changed, namely smaller dimensions, a more retracted cell body and more prominent nuclei, as seen at passage 10 in Figure 1. The literature describes protocols that also use the method without enzymatic digestion but emphasize certain cultivation conditions. To obtain a pure osteoblast culture, it seems that collagenase treatment is a necessary step to avoid contamination with bone marrow-derived cells. Supplementation of the cultivation medium with ascorbic acid is beneficial for both osteoblast isolation and for collagen's normal synthesis and secretion by osteoblast cells [27]. Other authors have reported that spontaneous isolation is also possible by defining osteoblasts with specific markers such as Bone Morphogenetic Protein-2,4 collagen type I and ALP activity. They found that the time needed in primary cultures to reach confluence depended on the age of the donors and on the method of cell harvesting, an advantage being aged under 65 years old and using the collagenase isolation method [28].

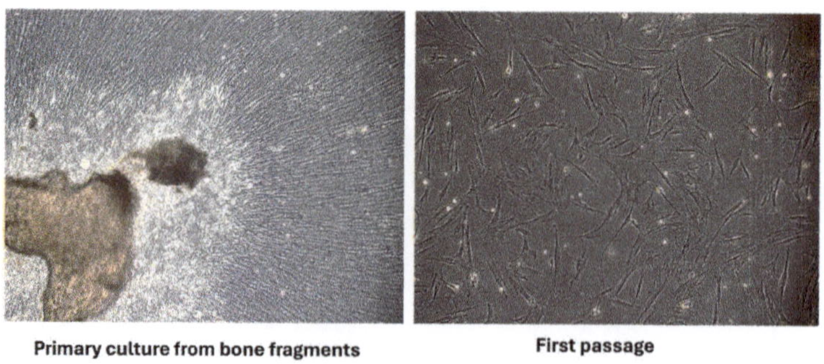

Figure 1. *Cont.*

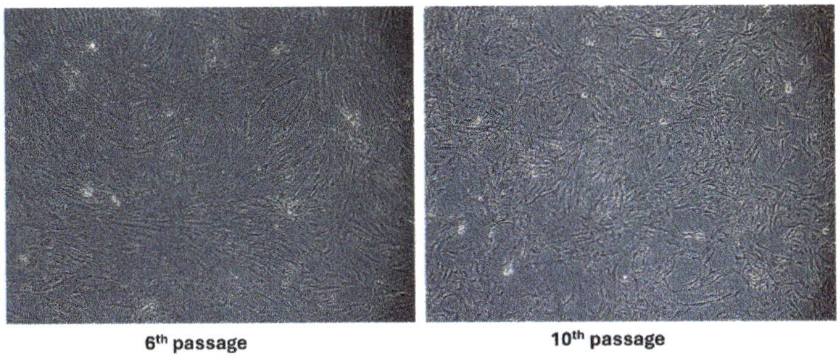

Figure 1. Phase-contrast optical microscopy appearance of osteoblast cells isolated from bone fragments in primary culture, at passage 1, 6 and 10 (objective magnification ×10).

3.2. Determining the Working Doses of JR Extract, EA and CAT by Using the MTT Assay

The effects of JR extract, EA and CAT on MSC, O6 and O10 cultured for 72 h in the presence of the treatments were assessed by using the cell viability test, MTT. The obtained results for MSCs are illustrated in Figure 2. High cytotoxicity was observed at doses between 1.5 mg mL^{-1} and 0.75 mg mL^{-1} of JR extract, with a percentage reduction in viable cells below 50% compared to the control untreated cells. The effects of EA were at the level of the untreated control values. CAT induced a cytotoxic effect of about 30% inhibition at higher doses from 1.5 mg mL^{-1} to 0.75 mg mL^{-1} (Figure 2).

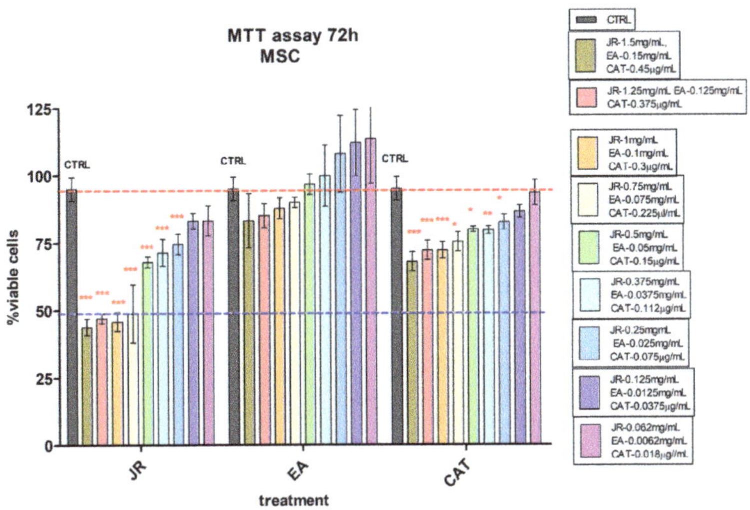

Figure 2. MTT viability test of adult mesenchymal cells (MSCs) from bone marrow, treated for 72 h with walnut leaf extract (JR), ellagic acid (EA) and catechin (CAT). Error bars represent the standard deviations of three separate measurements ($n = 3$). Two-way ANOVA followed by the Bonferroni post test was used for comparison of different treated groups with untreated control cells; * $p < 0.05$, ** $p < 0.01$ and *** $p < 0.001$.

Young osteoblastic cells (O6) behaved similarly to adult stem cells in terms of their response to JR, showing lower inhibition at high doses compared to MSCs. A 50% decrease in viability compared to control cells was observed at higher doses of 1.5–1.0 mg mL^{-1} of JR. Instead, they presented increased sensitivity to high concentrations of CAT (Figure 3).

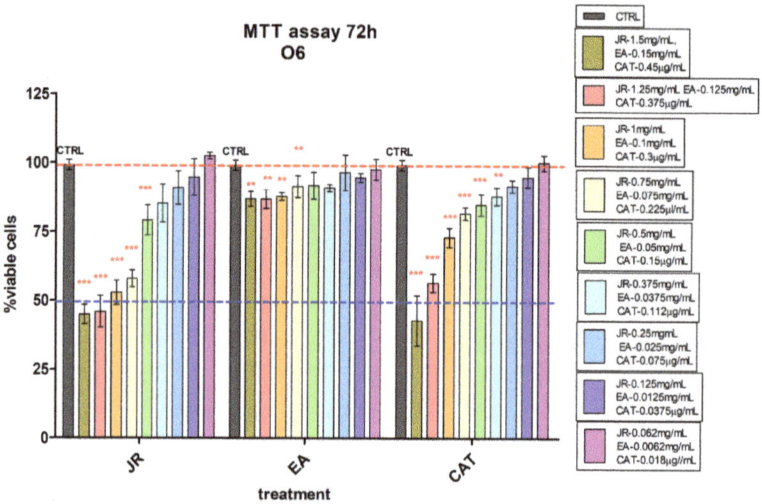

Figure 3. MTT viability test of osteoblastic cells—O6 treated with walnut leaf extract (JR), ellagic acid (EA) and catechin (CAT). Error bars represent the standard deviations of three separate measurements ($n = 3$). Two-way ANOVA followed by the Bonferroni post test was used for comparison of different treated groups with untreated control cells; ** $p < 0.01$ and *** $p < 0.001$.

Similar behavior was also observed in O10, but with higher sensitivity even at lower doses of JR up to 0.25 mg mL^{-1} and with a greatly increased cytotoxic response to high doses of CAT (Figure 4).

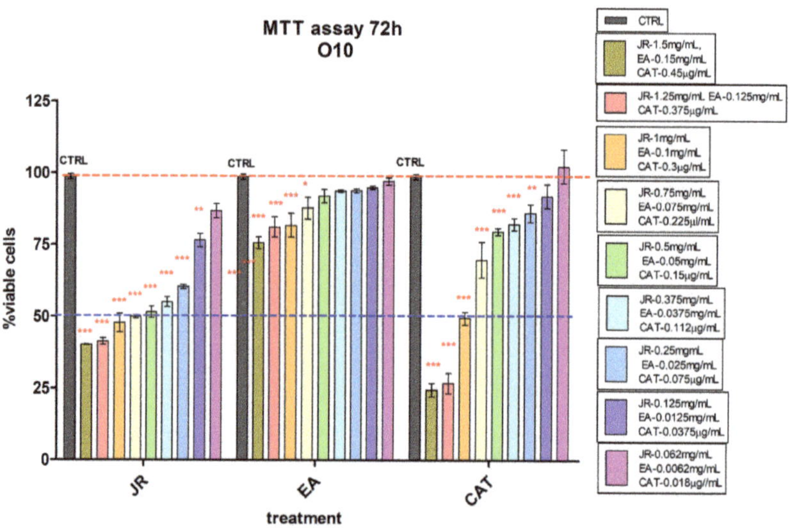

Figure 4. MTT viability test of differentiated osteoblastic cells (O10) treated with walnut leaf extract (JR), ellagic acid (EA) and catechin (CAT). Error bars represent the standard deviations of three separate measurements ($n = 3$). Two-way ANOVA followed by the Bonferroni post test was used for comparison of different treated groups with untreated control cells; * $p < 0.05$, ** $p < 0.01$ and *** $p < 0.001$.

3.3. Long-Term Effects of JR Extract, EA and CAT on Pre-Differentiated Cells' Viability and Proliferation

The results obtained in the MTT viability test enabled us to select a common dilution (×26.6) for the three treatments that would not be cytotoxic and that at the same time would exert effects on the osteogenesis process. The used concentrations were as follows: JR—0.5 mg mL^{-1}, EA—0.05 mg mL^{-1} and CAT—0.15 µg mL^{-1}. In the experiment using the alamarBlue test, this dose on cells cultured for a longer period of up to 6 days was followed with the administration of two treatments at 24 h and 3 days. The rationale for using a single dose for the three treatments derived from the viability assays was the observation of differences in the induction of bone differentiation of the three cell lines at different stages of maturity, in a manner similar to what occurs in a whole organism. This can be considered a limitation of this study, determined by the intention of simplification for the interpretation of the results.

Statistical analysis with comparison between the values at 3 and 6 days for each treatment (Figure 5) showed that control cells cultivated in osteogenic medium had a higher growth rate in this period, especially osteoblastic cells. All treatments induced a decrease in cell number after 3 days for cells with an early pre-differentiation phenotype—MSC and O6 cells. Treatment with JR at a dilution of ×26.2 (0.5 mg mL^{-1}) induced a decrease in MSC proliferation on day 6, unlike treatment with EA and CAT which supported the growth of these cells. Osteoblast cells were not influenced by JR treatment, but EA and CAT administration induced an increase in cell number (Figure 5).

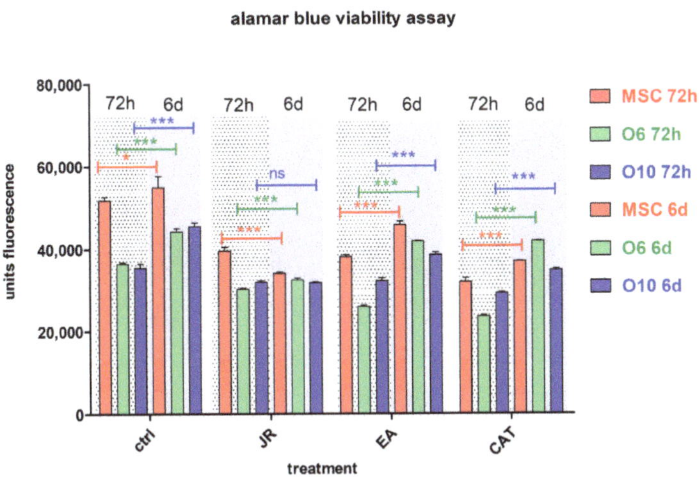

Figure 5. The fluorescence values emitted by cells in the alamarBlue test under the influence of two treatments with JR extract (0.5 mg mL^{-1}), EA (0.05 mg mL^{-1}) and CAT (0.15 µg mL^{-1}). The determinations were made at 3 and 6 days of cultivation after the two treatments. For multiple comparisons for each cell line of controls and treatments responses, two-way ANOVA analysis followed by the Bonferroni post test was performed, estimating the differences between values obtained at 72 h with those at 6 days. Error bars represent the standard deviations of measurements (n = 4). ^{ns} $p > 0.5$, * $p < 0.05$ and *** $p < 0.001$.

In another graphic representation of alamarBlue data, the effect at a certain time of the three treatments is revealed. In these graphs, each type of cell is compared with the untreated control and the obtained values after treatments for JR, EA and CAT at the same time (3 days or 6 days). (Figure 6).

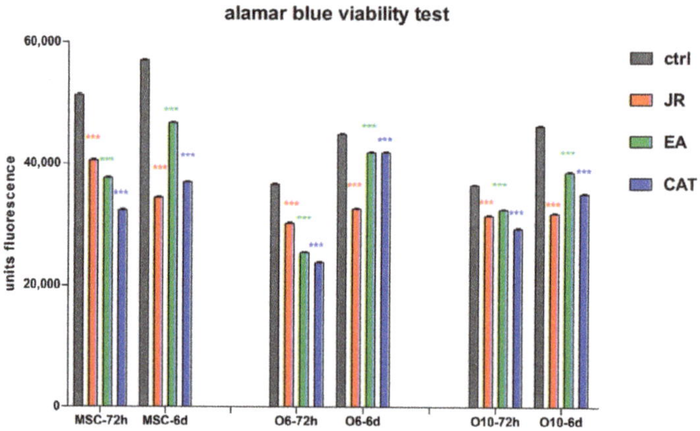

Figure 6. The fluorescence values emitted by cells in the alamarBlue test under the influence of two treatments with JR extract (0.5 mg mL^{-1}), EA (0.05 mg mL^{-1}) and CAT (0.15 μg mL^{-1}). The determinations were made at 3 and 6 days of cultivation after the two treatments. For comparisons of each cell line between controls (blue columns) and treatments responses at the same time of analysis (3 days or 6 days), two-way ANOVA analysis followed by the Bonferroni post test was performed. Error bars represent the standard deviations of measurements ($n = 4$). *** $p < 0.001$.

An obvious trend in all cell lines was observed in the first time interval of 3 days, where a similar decrease in cell proliferation was noticed for all three treatments, the most pronounced for CAT. After 6 days, the cells treated with EA showed a return to values like those of the untreated cells, especially for the O6 line. Mature osteoblasts O10 showed the lowest degree of variation across all treatments.

3.4. Alizarin Red Staining

The differences in behavior between the different stages of osteoblastic cell differentiation were also assessed by alizarin red staining. More visible staining of the differentiated control cells O6 and O10 was noticed compared to the adult MSCs. JR extract and EA induced slightly more distinct mineralization for pre-differentiated cells, while CAT had a cytotoxic effect at this dose.

By using phase-contrast microscopy, it was observed that the mineralization process after 14 days was almost similar between the control cells and those treated with JR, but slightly more pronounced for those treated with EA. CAT greatly decreased the cell viability rate at the dose used (Figure 7).

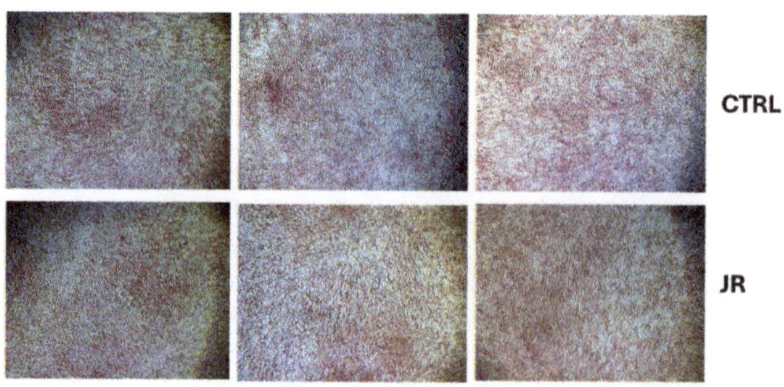

Figure 7. *Cont.*

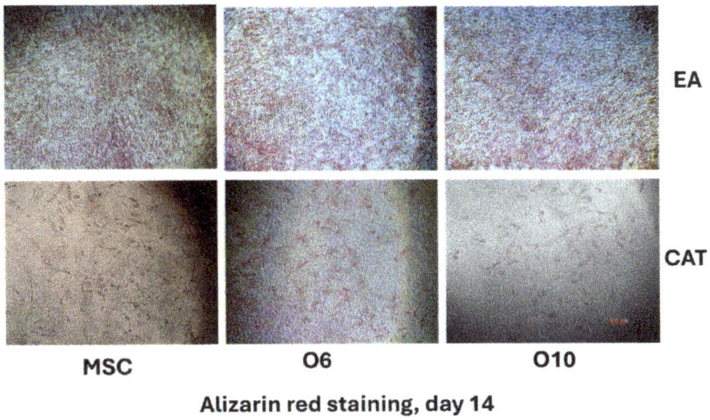

Figure 7. The microscopic appearance of alizarin red staining of MSC and osteoblastic cells after 14 days of cultivation with OS medium and 4 treatments with JR extract (0.5 mg mL^{-1}), EA (0.05 mg mL^{-1}) and CAT (0.15 μg mL^{-1}). (phase-contrast microscopy, ×10 objective). Scale bar = 100 μm.

By extracting the alizarin red stain with 10% CPC solution, the degree of mineralization induced by each treatment was quantified according to the degradation of bone differentiation of each cell line. It was observed that JR extract and EA induced the most significant increase in calcium deposition in the most differentiated cells O10, followed by O6 (Figure 8).

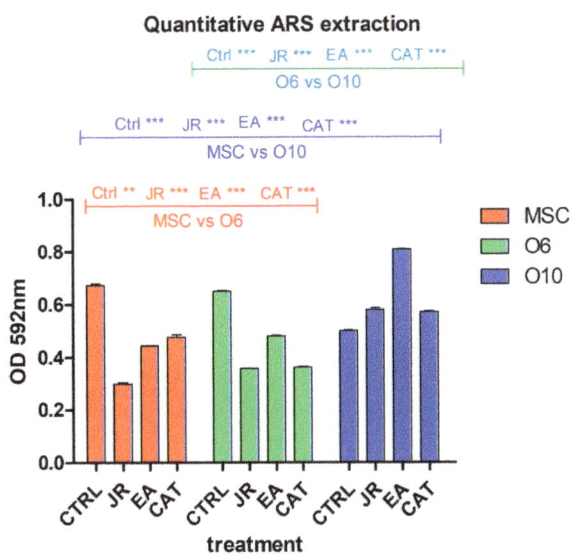

MSC vs O6		
Column Factor	P value	Summary
CTRL	P<0.01	**
JR	P<0.001	***
EA	P<0.001	***
CAT	P<0.001	***
MSC vs O10		
Column Factor	P value	Summary
CTRL	P<0.001	***
JR	P<0.001	***
EA	P<0.001	***
CAT	P<0.001	***
O6 vs O10		
Column Factor	P value	Summary
CTRL	P<0.001	***
JR	P<0.001	***
EA	P<0.001	***
CAT	P<0.001	***

Figure 8. A graphical representation of the results of the quantification of alizarin red staining via extraction with 10% cetylpyridinium chloride. The right panel illustrates the statistical analysis with two-way ANOVA followed by the Bonferroni post test. Experiments: $n = 3$. Values are expressed as the mean ± SD. Error bars represent the standard deviations of measurements. ** $p < 0.01$ and *** $p < 0.001$.

3.5. Evaluation of Osteogenic Differentiation by Immunocytochemistry

Osteogenic differentiation was evaluated by following the expression of bone markers characteristic of each stage of differentiation. For early differentiation towards osteoblasts, we evaluated OPN and collagen. For the phase of construction and maturation of the bone matrix, we evaluated collagen 1A. For the final phase of mineralization and final differentiation towards osteocytes, we evaluated OC, osteonectine and DMP1.

This study focused on the expression of osteogenesis markers, osteopontin, collagen 1A, osteocalcin and DMP1, depending on the differentiation stage of the three cell lines. In addition, the expression of NRF2 was also monitored, as the NRF2-Keap1 signaling pathway plays a role in cell defense and survival. The transcription factor NRF2 protects cells from toxic substances including xenobiotics and oxidative stress; therefore, NRF2 activators are currently being tested as chemoprotective compounds. The main role of NRF2 is the activation of the cellular antioxidant response by inducing the transcription of a wide range of genes that are able to combat the harmful effects of some extrinsic and intrinsic factors, especially those associated with oxidative stress. Additional tests on the antioxidative response were performed by determining the antioxidant profile of O6 cell lysates (malondialdehyde (MDA) and catalase activity). The obtained data are included in the Supplementary Materials.

The images in Figure 9 reveal several observations. The control cells expressed OPN with a moderate intensity, but the treatment with JR induced an increase in the fluorescence intensity (the highest for the younger cells—MSC and O6). The effect of EA was the opposite, with the increase in OPN expression in cells occurring in a more mature stage of differentiation. Cells exposed to CAT showed levels very similar to control cells.

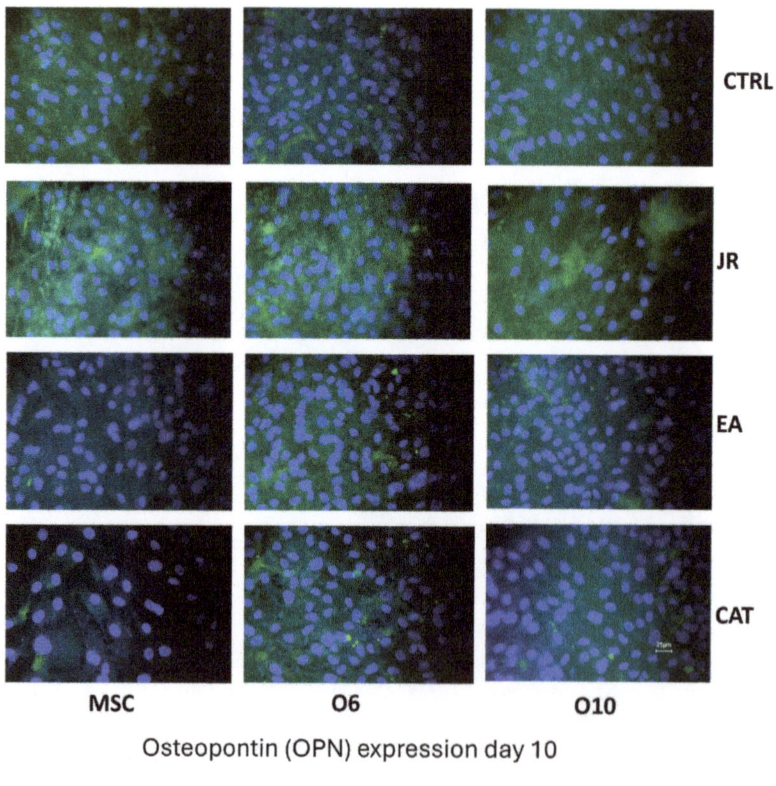

Figure 9. *Cont.*

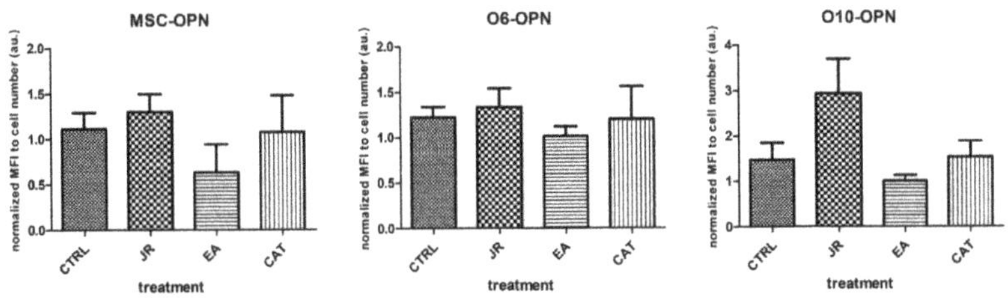

Figure 9. Fluorescence images for osteopontin staining (FITC) of MSC, O6 and O10 cells cultured for 7 days with osteogenic medium and the application of two treatments with JR extract (0.5 mg mL^{-1}), EA (0.05 mg mL^{-1}) and CAT (0.15 µg mL^{-1}). Magnification, ×40 objective. Scale bar = 25 µm. In the bottom panel, graphical representations of the average fluorescence intensity normalized to the number of DAPI-stained nuclei are shown. Data are presented as the mean ± SD (n = 3).

The effect of JR on inducing differentiation towards mature osteoblastic cells was also observed in the staining for the expression of OC. OC is a non-collagen protein that represents approximately 15% of the bone protein matrix. During the process of bone differentiation, OC secretion is very decreased and does not reach maximum levels until the mineralization's late stages. As shown in Figure 10, the untreated control cells expressed different levels of OC, the most intense being for the most advanced differentiated cells O10. JR extract substantially increased OC expression in MSCs, but also in O6 and O10 osteoblasts. EA augmented the production of OC, especially by younger osteoblasts O6, but also by MSCs, and inhibited OC expression in O10 cells. CAT amplified the expression of OC by MSCs and inhibited its expression by mature osteoblastic cells O6 and O10.

A similar effect was observed when staining for DMP1 (FITC). DMP1 is a critical protein for the proper mineralization process of bone and dentin; it is present in various cells of bone and dental tissues. In undifferentiated osteoblasts, it is primarily a nuclear protein regulating the expression of osteoblast-specific genes. During osteoblast maturation, the protein becomes phosphorylated and is exported to the extracellular matrix, where it initiates the formation of the mineralized matrix. In Figure 11, the intra-nuclear expression of DMP1 can be seen by young MSCs. This expression was much more intense and localized in the extracellular space for O6 and O10 cells, where the mineralization process was much more advanced. The effect of JR extract increases the synthesis of DMP1 by MSCs, with protein accumulation in the extracellular space. Mature O10 osteoblastic cells presented a slight decrease in DMP1 expression compared to the untreated control. EA acted only on young MSCs and induced a decrease in expression by mature cells. CAT also inhibited the expression of DMP1 in mature osteoblast lines.

To evaluate the protection-inducing effects of JR extract, EA and CAT against exogenous factors, especially against oxidative stress, immunocytochemical staining was employed to evaluate the expression of the NRF2 protein/transcription factor. It seems that the stage of osteogenic maturation also determined different expressions of NRF2, with the most intense expression found for mature O10 cells and the lowest level for O6 cells. Both JR extract and EA determined a significant increase in this protein in all three cell lines, meanwhile CAT inhibited the expression of NRF2 by MSCs (Figure 12).

The expression of collagen 1A was investigated in parallel with collagen dosing from the supernatants after 10 days of cultivation of the three cell lines in osteogenic medium and after the application of three treatments with JR, EA and CAT performed at each medium change. Collagen was expressed intracellularly and partially extracellularly by MSCs, and this expression increased in more mature osteoblasts (O6 and O10). In immunocytochemical staining, this expression was detected in untreated cells, more intense in the case of O10 cells. JR extract increased collagen expression in the supernatants after 10 days of cultivation which were collected from MSC cultures (average 63 μg mL^{-1}) and to a lesser extent from O6 (average 43.6 μg mL^{-1}) and O10 cells (average 40.9 μg mL^{-1}). The effect of EA was demonstrated especially in mature osteoblasts (O10), with increased collagen expression both intracellularly and in the supernatants (average 52 μg mL^{-1}). CAT had an effect of increasing intracellular expression only in O6 osteoblasts, but in the supernatants the level decreased for all three cell lines at a level of around 25 μg mL^{-1} (Figure 13).

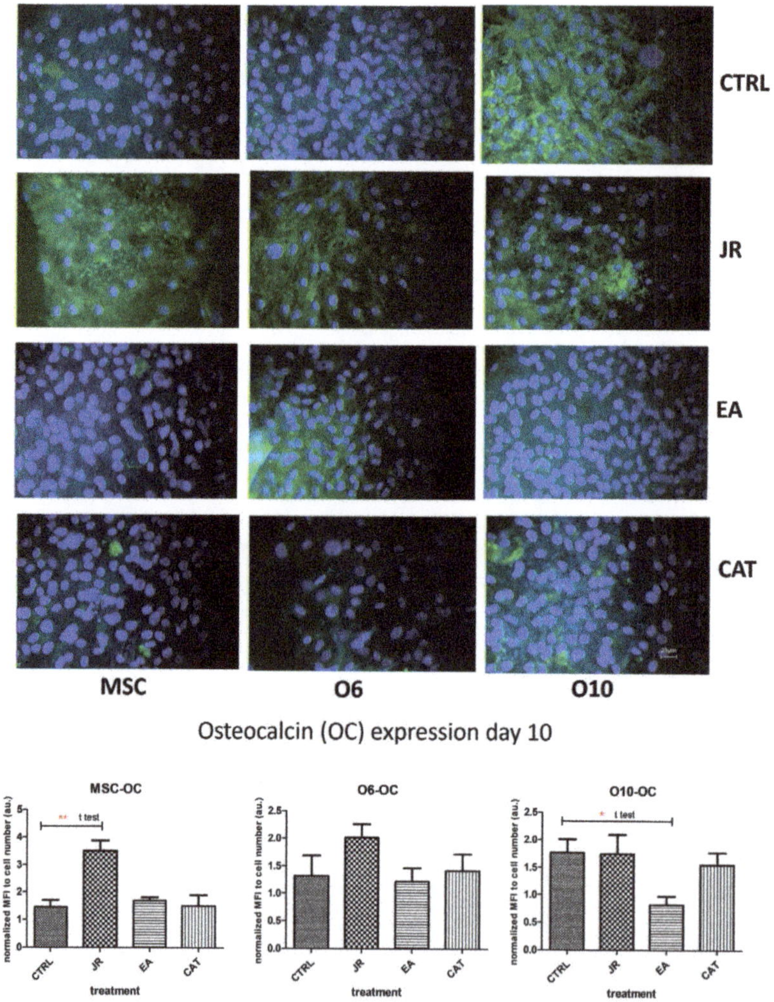

Figure 10. Fluorescence images for osteocalcin staining (FITC) of MSC, O6 and O10 cells cultured for 7 days with osteogenic medium and the application of two treatments with JR extract (0.5 mg mL^{-1}), EA (0.05 mg mL^{-1}) and CAT (0.15 μg mL^{-1}). Magnification, ×40 objective. Scale bar = 25 μm. In the bottom panel, graphical representations of the average fluorescence intensity normalized to the number of DAPI-stained nuclei are shown. Statistics were assessed using a paired *t*-test. Values are expressed as the mean ± SD). * $p < 0.05$, ** $p < 0.01$.

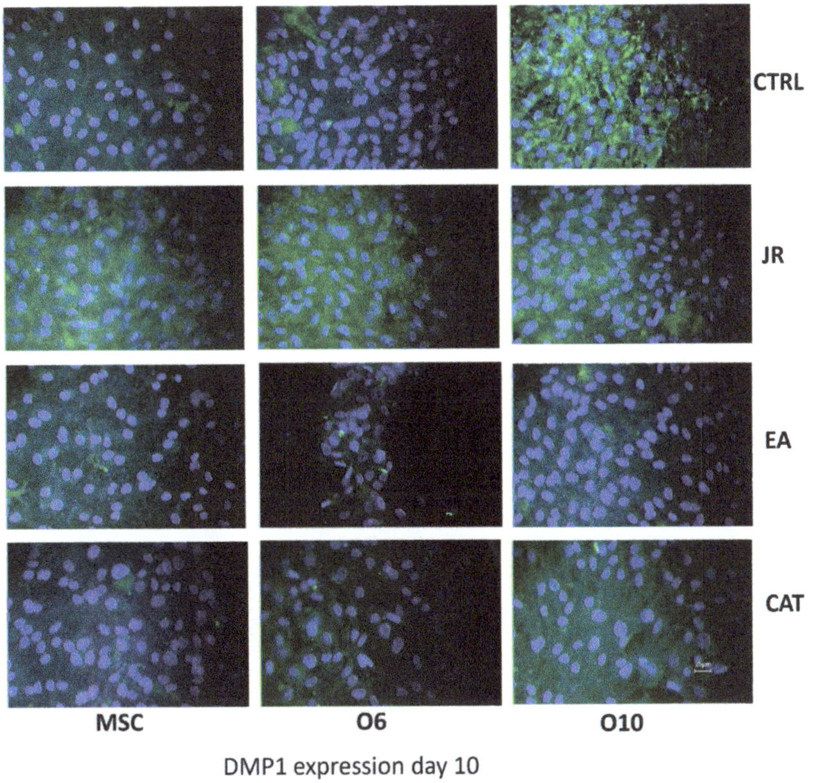

Figure 11. Fluorescence images for DMP1 staining (FITC) of MSC, O6 and O10 cells cultured for 7 days with osteogenic medium and the application of two treatments with JR extract (0.5 mg mL^{-1}), EA (0.05 mg mL^{-1}) and CAT (0.15 μg mL^{-1}). Magnification, ×40 objective. Scale bar = 25 μm.

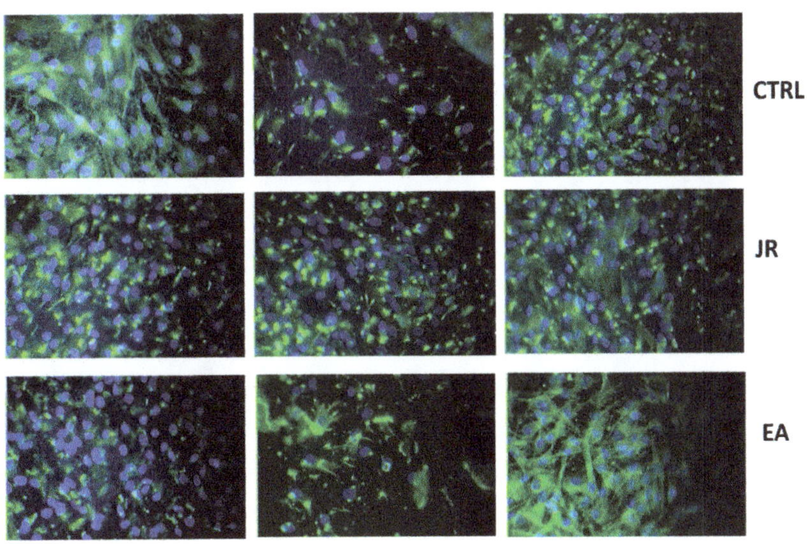

Figure 12. *Cont.*

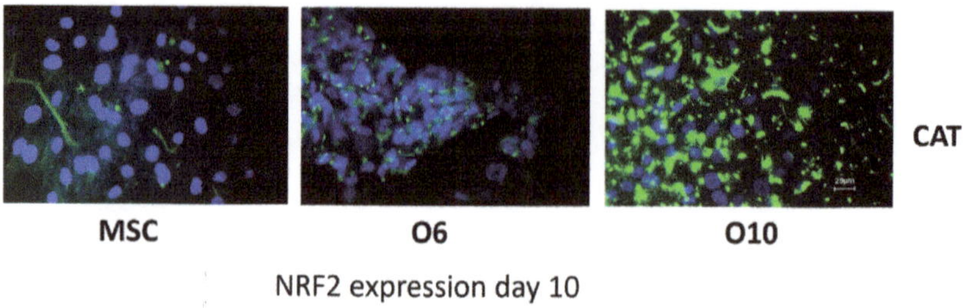

Figure 12. Fluorescence images for NRF2 staining (FITC) of MSC, O6 and O10 cells cultured for 7 days with osteogenic medium and the application of two treatments with JR extract (0.5 mg mL^{-1}), EA (0.05 mg mL^{-1}) and CAT (0.15 µg mL^{-1}). Magnification, ×40 objective. Scale bar = 25 µm.

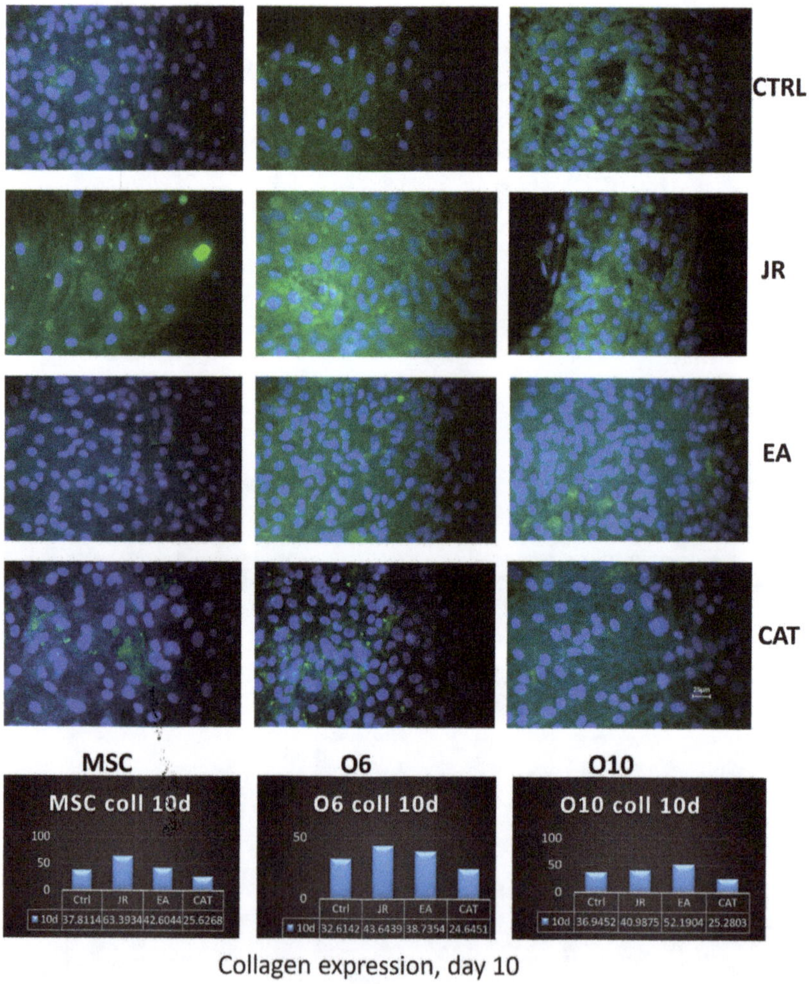

Figure 13. Fluorescence images for collagen 1A staining (FITC) of MSC, O6 and O10 cells cultured for 7 days with osteogenic medium and the application of two treatments with JR extract (0.5 mg mL^{-1}), EA (0.05 mg mL^{-1}) and CAT (0.15 µg mL^{-1}) (×40 objective). Scale bar = 25 µm. The bottom panel corresponds to the collagen measurements (µg mL^{-1}) performed after 10 days in the osteogenic cell culture medium.

3.6. Collagen Levels in Supernatants and Cell Lysates

For the MSC cell line, we observed a linear evolution of the amount of collagen for the control samples in the first 3–7 days (average 31 μg mL^{-1}), with a slight increase to 37 μg mL^{-1} at 10 days. Treatment with JR led to a significant increase in collagen in the culture medium progressively up to 10 days (average 63 μg mL^{-1}). EA induced an increase in collagen to values similar to those induced by JR at 7 days (54 μg mL^{-1}), while CAT triggered a slight increase in collagen on the 7th day. The younger control osteoblastic cells (O6) showed the highest level of collagen in the medium on day 7. The response to JR and EA was the highest after 3 days, and then the collagen level decreased gradually. CAT induced the maximum increase in collagen on day 7 (60 μg mL^{-1}). Mature osteoblastic cells (O10) showed a more moderate response to JR with an increase up to 40 μg mL^{-1} at 7 and 10 days. EA also induced an increase in collagen levels from day 3, while CAT inhibited collagen secretion in O10 cells (Figure 14).

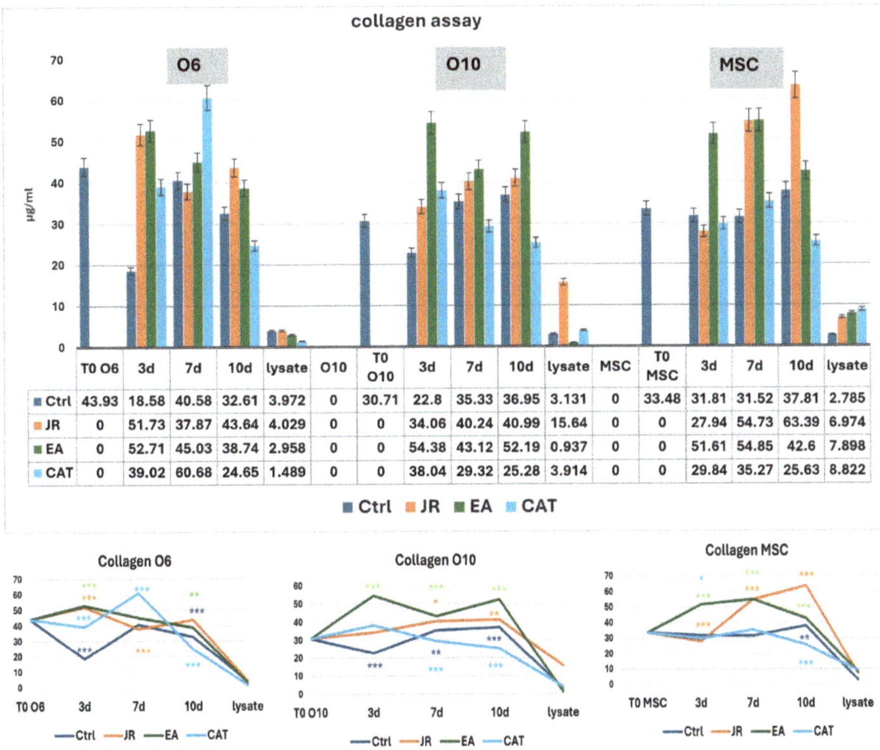

	T0 O6	3d	7d	10d	lysate	O10	T0 O10	3d	7d	10d	lysate	MSC	T0 MSC	3d	7d	10d	lysate
Ctrl	43.93	18.58	40.58	32.61	3.972	0	30.71	22.8	35.33	36.95	3.131	0	33.48	31.81	31.52	37.81	2.785
JR	0	51.73	37.87	43.64	4.029	0	0	34.06	40.24	40.99	15.64	0	0	27.94	54.73	63.39	6.974
EA	0	52.71	45.03	38.74	2.958	0	0	54.38	43.12	52.19	0.937	0	0	51.61	54.85	42.6	7.898
CAT	0	39.02	60.68	24.65	1.489	0	0	38.04	29.32	25.28	3.914	0	0	29.84	35.27	25.63	8.822

Figure 14. Collagen levels (μg mL^{-1}) at time 0 and at 3, 7 and 10 days in the supernatants of MSC, O6 and O10 cell cultures in osteogenic medium and after 3 treatments with JR extract (0.5 mg mL^{-1}), EA (0.05 mg mL^{-1}) and CAT (0.15 μg mL^{-1}). The cellular lysate was harvested 10 days after the culture. The values are expressed in μg mL^{-1}. Error bars represent the standard deviations of three measurements (n = 3). One-way ANOVA followed by the Bonferroni post test demonstrating the statistically significant effects for each cell line in response to JR, EA and CAT treatments compared to control samples is illustrated in the bottom panel. The p-values are set at * $p < 0.05$, ** $p < 0.01$, or *** $p < 0.001$.

3.7. Evaluation of ALP Activity

ALP activity in control samples was the highest in the undifferentiated stem cells, then decreased as MSC differentiation was initiated. Increased ALP activity is described in pluripotent stem cells and related cells as a key marker. Tissue non-specific alkaline phosphatase (TNAP) is an isoenzyme implicated in the initiation of bone mineralization by providing free inorganic phosphate for local hydroxyapatite formation with further deposition in the extracellular space between collagen fibrils. It can also be an inhibitor of the bone matrix via hydrolysis of phosphates [29–31].

Untreated O6 cells had the lowest activity at the initiation of cultures (T0), but ALP activity increased progressively until day 10. Control O10 cells showed relatively increased levels of ALP activity compared to O6, but these levels remained relatively constant throughout the 10 days of culture. JR compared to untreated samples did not induce an increase in ALP activity. It was observed that JR on day 10 was more active on O10 cells. EA was the most active on O10 cells, with an increase in values above those measured in untreated control cells. MSCs also showed increased activity 10 days after EA treatment, above the levels in controls at the same time. CAT also had stimulatory effects on O10 cells and MSCs, especially on day 7 (Figures 15 and 16).

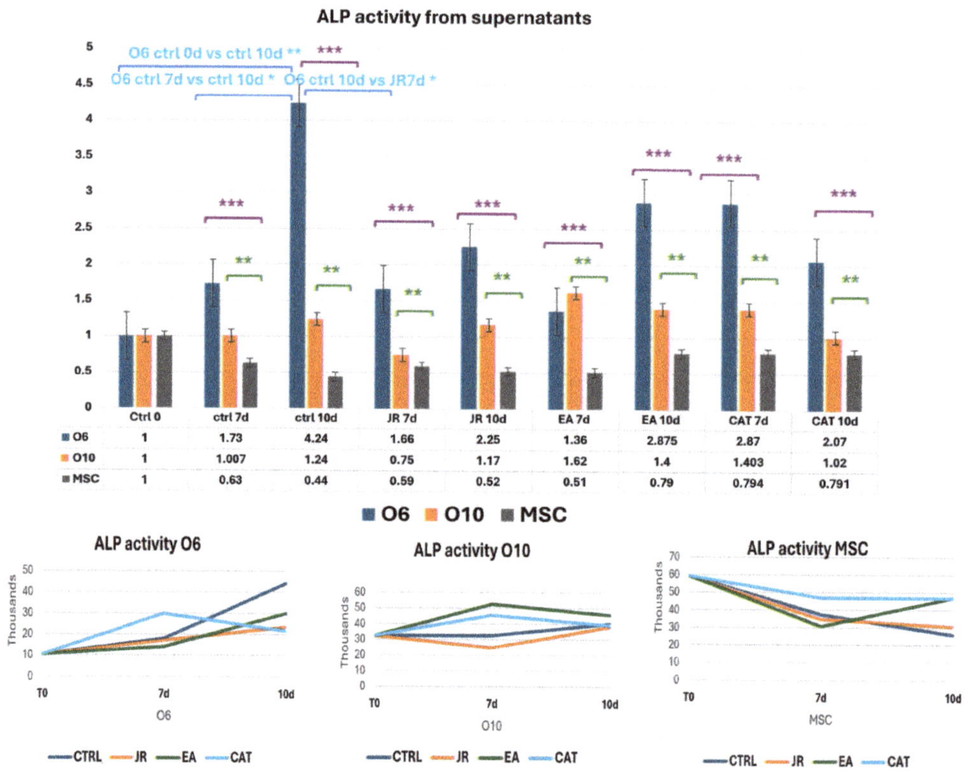

Figure 15. Graphics of induced alkaline phosphatase activity represented as fold-change versus control from supernatants of MSC, O6 and O10 cell cultures in osteogenic medium and after 3 treatments of JR, EA and CAT; samples were collected at time 0, 7 and 10 days. Error bars represent the standard deviations of three measurements ($n = 3$). Statistical analysis representation: One-way ANOVA (Kruskal–Wallis test followed by Dunn's Multiple Comparison Test) results are illustrated with comparison of cell groups: O6 vs. MSCs (purple brackets and stars *** $p < 0.001$); O10 vs. MSCs (green brackets and stars ** $p < 0.01$). Two-way ANOVA analysis is represented with blue brackets and stars * $p < 0.05$).

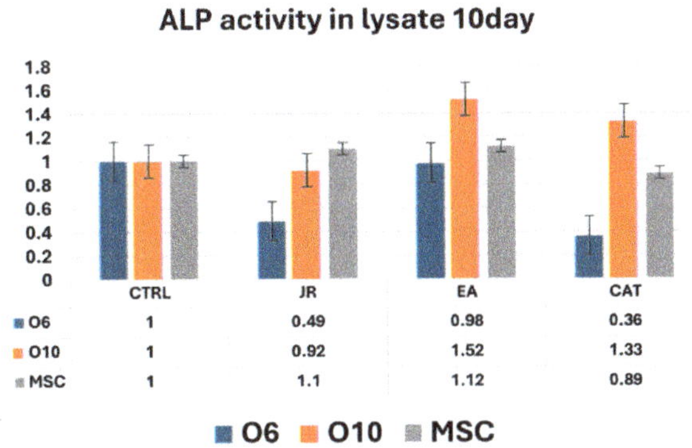

Figure 16. A graphic representation of alkaline phosphatase activity (as fold-change versus control) measured from cell lysate samples of MSC, O6 and O10 cell cultures in osteogenic medium and after 3 treatments of JR, EA and CAT; samples were collected at 10 days. Error bars represent the standard deviations of three measurements ($n = 3$).

Intracellular ALP activity observed in cell lysates harvested after 10 days of culture was increased in control MSCs compared to O6 and O10 cells. JR induced a decrease in ALP values for O6 cells without affecting those of MSC and O10. EA and CAT induced an increase in ALP activity in O10 and MSCs and a decrease in ALP values for O6 (Figure 16).

3.8. Evaluation of NF-kB Levels

NF-kB values showed variations without significant differences in almost all the samples analyzed. CAT, through its cytotoxic action, reduced the number of cells to much lower values at 10 days for MSCs and O10 cells. Statistical differences in NF-kB (two-way ANOVA followed by Bonferroni posttest) were only registered for CAT at 10 days of cultivation compared to the control samples (Figure 17).

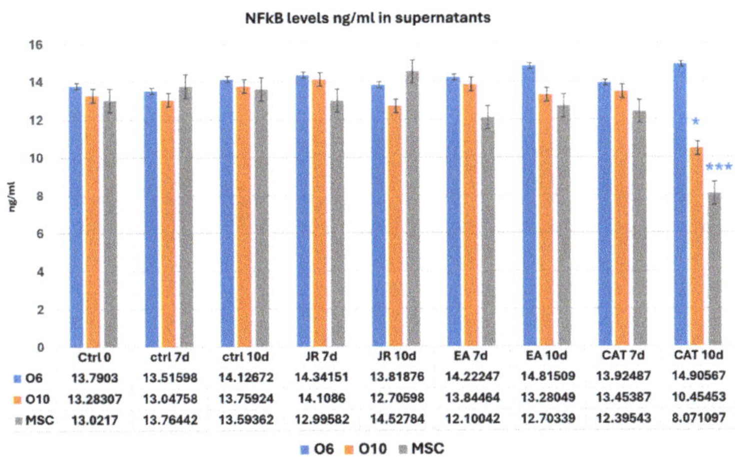

Figure 17. A graphic representation of NF-kB values (ng mL^{-1}) evaluated from cell supernatants of MSC, O6 and O10 cell cultures in osteogenic medium and after 3 treatments of JR, EA and CAT; samples were collected at 0, 3 and 10 days. Error bars represent the standard deviations of three measurements ($n = 3$). Two-way ANOVA followed by the Bonferroni post test for evaluation of significant differences between cell lines in response to JR, EA and CAT treatments. The p-values are set as * $p < 0.05$ or *** $p < 0.001$.

It was observed, however, that in the lysates harvested at 10 days of culture, that young osteoblastic cells responded with an increase in NF-kB expression to the treatment with EA, in contrast to much lower values obtained in mature osteoblasts and MSCs (Figure 18).

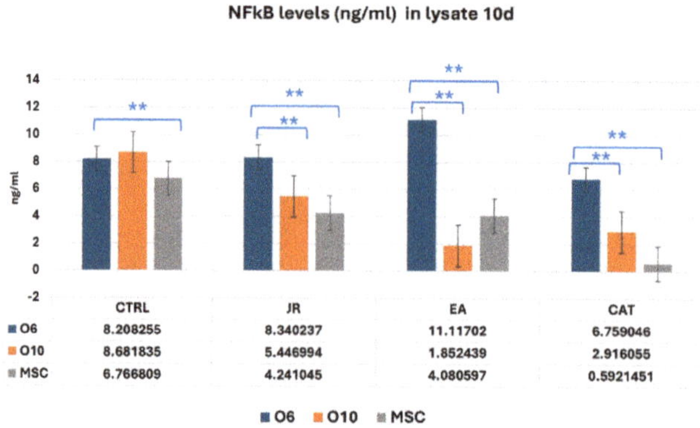

Figure 18. A graphic representation of NF-kB values (ng mL^{-1}) evaluated from cell lysate samples of MSC, O6 and O10 cell cultures in osteogenic medium and after 3 treatments of JR, EA and CAT; samples were collected at 10 days. Error bars represent the standard deviations of three measurements ($n = 3$). One-way ANOVA followed by the Bonferroni post test for evaluation of significant differences between cell lines in response to JR, EA and CAT treatments. The p-values is set as ** $p < 0.05$.

4. Discussion

In the present study, we analyzed the effects of treatment with JR extract compared to EA and CAT on cells at different stages of bone differentiation to understand at which stage of bone differentiation these bio-active agents act. This pilot study investigated the treatment response of osteoblasts derived from osteoporotic tissue. The bone tissue was collected from a 77-year-old donor who had a femoral neck fracture due to osteoporosis. These results cannot be extended to the general population (younger or male persons). Therefore, we plan to enlarge our studies by involving multiple donors for future validation.

Our results show a greater sensitivity of mature osteoblasts and to a lesser extent osteoblastic progenitors (MSCs and pre-osteoblasts) to JR, as shown by the MTT viability tests after 72 h of treatment with different doses of JR. In the evaluation of bone differentiation experiments using the OS differentiation medium, the cells were exposed to successive treatments with a predetermined dose as follows: JR extract (0.5 mg mL^{-1}), EA (0.05 mg mL^{-1}) and CAT (0.15 μg mL^{-1}). The rationale for using a single concentration for the three kinds of treatments was to create a comparative parallel with in vivo conditions where the osteoblastic cells are present at different stages of differentiation, so their response to therapy with a predetermined dose will probably determine different responses depending on the cells' maturity phase. The approach adopted in our experiments represents a limitation that was determined by simplifying the interpretation of the results due to numerous variables. In the experiment in which the cells were exposed to multiple administrations of a non-toxic dose of JR, EA and CAT, it was observed that JR did not induce the proliferation of any of the three cell lines, in contrast to EA and CAT. This observation suggests cytostasis induced by the initiation of a differentiation process often associated with decreased cell proliferation. The mineralization process, assessed with alizarin red staining, was more pronounced for osteoblasts (O6 and O10 cells) treated with EA compared to JR and CAT treatments, the latter leading to an inhibition of calcium deposition.

The intracellular expression of osteogenic markers, collagen, OPN, OC, DMP1 and NRF2 was increased mainly in the young cells (MSC and O6) treated with JR extract. A particularity was the expression of NRF2, which was increased in all cell lines after EA administration. CAT had this effect only on osteoblastic cells. NRF2 is widely present as an inactive form in the cytoplasm of many cell types such as osteoblasts, osteocytes and osteoclasts. NRF2-deficient mice presented a reduced resistance to oxidative stress with increased predisposition to osteoporosis [32]. Oxidative stress exposure determines NRF2 release from its Keap1 inhibitory protein and its translocation into the nucleus, activating various antioxidant and detoxification enzymes. Cytoplasmic expression of NRF2 keeps the bone cells safe from damage induced by oxidative stress, inhibiting the imbalance of bone remodeling and improving fracture healing. Oxidative stress is incriminated to be a well-known risk factor for osteoporosis development. A multitude of natural compounds such as resveratrol, schisandrin A, lutein and psoralen target the NRF2 pathway and consequently reduce osteoporosis prevalence due to oxidative stress regulation [33]. Polyphenolic compounds found in the composition of JR leaf extract, such as caffeic acid, resveratrol, ellagic acid and quercetin, act as indirect antioxidants, increasing NRF2 activity by inducing its target genes [34,35].

Collagen levels quantified in supernatants, as an indicator of collagen secretion into the extracellular space, revealed a significant increase in MSCs in response to JR extract and EA and in mature osteoblasts treated with EA. The cellular response to JR, EA and CAT treatments in terms of ALP activity was primarily dependent on the stage of cellular differentiation. JR extract and especially EA increased ALP activity in extracellular and intracellular compartments in MSC cultures after 10 days. The highest increase in ALP activity was observed in mature osteoblasts (O10), and its activity was inhibited in pre-osteoblasts (O6) with all treatments after 10 days. As can be seen in the schematic diagram in Figure 19, JR extract did not have a significant effect in inducing mineralization, but did have a significant effect on the construction of the extracellular protein matrix (collagen, OPN and OC). This behavior can be explained by higher bone remodeling, in which average tissue mineralization is reduced and cellular heterogeneity is increased. This fact is described by Allen M.R et al. (2019), who state that older, more mineralized bones are replaced by newer bone tissue, which has lower mineralization [36].

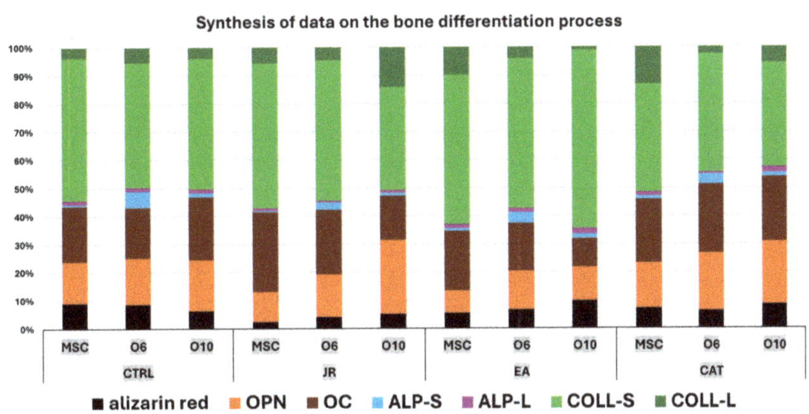

Figure 19. A schematic diagram of the majority of the main elements of bone differentiation with a synthesis of quantifiable results reflecting the response of MSCs and osteoblasts to treatments with JR, EA and CAT (OPN—osteopontin, OC—osteocalcin, ALP-S—alkaline phosphatase in supernatants, ALP-L—alkaline phosphatase in cell lysates, COLL-S—collagen levels in supernatants, COLL-l—collagen levels in cell lysates).

The significance of the values obtained for NF-kB in our experiments is given by the type of treatment applied. EA and JR did not influence NF-kB levels. However, CAT, which proved to be slightly cytotoxic at the doses used in this study, induced significant decreases in all three cell lines, with MSCs being more sensitive.

NF-kB is a transcription factor implicated in inflammation, apoptosis and cellular degradation processes. Inflammation induced by TNF-β on MSCs induced suppression of osteogenesis with downregulation of bone-specific matrix β1-integrin and Runx2 genes and upregulation of NF-κB phosphorylation. p105 is an inhibitory protein that maintains NF-κB heterodimers, p50 and p65, in the cytoplasm. NF-κB canonical activation by cytokines (such as TNF-α and RANKL) results in p105 phosphorylation, leading to its degradation and allowing for nuclear translocation of NF-κB. Therefore, it highlights the formation and activity of osteoclasts induced by inflammation in bone diseases [37,38]. In a previous study, resveratrol reversed TNF-β through NF-κB downregulation by activating Sirt1 (Sirtuin 1) and Runx2, thus demonstrating MSC differentiation into osteoblasts [39]. Hess et al. in a study on hMSCs treated with TNF-α showed activation of NF-kB, which in turn enhanced BMP-2 and ALP expression concomitant with increased matrix mineralization. A decrease in NF-kB signaling did not result in a decrease in matrix mineralization [14].

Caro et al. demonstrated in various in vitro models of oxidative stress at the hepatic level that catechin can lead to an antioxidant or pro-oxidant profile depending on various factors not well described [40]. Such findings may guide us towards an interpretation of our results, in which free CAT, although in concentrations similar to those detected in JR extract, did not have similar effects on osteogenic differentiation as those of JR and EA treatments. NRF2 studied by immunocytochemistry showed an increase in expression after treatments with JR and EA, especially in younger cells, MSCs and O6 cells. Mature osteoblasts O10 expressed higher positivity in control samples and CAT-treated cells. An explanation for this increase in NRF2 expression is given by a different mechanism independent of the oxidative stress response by phytochemical compounds such as curcumin, EGCG and resveratrol. It was proposed that the NRF2 signaling pathway can be activated by phytochemicals by inducing phosphorylation of NRF2 via activation of upstream protein kinases and/or through direct interaction with Keap1 cysteine thiols. Subsequent synthesis of cytoprotective enzymes can suppress oxidative stress and the inflammatory response [41].

Phytotherapy approaches have been studied particularly in ovariectomized animal models for the induction of osteoporosis. There are numerous such studies in the literature, from which a few have been selected in this paper. *Piper sarmentosum* extract, a Chinese plant from the barberry family, was used for fractures of osteoporotic status and improved fracture healing [42]. Another plant originally from North America belonging to the family *Ranuculaceae*, *Cimicifuga racemosa*, is currently used in complementary and alternative therapies to improve postmenopausal syndrome [43]. *Cimicifuga racemosa* rhizomes contain some suggestive phytochemicals such as triterpene glycosides, phenols, flavonoids and alkaloids, ferulic acid, caffeic acid and isoflavones phytoestrogen (formononetin), the latter being responsible for the estrogenic activity [44,45].

Plant extracts obtained from *Eucommiae cortex*, *Salvia miltiorrhiza* and *Curcuma longa* have been successfully applied in the treatment of fractures because of their strong anti-inflammatory effects. These properties are not due to a single component, but rather a complex mixture of bioactive compounds with synergistic action on the bone microstructure [2]. QEP, a traditional Chinese medicine formula composed of *E. ulmoides* Oliv., *P. corylifolia* L., *A. sativum* L. and *J. regia* L., has shown beneficial effects on osteoporosis in ovariectomized rats and in vitro on human osteoblastic cell cultures hFOB 1.19. Alizarin red quantification also showed mineralization of hFOB 1.19 cells, and ALP determination supported osteogenic differentiation [46]. Pang et al. promoted osteoblast differentiation of human bone marrow-derived mesenchymal stromal cells (hBMSCs) by using JR leaf extract, indicating improved expression of pro-osteogenic biomarkers such as ALP, OCN or OPG. However, the extract did not present a significant effect on cell proliferation, while ALP determination revealed that walnut leaf extract improved osteogenic activity depending on the dose [24].

Papoutsi et al. showed that EA and methanolic JR extract significantly induced nodule formation on KS483 osteoblasts, promoting mineralization, meanwhile also providing a high anti-atherogenic potential. The study was also conducted on human aorta endothelial cells. Ellagic acid at a low concentration induced ossification, while JR methanolic extract also activated the mineralized nodules' formation [47]. In our study in an osteogenic environment, the treatments with JR and EA, as well as the control group, initiated the formation of osteogenic nodules, while the treatment with CAT had no mineralization effect. Similar findings were reported by Dong et al. In this in vitro study, EA promoted the osteoblastic differentiation of BMSCs and induced osteogenesis in ovariectomized mice by increasing bone markers such as ALP, collagen 1 alpha 1, osteopontin and osteocalcin. On the seventh day of treatment, the ALP and alizarin red staining results indicated significant osteogenic differentiation in the EA groups compared to the control [48].

In our previous study, several polyphenolic compounds were identified in JR leaf extract such as ellagic acid, caffeic acid, ferulic acid, quercetin, resveratrol and catechin [19,25]. Each biocompound in JR extract can exert its antioxidative and anti-inflammatory action in a synergistic manner. *Juglans regia L.* extract compared to ellagic acid on bone neoformation in rats has also been studied individually in relation to its effect on osteogenesis (RANKL and hydroxyproline serum levels), oxidative stress (MDA levels, and SOD and CAT activities) and the anti-inflammatory effect (blood levels of TNFα and IL-6) in our previous in vivo study. The conclusion of the study was that oxidative stress was diminished in the presence of JR extract (MDA levels were lowest in the JR group), while the proinflammatory mediators were not significantly modified. Also, regarding bone regeneration and reabsorption markers, RANKL levels were low in the JR-treated group, and hydroxyproline concentration was not significantly improved [25].

Santos et al. in an in vitro study on osteoblast-like cell cultures treated with caffeic acid phenethyl ester showed that caffeic acid at low concentrations induced an increase in Rux2, Alp and SPP1 gene expression after 5 days of treatment and an enhancement in mineralization after 14 days [49]. The effects of caffeic acid were also presented in an in vivo study using a rat cranial defect model, demonstrating significantly improved bone defect healing [50]. Lotfi et al. showed that quercetin expressed outstanding osteogenic activity, anti-inflammatory effects and antioxidant capacity, all of which can be used for improving the bone regeneration process [18]. Liang et al. proved the potential of ferulic acid in promoting the proliferation and differentiation of skeletal stem cells after irradiation [51].

Catechins, an important class of natural polyphenolic phytochemicals, have multiple therapeutics effects due to their osteoprotective, antioxidant, and anti-inflammatory properties. Catechin has shown a stimulatory effect on osteoblast growth in an in vitro study on osteoblastic MC3T3-E1 cells due to inhibition of TNF-α and IL-6, while increasing cell viability and ALP activity [52].

Epigallocatechin gallate was demonstrated to augment calcium and phosphate levels and to reduce IL-6, TNF and RANKL. EGCG showed biocompatible and osteoinductive properties; it also stimulated the expression of osteogenic growth factors and increased osteocalcin and ALP levels [53]. In another study conducted on osteoblast-like MC3T3-E1 cells, Kawabata et al. showed that EGCG repressed the effect of insulin-like growth factor-I in inducing the migration of osteoblasts [54]. One beneficial green tea compound is epicatechin gallate (ECG), which in a study on C3H10T1/2 cells and hMSCs stimulated the differentiation of osteoblasts [55]. ECG showed protective effects on MC3T3-E1 cells against cadmium-induced osteoblast apoptosis [56].

All of these studies demonstrate the complex effects induced by each of these bio-compounds, mainly with anti-inflammatory, antioxidative, osteoinductive and protective activity. The presence of these compounds in adequate proportions in the JR leaf extract allows for their synergistic action, thus explaining the differences in cellular response compared to EA and CAT treatments.

Our study presents several limitations in the interpretation of the results, including the use of a single donor and the assessment of indirect biomarkers of oxidative stress, which only partially reflect the effects of reactive oxygen species. Interindividual variability could influence cellular responses to treatment, an aspect that should be considered when interpreting the data. Another limitation is using a single concentration for each treatment in differentiation experiments. Nevertheless, we consider the obtained results to be a first, using JR extract, bringing new insights that support its osteoregenerative potential, as evaluated on multiple cell lines. Thus, a natural product such as JR extract, with its complex phytochemical composition, may open promising perspectives for the development of new phytotherapeutic solutions with clinical applicability. In future research, we aim to use a more robust experimental design with larger sample sizes to validate the current findings.

5. Conclusions

The complex process of bone regeneration is influenced by both cells (mesenchymal stem cells and osteoblasts) and the presence of osteoregenerative factors. In vitro studies performed on cells in different stages of development, such as adult mesenchymal stem cells, pre-osteoblasts and late-stage osteoblasts, showed that the cells responded differently to treatments with *Juglans regia* L. extract in comparison with EA and CAT. The obtained results highlight the protective effects and osteogenic differentiation generated by JR extract in a manner comparable to EA treatment, while a cytotoxic effect and increased sensitivity was observed under CAT administration. Our study provides new insights into understanding the function of JR leaves and presents new potential treatment options for osteoporosis.

Supplementary Materials: The following supporting information can be downloaded at: https://www.mdpi.com/article/10.3390/jfb16070268/s1.

Author Contributions: Conceptualization, A.H.-F. and S.V.C.; methodology, O.Ş. and O.A.; software, O.Ş.; validation, G.A.F. and L.G.V.; formal analysis, L.F. and L.G.V.; investigation, L.F.; resources, A.H.-F.; data curation, L.F.; writing—original draft preparation, O.Ş.; writing—review and editing, G.A.F.; visualization, M.E.M.; supervision, M.E.M. All authors have read and agreed to the published version of the manuscript.

Funding: The APC was funded by the University of Oradea.

Institutional Review Board Statement: The study was conducted in accordance with the Declaration of Helsinki, and approved by the Ethical Committee of "Iuliu Hatieganu" University (16/9.11.2021) and the Ethic Council of the County Emergency Clinical Hospital of Oradea, Romania (no. 7700/02.03.2023).

Informed Consent Statement: Informed consent was obtained from all subjects involved in the study.

Data Availability Statement: The original contributions presented in this study are included in the article/Supplementary Material. Further inquiries can be directed to the corresponding authors.

Conflicts of Interest: The authors declare no conflicts of interest.

References

1. Miranda, L.L.; Guimarães-Lopes, V.D.P.; Altoé, L.S.; Sarandy, M.M.; Melo, F.C.S.A.; Novaes, R.D.; Gonçalves, R.V. Plant Extracts in the Bone Repair Process: A Systematic Review. *Mediat. Inflamm.* **2019**, *2019*, 1–22. [CrossRef] [PubMed]
2. Ilyas, S.; Lee, J.; Lee, D. Emerging Roles of Natural Compounds in Osteoporosis: Regulation, Molecular Mechanisms and Bone Regeneration. *Pharmaceuticals* **2024**, *17*, 984. [CrossRef] [PubMed]
3. Safadi, F.F.; Barbe, M.F.; Abdelmagid, S.M.; Rico, M.C.; Aswad, R.A.; Litvin, J.; Popoff, S.N. Bone Structure, Development and Bone Biology. In *Bone Pathology*; Khurana, J.S., Ed.; Humana Press: Totowa, NJ, USA, 2009; pp. 1–50, ISBN 978-1-58829-766-2.
4. Pitkänen, S. In Vitro and in Vivo Osteogenesis and Vasculogenesis in Synthetic Bone Grafts. Ph.D. Thesis, Tampere University, Tampere, Finland, 2020; Volume 200.
5. Miron, R.J.; Zhang, Y.F. Osteoinduction: A Review of Old Concepts with New Standards. *J. Dent. Res.* **2012**, *91*, 736–744. [CrossRef]
6. Si, J.; Wang, C.; Zhang, D.; Wang, B.; Hou, W.; Zhou, Y. Osteopontin in Bone Metabolism and Bone Diseases. *Med. Sci. Monit.* **2020**, *26*, e919159. [CrossRef]
7. Toyosawa, S.; Oya, K.; Sato, S.; Ishida, K. [Osteocyte and DMP1]. *Clin. Calcium* **2012**, *22*, 713–720.
8. Rios, H.F.; Ye, L.; Dusevich, V.; Eick, D.; Bonewald, L.F.; Feng, J.Q. DMP1 Is Essential for Osteocyte Formation and Function. *J. Musculoskelet. Neuronal Interact.* **2005**, *5*, 325–327.
9. Han, J.; Yang, K.; An, J.; Jiang, N.; Fu, S.; Tang, X. The Role of NRF2 in Bone Metabolism—Friend or Foe? *Front. Endocrinol.* **2022**, *13*, 813057. [CrossRef]
10. Shaw, A.T.; Gravallese, E.M. Mediators of Inflammation and Bone Remodeling in Rheumatic Disease. *Semin. Cell Dev. Biol.* **2016**, *49*, 2–10. [CrossRef]
11. Chang, J.; Liu, F.; Lee, M.; Wu, B.; Ting, K.; Zara, J.N.; Soo, C.; Al Hezaimi, K.; Zou, W.; Chen, X.; et al. NF-κB Inhibits Osteogenic Differentiation of Mesenchymal Stem Cells by Promoting β-Catenin Degradation. *Proc. Natl. Acad. Sci. USA* **2013**, *110*, 9469–9474. [CrossRef]
12. Chen, X.; Hu, C.; Wang, G.; Li, L.; Kong, X.; Ding, Y.; Jin, Y. Nuclear Factor-κB Modulates Osteogenesis of Periodontal Ligament Stem Cells through Competition with β-Catenin Signaling in Inflammatory Microenvironments. *Cell Death Dis.* **2013**, *4*, e510. [CrossRef]
13. Deng, P.; Zhou, C.; Alvarez, R.; Hong, C.; Wang, C.-Y. Inhibition of IKK/NF-κB Signaling Enhances Differentiation of Mesenchymal Stromal Cells from Human Embryonic Stem Cells. *Stem Cell Rep.* **2016**, *6*, 456–465. [CrossRef]
14. Hess, K.; Ushmorov, A.; Fiedler, J.; Brenner, R.E.; Wirth, T. TNFα Promotes Osteogenic Differentiation of Human Mesenchymal Stem Cells by Triggering the NF-κB Signaling Pathway. *Bone* **2009**, *45*, 367–376. [CrossRef] [PubMed]
15. Bord, S.; Ireland, D.C.; Beavan, S.R.; Compston, J.E. The Effects of Estrogen on Osteoprotegerin, RANKL, and Estrogen Receptor Expression in Human Osteoblasts. *Bone* **2003**, *32*, 136–141. [CrossRef] [PubMed]
16. Cheng, C.-H.; Chen, L.-R.; Chen, K.-H. Osteoporosis Due to Hormone Imbalance: An Overview of the Effects of Estrogen Deficiency and Glucocorticoid Overuse on Bone Turnover. *Int. J. Mol. Sci.* **2022**, *23*, 1376. [CrossRef] [PubMed]
17. El-Ma'aita, A.; Da'as, N.; Al-Hattab, M.; Hassona, Y.; Al-Rabab'ah, M.; Al-Kayed, M.-A. Awareness of the Risk of Developing Medication-Related Osteonecrosis of the Jaw among Bisphosphonate Users. *J. Int. Med. Res.* **2020**, *48*, 0300060520955066. [CrossRef] [PubMed]
18. Lotfi, M.-S.; Sheibani, M.; Jafari-Sabet, M. Quercetin-Based Biomaterials for Enhanced Bone Regeneration and Tissue Engineering. *Tissue Cell* **2024**, *91*, 102626. [CrossRef] [PubMed]
19. Hanga-Farcaș, A.; Miere (Groza), F.; Filip, G.A.; Clichici, S.; Fritea, L.; Vicaș, L.G.; Marian, E.; Pallag, A.; Jurca, T.; Filip, S.M.; et al. Phytochemical Compounds Involved in the Bone Regeneration Process and Their Innovative Administration: A Systematic Review. *Plants* **2023**, *12*, 2055. [CrossRef]
20. Chircov, C.; Miclea, I.I.; Grumezescu, V.; Grumezescu, A.M. Essential Oils for Bone Repair and Regeneration—Mechanisms and Applications. *Materials* **2021**, *14*, 1867. [CrossRef]
21. Zhou, C.; Shen, S.; Zhang, M.; Luo, H.; Zhang, Y.; Wu, C.; Zeng, L.; Ruan, H. Mechanisms of Action and Synergetic Formulas of Plant-Based Natural Compounds from Traditional Chinese Medicine for Managing Osteoporosis: A Literature Review. *Front. Med.* **2023**, *10*, 1235081. [CrossRef]
22. Bhat, A.A.; Shakeel, A.; Rafiq, S.; Farooq, I.; Malik, A.Q.; Alghuthami, M.E.; Alharthi, S.; Qanash, H.; Alharthy, S.A. *Juglans regia* Linn.: A Natural Repository of Vital Phytochemical and Pharmacological Compounds. *Life* **2023**, *13*, 380. [CrossRef]
23. Al-Nadaf, A.H.; Thiab, S.; Obidat, R.; AL-Arman, S.; Shahin, N.A. Pharmacological Evaluation of the Protective Effect of Hydroalcoholic Extract of Walnut (*Juglans regia* L.) Leaf on 6-Mercaptopurine-Induced Hepatotoxicity in Rats. *Phytomed. Plus* **2024**, *4*, 100617. [CrossRef]
24. Pang, X.; Zhong, Z.; Jiang, F.; Yang, J.; Nie, H. *Juglans regia* L. Extract Promotes Osteogenesis of Human Bone Marrow Mesenchymal Stem Cells through BMP2/Smad/Runx2 and Wnt/β-Catenin Pathways. *J. Orthop. Surg. Res.* **2022**, *17*, 88. [CrossRef] [PubMed]

25. Hanga-Farcas, A.; Fritea, L.; Filip, G.A.; Clichici, S.; Vicas, L.G.; Toma, V.-A.; Marian, E.; Gligor, F.G.; Abu Dayyih, W.; Muresan, M.E. The Influence of *Juglans regia* L. Extract and Ellagic Acid on Oxidative Stress, Inflammation, and Bone Regeneration Biomarkers. *Int. J. Mol. Sci.* **2024**, *25*, 12577. [CrossRef] [PubMed]
26. Tomuleasa, C.I.; Foris, V.; Soriţău, O.; Páll, E.; Fischer-Fodor, E.; Lung-Illes, V.; Brie, I.; Virág, P.; Perde-Schrepler, M.; Postescu, I.D.; et al. Effects of 60Co Gamma-Rays on Human Osteoprogenitor Cells. *Rom. J. Morphol. Embryol.* **2009**, *50*, 349–355. [PubMed]
27. Kostina, D.; Lobov, A.; Klausen, P.; Karelkin, V.; Tikhilov, R.; Bozhkova, S.; Sereda, A.; Ryumina, N.; Enukashvily, N.; Malashicheva, A. Isolation of Human Osteoblast Cells Capable for Mineralization and Synthetizing Bone-Related Proteins In Vitro from Adult Bone. *Cells* **2022**, *11*, 3356. [CrossRef]
28. Voegele, T.J.; Voegele-Kadletz, M.; Esposito, V.; Macfelda, K.; Oberndorfer, U.; Vecsei, V.; Schabus, R. The Effect of Different Isolation Techniques on Human Osteoblast-like Cell Growth. *Anticancer. Res.* **2000**, *20*, 3575–3581.
29. Vimalraj, S. Alkaline Phosphatase: Structure, Expression and Its Function in Bone Mineralization. *Gene* **2020**, *754*, 144855. [CrossRef]
30. Štefková, K.; Procházková, J.; Pacherník, J. Alkaline Phosphatase in Stem Cells. *Stem Cells Int.* **2015**, *2015*, 1–11. [CrossRef]
31. Orimo, H. The Mechanism of Mineralization and the Role of Alkaline Phosphatase in Health and Disease. *J. Nippon. Med. Sch.* **2010**, *77*, 4–12. [CrossRef]
32. Ibáñez, L.; Ferrándiz, M.L.; Brines, R.; Guede, D.; Cuadrado, A.; Alcaraz, M.J. Effects of NRF2 Deficiency on Bone Microarchitecture in an Experimental Model of Osteoporosis. *Oxidative Med. Cell. Longev.* **2014**, *2014*, 1–9. [CrossRef]
33. Che, J.; Yang, X.; Jin, Z.; Xu, C. NRF2: A Promising Therapeutic Target in Bone-Related Diseases. *Biomed. Pharmacother.* **2023**, *168*, 115748. [CrossRef] [PubMed]
34. Tebay, L.E.; Robertson, H.; Durant, S.T.; Vitale, S.R.; Penning, T.M.; Dinkova-Kostova, A.T.; Hayes, J.D. Mechanisms of Activation of the Transcription Factor NRF2 by Redox Stressors, Nutrient Cues, and Energy Status and the Pathways through Which It Attenuates Degenerative Disease. *Free Radic. Biol. Med.* **2015**, *88*, 108–146. [CrossRef] [PubMed]
35. Zhou, Y.; Jiang, Z.; Lu, H.; Xu, Z.; Tong, R.; Shi, J.; Jia, G. Recent Advances of Natural Polyphenols Activators for Keap1-NRF2 Signaling Pathway. *Chem. Biodivers.* **2019**, *16*, e1900400. [CrossRef] [PubMed]
36. Allen, M.R.; Burr, D.B. Bone Growth, Modeling, and Remodeling. In *Basic and Applied Bone Biology*; Elsevier: Amsterdam, The Netherlands, 2019; pp. 85–100, ISBN 978-0-12-813259-3.
37. Boyce, B.F.; Xiu, Y.; Li, J.; Xing, L.; Yao, Z. NF-κB-Mediated Regulation of Osteoclastogenesis. *Endocrinol. Metab.* **2015**, *30*, 35. [CrossRef]
38. Wangsanut, T.; Brann, K.R.; Adcox, H.E.; Carlyon, J.A. Orientia Tsutsugamushi Modulates Cellular Levels of NF-κB Inhibitor P105. *PLoS Negl. Trop. Dis.* **2021**, *15*, e0009339. [CrossRef]
39. Constanze, B.; Popper, B.; Aggarwal, B.B.; Shakibaei, M. Evidence That TNF-β Suppresses Osteoblast Differentiation of Mesenchymal Stem Cells and Resveratrol Reverses It through Modulation of NF-κB, Sirt1 and Runx2. *Cell Tissue Res.* **2020**, *381*, 83–98. [CrossRef]
40. Caro, A.A.; Davis, A.; Fobare, S.; Horan, N.; Ryan, C.; Schwab, C. Antioxidant and Pro-Oxidant Mechanisms of (+) Catechin in Microsomal CYP2E1-Dependent Oxidative Stress. *Toxicol. Vitr.* **2019**, *54*, 1–9. [CrossRef]
41. Surh, Y.-J.; Kundu, J.; Na, H.-K. NRF2 as a Master Redox Switch in Turning on the Cellular Signaling Involved in the Induction of Cytoprotective Genes by Some Chemopreventive Phytochemicals. *Planta Med.* **2008**, *74*, 1526–1539. [CrossRef]
42. Estai, M.A.; Suhaimi, F.H.; Das, S.; Fadzilah, F.M.; Majedah Idrus Alhabshi, S.; Shuid, A.N.; Soelaiman, I.-N. Piper Sarmentosum Enhances Fracture Healing in Ovariectomized Osteoporotic Rats: A Radiological Study. *Clinics* **2011**, *66*, 865–872. [CrossRef]
43. Kolios, L.; Schumann, J.; Sehmisch, S.; Rack, T.; Tezval, M.; Seidlova-Wuttke, D.; Frosch, K.-H.; Stuermer, K.; Stuermer, E. Effects of Black Cohosh (*Cimicifuga racemosa*) and Estrogen on Metaphyseal Fracture Healing in the Early Stage of Osteoporosis in Ovariectomized Rats. *Planta Med.* **2010**, *76*, 850–857. [CrossRef]
44. Mohapatra, S.; Iqubal, A.; Ansari, M.J.; Jan, B.; Zahiruddin, S.; Mirza, M.A.; Ahmad, S.; Iqbal, Z. Benefits of Black Cohosh (*Cimicifuga racemosa*) for Women Health: An Up-Close and In-Depth Review. *Pharmaceuticals* **2022**, *15*, 278. [CrossRef]
45. Qin, Z.; Dong, Z.; Liu, J.; Zhong, A.; Bao, M.; Wang, H.; Yu, H.; Zhang, S.; Zhang, W.; Shen, L.; et al. A Preliminary Study on the Effects of Black Cohosh Preparations on Bone Metabolism of Rat Models with GnRH-a-Induced Peri-Menopausal Symptoms. *Front. Endocrinol.* **2022**, *13*, 854345. [CrossRef] [PubMed]
46. Hao, J.; Bei, J.; Li, Z.; Han, M.; Ma, B.; Ma, P.; Zhou, X. Qing'e Pill Inhibits Osteoblast Ferroptosis via ATM Serine/Threonine Kinase (ATM) and the PI3K/AKT Pathway in Primary Osteoporosis. *Front. Pharmacol.* **2022**, *13*, 902102. [CrossRef] [PubMed]
47. Papoutsi, Z.; Kassi, E.; Chinou, I.; Halabalaki, M.; Skaltsounis, L.A.; Moutsatsou, P. Walnut Extract (*Juglans regia* L.) and Its Component Ellagic Acid Exhibit Anti-Inflammatory Activity in Human Aorta Endothelial Cells and Osteoblastic Activity in the Cell Line KS483. *Br. J. Nutr.* **2008**, *99*, 715–722. [CrossRef]
48. Dong, H.; Cao, Y.; Zou, K.; Shao, Q.; Liu, R.; Zhang, Y.; Pan, L.; Ning, B. Ellagic Acid Promotes Osteoblasts Differentiation via Activating SMAD2/3 Pathway and Alleviates Bone Mass Loss in OVX Mice. *Chem. Biol. Interact.* **2024**, *388*, 110852. [CrossRef]

49. Santos, P.H.N.; Silva, H.L.; Martinez, E.F.; Joly, J.C.; Demasi, A.P.D.; De Castro Raucci, L.M.S.; Teixeira, L.N. Low Concentrations of Caffeic Acid Phenethyl Ester Stimulate Osteogenesis in Vitro. *Tissue Cell* **2021**, *73*, 101618. [CrossRef]
50. Uçan, M.; Koparal, M.; Ağaçayak, S.; Gunay, A.; Ozgoz, M.; Atilgan, S.; Yaman, F. Influence of Caffeic Acid Phenethyl Ester on Bone Healing in a Rat Model. *J. Int. Med. Res.* **2013**, *41*, 1648–1654. [CrossRef]
51. Liang, J.-W.; Li, P.-L.; Wang, Q.; Liao, S.; Hu, W.; Zhao, Z.-D.; Li, Z.-L.; Yin, B.-F.; Mao, N.; Ding, L.; et al. Ferulic Acid Promotes Bone Defect Repair After Radiation by Maintaining the Stemness of Skeletal Stem Cells. *Stem Cells Transl. Med.* **2021**, *10*, 1217–1231. [CrossRef]
52. Choi, E.-M.; Hwang, J.-K. Effects of (+)-Catechin on the Function of Osteoblastic Cells. *Biol. Pharm. Bull.* **2003**, *26*, 523–526. [CrossRef]
53. German, I.J.S.; Barbalho, S.M.; Andreo, J.C.; Zutin, T.L.M.; Laurindo, L.F.; Rodrigues, V.D.; Araújo, A.C.; Guiguer, E.L.; Direito, R.; Pomini, K.T.; et al. Exploring the Impact of Catechins on Bone Metabolism: A Comprehensive Review of Current Research and Future Directions. *Metabolites* **2024**, *14*, 560. [CrossRef]
54. Kawabata, T.; Tokuda, H.; Sakai, G.; Fujita, K.; Matsushima-Nishiwaki, R.; Otsuka, T.; Kozawa, O. Repression of IGF-I-Induced Osteoblast Migration by (-)-Epigallocatechin Gallate through P44/P42 MAP Kinase Signaling. *Biomed. Rep.* **2018**, *9*, 318–326. [CrossRef] [PubMed]
55. Byun, M.R.; Sung, M.K.; Kim, A.R.; Lee, C.H.; Jang, E.J.; Jeong, M.G.; Noh, M.; Hwang, E.S.; Hong, J.-H. (−)-Epicatechin Gallate (ECG) Stimulates Osteoblast Differentiation via Runt-Related Transcription Factor 2 (RUNX2) and Transcriptional Coactivator with PDZ-Binding Motif (TAZ)-Mediated Transcriptional Activation. *J. Biol. Chem.* **2014**, *289*, 9926–9935. [CrossRef]
56. Wei, F.; Lin, K.; Ruan, B.; Wang, C.; Yang, L.; Wang, H.; Wang, Y. Epigallocatechin Gallate Protects MC3T3-E1 Cells from Cadmium-Induced Apoptosis and Dysfunction via Modulating PI3K/AKT/mTOR and NRF2/HO-1 Pathways. *PeerJ* **2024**, *12*, e17488. [CrossRef] [PubMed]

Journal of
Functional Biomaterials

MDPI

Article

Three-Dimensional-Printed Biomimetic Scaffolds for Investigating Osteoblast-Like Cell Interactions in Simulated Microgravity: An In Vitro Platform for Bone Tissue Engineering Research

Eleonora Zenobi [1,2,†], Giulia Gramigna [3,†], Elisa Scatena [1,2], Luca Panizza [4], Carlotta Achille [1], Raffaella Pecci [5], Annalisa Convertino [6], Costantino Del Gaudio [7,*], Antonella Lisi [3,‡] and Mario Ledda [3,*,‡]

1 E. Amaldi Foundation, Via del Politecnico snc, 00133 Rome, Italy; eleonora.zenobi@fondazioneamaldi.it (E.Z.); elisa.scatena@fondazioneamaldi.it (E.S.); carlotta.achille@fondazioneamaldi.it (C.A.)
2 Hypatia Research Consortium, Viale I Maggio, 156, Grottaferrata, 00046 Rome, Italy
3 Institute of Translational Pharmacology, National Research Council, Via Fosso del Cavaliere 100, 00133 Rome, Italy; giulia.gramigna@ift.cnr.it (G.G.); antonella.lisi@ift.cnr.it (A.L.)
4 Violatech srl, Via Kenia 74, 00144 Rome, Italy; luca.panizza@violatech.it
5 National Centre for Innovative Technologies in Public Health, Istituto Superiore di Sanità, Viale Regina Elena, 00161 Rome, Italy; raffaella.pecci@iss.it
6 Institute for Microelectronics and Microsystems, National Research Council, Via Fosso del Cavaliere 100, 00133 Rome, Italy; annalisa.convertino@cnr.it
7 Italian Space Agency, Via del Politecnico snc, 00133 Rome, Italy
* Correspondence: costantino.delgaudio@asi.it (C.D.G.); mario.ledda@ift.cnr.it (M.L.)
† These authors contributed equally to this work.
‡ These authors shared senior author position contributing equally to this work.

Abstract: Three-dimensional cell culture systems are relevant in vitro models for studying cellular behavior. In this regard, this present study investigates the interaction between human osteoblast-like cells and 3D-printed scaffolds mimicking physiological and osteoporotic bone structures under simulated microgravity conditions. The objective is to assess the effects of scaffold architecture and dynamic culture conditions on cell adhesion, proliferation, and metabolic activity, with implications for osteoporosis research. Polylactic acid scaffolds with physiological (P) and osteoporotic-like (O) trabecular architectures were 3D-printed by means of fused deposition modeling technology. Morphometric characterization was performed using micro-computed tomography. Human osteoblast-like SAOS-2 and U2OS cells were cultured on the scaffolds under static and dynamic simulated microgravity conditions using a rotary cell culture system (RCCS). Scaffold biocompatibility, cell viability, adhesion, and metabolic activity were evaluated through Bromodeoxyuridine incorporation assays, a water-soluble tetrazolium salt assay, and an enzyme-linked immunosorbent assay of tumor necrosis factor-α secretion. Both scaffold models supported osteoblast-like cell adhesion and growth, with an approximately threefold increase in colonization observed on the high-porosity O scaffolds under dynamic conditions. The dynamic environment facilitated increased surface interaction, amplifying the effects of scaffold architecture on cell behavior. Overall, sustained cell growth and metabolic activity, together with the absence of detectable inflammatory responses, confirmed the biocompatibility of the system. Scaffold microstructure and dynamic culture conditions significantly influence osteoblast-like cell behavior. The combination of 3D-printed scaffolds and a RCCS bioreactor provides a promising platform for studying bone remodeling in osteoporosis and microgravity-induced bone loss. These findings may contribute to the development of advanced in vitro models for biomedical research and potential countermeasures for bone degeneration.

 https://doi.org/10.3390/jfb16080271

Keywords: 3D-printed bone-like scaffolds; biomimetics; microgravity conditions

1. Introduction

Three-dimensional cell culture systems have transformed biological research, particularly in the field of human health, by offering a more realistic representation of the native tissue microenvironment compared to traditional two-dimensional in vitro cultures [1]. These systems enhance cell–cell and cell–matrix interactions thanks to their spatial organization and biomimetic scaffold surfaces, making them invaluable tools for applications in tissue engineering and regenerative medicine.

In the context of bone tissue engineering, the development of advanced devices is essential for overcoming the limitations of conventional medical prostheses, which can significantly impact the recipient's quality of life. The possibility to deal with ad hoc structures that are capable of replicating the natural extracellular matrix, a highly complex and functional structure, can effectively support cellular growth, differentiation, and integration with surrounding tissues [2]. Recent advancements in 3D-printing technologies, such as fused deposition modeling (FDM), have enabled the fabrication of scaffolds that mimic the intricate microarchitecture of native bone, including healthy and pathological conditions like osteoporosis [3].

A comprehensive research strategy can then be planned by integrating biomimetic, biocompatible, and bioresorbable scaffolds with dynamic bioreactors, such as rotary cell culture systems (RCCSs), which represent a cutting-edge 3D in vitro set-up to closely reproduce the native bone tissue microenvironment [4]. This experimental approach can be regarded as a promising platform for advanced bone tissue engineering investigations, also supporting the development of innovative therapies for improved treatment strategies and pre-clinical research models.

Bone is a dynamic and hierarchically structured tissue that provides structural support, facilitates movement, and protects internal organs [5,6]. Two different types can be mainly observed: trabecular and cortical bone. Trabecular bone, which accounts for approximately 20% of the skeleton, is characterized by a sponge-like structure with interconnected trabeculae. The porosity ranges from 50% to 90%, significantly influencing mechanical properties, such as modulus and compressive strength [7]. Cortical bone, in contrast, is denser, with a porosity of about 10%, forming the outer layer of bones and providing resistance to bending and torsion. This structure has evolved to respond to a number of different conditions, including mechanical loading, which plays a pivotal role to maintain bone strength through a dynamic remodeling process involving osteocytes, osteoblasts, and osteoclasts.

A substantial modification of external inputs may induce osteoporotic bone condition, a progressive skeletal disorder characterized by bone mass and density loss, compromised bone strength, and microarchitectural deterioration, which significantly increases fracture risk. The disease primarily affects trabecular bone due to its high surface area and metabolic activity. Osteoporotic trabecular bone exhibits notable structural changes, including increased porosity, reduced trabecular thickness, and a loss of connectivity density. These alterations lead to a more fragile, impaired network, reducing its ability to withstand mechanical loads and resist deformation [8]. Replicating such structural features in vitro is essential for creating realistic models to study osteoporotic bone-like tissue behavior and assess scaffold performance under pathological-like conditions.

To better simulate the in vivo bone microenvironment and overcome the limitations of static 2D cultures, dynamic bioreactor systems have been developed. Among them, RCCSs are particularly advantageous due to their ability to generate a low-shear, three-dimensional continuous suspension of bioengineered constructs, enhance nutrient and oxygen diffusion, and mimic in vivo conditions, fostering cell growth, differentiation, and metabolic activity within 3D scaffolds [9]. In a RCCS, scaffolds and cells are maintained in a state of constant free-fall, effectively reducing sedimentation and mimicking some aspects of microgravity. This unique environment supports enhanced cell–scaffold interactions, improved mass transport, and mechanical stimulation, all of which are crucial factors in tissue development, especially in mechanically responsive tissues like bone [10].

Dynamic culture also contributes to mechanotransduction, a process in which cells sense and respond to mechanical cues in their environment, influencing cell proliferation, morphology, gene expression, and matrix production. In bone tissue engineering, such mechanical inputs are especially important, as osteoblast-lineage cells are naturally responsive to mechanical stress. Moreover, when applied to scaffolds with tailored porosity and architecture, such as those designed to mimic osteoporotic traits, dynamic systems allow for a more refined evaluation of how architecture and mechanical cues interact to guide cellular responses.

By integrating these principles, our study aims to validate a platform, combining biomimetic scaffold design and rotary dynamic culture as a realistic and tunable environment for investigating cellular behavior under conditions relevant to osteoporotic bone tissue. Starting from this assumption, we designed and fabricated 3D-printed polylactic acid (PLA) scaffolds using FDM to replicate both physiological and osteoporotic bone architectures. The selected polymer is a common material for bone scaffold investigations, being a biocompatible and biodegradable aliphatic polyester approved by the Food and Drug Administration [11,12]. In addition, PLA exhibits several advantages including the possibility of being obtained from renewable sources and slight piezoelectric properties, making it suitable to design scaffolds for bone tissue applications. The fabricated bone-like scaffolds were thus located within a RCCS bioreactor to be seeded and cultured with osteoblast-like cells (i.e., SAOS-2 and U2OS) in dynamic conditions to be compared with static cultures.

This study's primary objectives were (i) to evaluate the biocompatibility and functionality of the scaffolds using assays to measure cell proliferation, adhesion, and inflammatory response; (ii) assess how scaffold internal microarchitecture, porosity, and fluid dynamics in the RCCS influence cellular behavior, particularly growth and metabolic activity; and (iii) optimize scaffold design for simulated microgravity conditions, hypothesizing that higher porosity enhances cell growth and metabolic activity by increasing surface area and improving fluid dynamics. The collected findings are expected to highlight the transformative potential of combining advanced 3D cell culture technologies with dynamic bioreactor systems, bridging the gap between in vitro models and in vivo applications. This investigation provides valuable insights into bone remodeling induced by dynamic conditions, paving the way for innovative therapeutic strategies that may address a range of challenges, laying the groundwork for more advanced studies targeting long-term bone cells and scaffolds to treat conditions like osteoporosis more effectively. This integrated platform may serve as a step forward in creating next-generation tissue engineering solutions for both research and clinical applications.

2. Materials and Methods

2.1. Biomimetic Scaffold Design

The design and preliminary analysis of the proposed scaffolds were performed using Meshmixer software (Meshmixer version 2018, Autodesk, San Rafael, CA, USA). Starting from a previous approach [13], an updated model was prepared to simulate osteoporotic-like trabecular bone tissue. Briefly, biomimetic bone scaffolds were realized to deal with a physiological (P) and osteoporotic (O) model by first subtracting a three-dimensional random cluster of spheres from a box-shaped solid to obtain a porous pattern. Subsequently, specific parameters were set for both the proposed models in terms of pore dimension, pore spacing, trabecular thickness, and trabecular spacing, as summarized in Table 1. The trabecular network was generated using an algorithm based on Delaunay triangulation.

Table 1. Parameters used for CAD modeling: pore dimension (PD), pore spacing (PS), trabecular thickness (TT), and trabecular spacing (TS).

Sample	PD (μm)	PS (μm)	TT (μm)	TS (μm)
P	700	700	200	600
O	800	500	150	800

2.2. Three-Dimensional Printing Set-Up and Scaffold Evaluation

Each scaffold file (Figure 1) was imported in IdeaMaker software (IdeaMaker version 3.6.1, Raise 3D Inc., Irvine, CA, USA) to slice the model and drive the Raise 3D N2 printer (Raise 3D Inc., Irvine, CA, USA). A commercial PLA filament (FILOALFA, Turin, Italy) was extruded at 205 °C through a 0.4 mm-diameter nozzle, the build platform temperature was set at 60 °C, and the positioning accuracy of the 3D printer was 0.0125 mm (XY axis) and 0.00125 mm (Z axis).

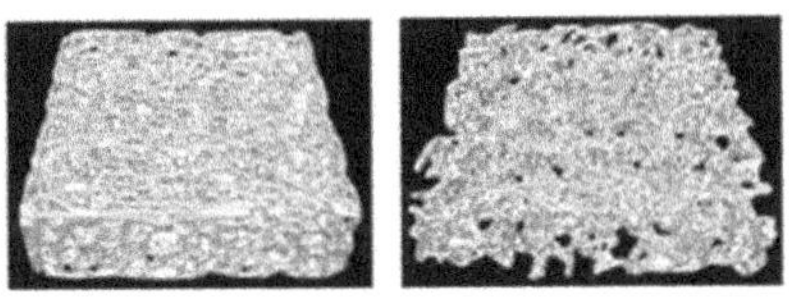

Figure 1. Preview printing of the P scaffold model (**left** panel) and the O scaffold model (**right** panel).

Microtomographic analysis (μ-CT) is a non-destructive technique, which allows for obtaining qualitative and quantitative information on the internal structure of the investigated samples. Each scaffold model was therefore analyzed to evaluate the printed microstructure and verify the conformity to the design criteria. For this aim, the microtomograph Skyscan 1072 (Bruker microCT, Kontich, Belgium) was used, powering the X-ray tube at 40 kV and 248 μA, and setting pixel dimensions at 14.65 × 14.65 μm (corresponding to 20× magnification and 180° rotation with a step of 0.45°). The 3D reconstruction of the sample was performed using the NRecon software (Version 1.7.0; Bruker microCT, Kontich, Belgium). The CT Analyzer software (Version 1.16.9.0; Bruker microCT, Kontich, Belgium) was employed to elaborate the collected data and calculate the histomorphometric parameters.

2.3. Cell Culture

The human osteosarcoma cell lineages SAOS-2 and U2OS (purchased from the American Type Culture Collection, ATCC, HTB-85 and HTB-96, Rockville, MD, USA) were selected to assess in vitro scaffold biocompatibility, cell growth, and adhesion. Before starting the experiments, cells were seeded on a plastic Petri dish and placed in a humidified incubator at 37 °C and 5% CO_2. Cells were cultured in high-glucose Dulbecco's modified Eagle's Medium (DMEM; Euroclone, Milan, Italy), completed with 10% heat-inactivated fetal bovine serum (FBS; Euroclone, Milan, Italy), 2 mM L-glutamine (Sigma, Darmstadt, Germany), 1.0 unit mL^{-1} penicillin (Sigma, Darmstadt, Germany), and 1.0 mg mL^{-1} streptomycin (Sigma, Darmstadt, Germany). Scaffolds were sterilized with 70% ethanol solution and accurately washed before cell seeding.

2.4. RCCS Bioreactor

Dynamic cell cultures in microgravity conditions were carried out using the Rotary Cell Culture System™ 4SCQ (Synthecon Incorporated, 8044 El Rio, Houston, TX 77054, USA). The RCCS bioreactor is shown in Figure 2; the cell culture chamber is horizontally rotated to constantly suspend the inoculated cells and any support (e.g., scaffolds) in the culture medium. Scaffolds of three different sizes ($4 \times 4 \times 4$ mm^3, $6 \times 6 \times 6$ mm^3, $8 \times 8 \times 8$ mm^3) were tested using increasing rotational speeds (from 25 to 45 rotations per minute) to identify the optimal floating condition of the scaffolds in the culture bioreactor vessel. Proper suspension of the scaffolds is essential for establishing an effective 3D dynamic culture environment, as it allows for uniform exposure of the scaffold surfaces to nutrient flow and shear forces, which are key to promoting effective cell adhesion, proliferation, and viability. The rotational speed that ensured their optimal suspension in the medium was found to be 45 rpm, while the most suitable scaffold size for proper floating conditions was $4 \times 4 \times 4$ mm^3. All tests were performed using these parameters.

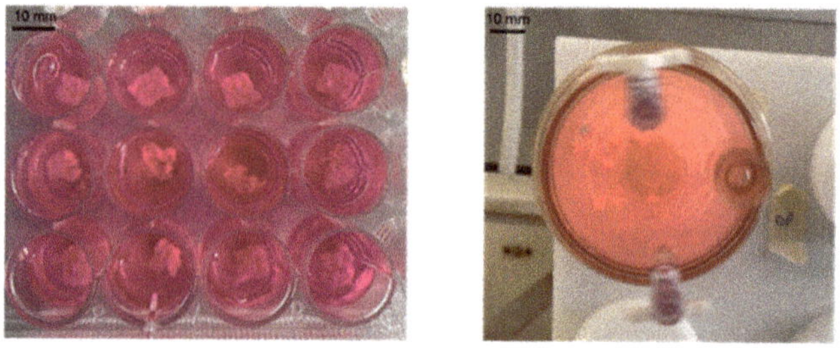

Figure 2. RCCS bioreactor model used in this study. Static conditions (**left** panel) and dynamic conditions (**right** panel).

2.5. Biological Assays

For the static and dynamic cultures, SAOS-2 and U2OS were seeded on the 3D-printed scaffolds at a density of 2×10^4 cell/cm^2 and cultured in the same humidified incubator at 37 °C and 5% CO_2, with the CO_2 being passively maintained within the bioreactor by means of an ad hoc filter that allows for gas exchange. The scaffolds were specifically sized ($4 \times 4 \times 4$ mm^3) to be cultured in a suspension mode into the RCCS bioreactor to balance the force acting on the scaffolds themselves (i.e., gravity, buoyancy, and centrifugal and drag forces). For the SAOS-2 and U2OS cultures in simulated microgravity, P and O scaffolds were inserted into the RCCS bioreactor chambers and the cells were directly seeded into the chamber filled with culture medium at a density of 2.0×10^4 cell/cm^2. Bubbles were removed using a syringe. On day 4, scaffolds were removed for the characterization.

2.5.1. Cell Proliferation and Metabolic Activity Analysis

Cell proliferation was assessed using the Bromodeoxyuridine (BrdU) incorporation assay (Cell Proliferation Kit, Roche Diagnostics), and metabolic activity was evaluated with a colorimetric assay based on the oxidation of water-soluble tetrazolium salts (WST-1 reagent, Roche Diagnostics).

On day 4, scaffolds were removed by the RCCS bioreactor and located in a multiwall dish for assay analysis. This step ensured that the assays (BrdU, WST-1, and ELISA) specifically measured the cellular activity of scaffold-adherent cells, minimizing interference from any free-floating cells.

For the metabolic activity assay, WST-1 reagent (diluted 1:10) was added directly to the culture medium, and cells were incubated for 2 h under standard humidified conditions. Following incubation, 100 µL of the resulting supernatant was transferred to a 96-well plate, and the formation of the formazan product was quantified spectrophotometrically. For the BrdU incorporation assay, cells were exposed to 10 mM BrdU for 18 h. After incubation, the cells were fixed and treated with an anti-BrdU antibody (dilution 1:100) for 30 min at 37 °C. Subsequently, the chromogenic substrate 2,2′-azino-bis(3-ethylbenzothiazoline-6-sulfonic acid) was added and incubated for an additional 30 min. Absorbance was then measured at 450 nm using a VICTOR3 multilabel plate reader (PerkinElmer, Waltham, MA, USA).

2.5.2. Quantification of TNF-α

Quantification of tumor necrosis factor-α (TNF-α) released by the cells was obtained using an ELISA development kit (PeproTech® EC Ltd., London, UK) at 405 nm. Supernatants from SAOS-2 and U2OS cells grown on the scaffolds were collected at 96 h of culture, centrifuged at 1200 rpm for 5 min, and stored at −80 °C until use. The release of TNF-α was quantified in the culture medium using a human ELISA development kit (PeproTech® EC Ltd., London, UK), following the manufacturer's protocol. Cell culture supernatants and recombinant human TNF-α standards were serially diluted in PBS containing 0.05% Tween-20 and 0.1% BSA (Sigma-Aldrich®, Dorset, UK). Cytokine detection was carried out using a biotin–avidin system, followed by incubation with the chromogenic substrate 2,2′-azino-bis(3-ethylbenzothiazoline-6-sulfonic acid) (ABTS; Sigma-Aldrich®, Dorset, UK).

2.5.3. Scanning Electron Microscopy (SEM) Analysis

Scanning electron microscopy (SEM) analysis was carried out to confirm the results obtained by cell growth assays. Cells cultured on the scaffolds were fixed with 4% paraformaldehyde for 10 min. After dehydration, the samples were coated with a 10 nm gold layer using an evaporation technique. SEM imaging was performed using a ZEISS SIGMA 300 field emission microscope (ZEISS, Oberkochen, Germany). Morphological evaluation was conducted at an accelerating voltage of 5 kV with secondary electron detection to visualize the cell–scaffold interactions in detail.

2.6. Statistical Analysis

Data are expressed as mean ± standard deviation, and MedCalc version 23.0.9 software was used for statistical analysis. Each experiment was performed in triplicate, data were analyzed with Student's *t*-tests, and the significance level adopted for all analyses was $p < 0.05$.

3. Results

3.1. Scaffold Design Assessment

The nominal porosity of the scaffold models was measured by means of the "analysis" tool of the 3D modeling software Meshmixer 3.5. The P scaffold porosity was 53.77%, while the O model was characterized by a porosity level of 82.96%. The bone porosity may change depending on the skeletal site and can increase up to 90% in osteoporotic conditions [14]; in this study, the design of the O scaffolds allowed us to reach the highest porosity with thin trabeculae and an interconnected bone-like structure in the required osteoporotic range, as previously reported [15–17].

3.2. Three-Dimensional-Printed Scaffolds and Morphometric Analysis

The 3D-printed PLA scaffolds are shown in Figure 3, where it is possible to observe the different microarchitectures of both the physiological and osteoporotic-like constructs.

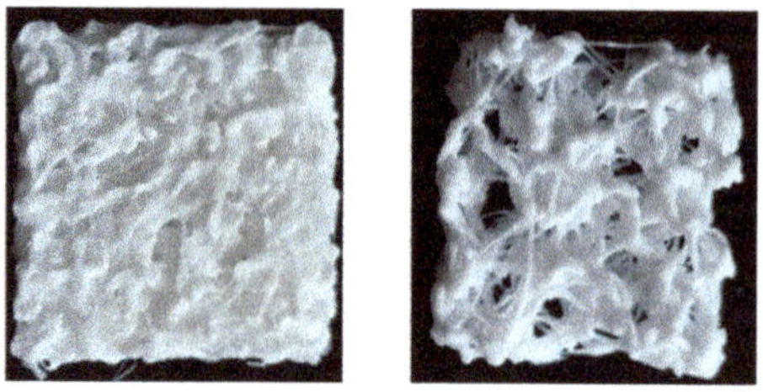

Figure 3. Three-dimensional-printed PLA scaffolds: physiological P model (**left** panel) and osteoporotic O model (**right** panel).

Each scaffold was analyzed using μ-CT to study the internal microstructure and quantify the compliance of the design criteria to the printed cases. A quantitative analysis was performed analyzing μ-CT data to extract the main bone histomorphometry parameters relevant to this study (Tables 2 and 3), which were then compared with corresponding values reported in the literature.

Table 2. Morphometric parameters for the physiological P scaffold.

Parameters	μ-CT Data	Literature Range	Reference
Trabecular thickness [μm]	258	310–490	[18]
Trabecular spacing [μm]	284	410–850	[18]
Porosity (%)	44.7	34–78	[18]

Table 3. Morphometric parameters for the pathological O scaffold.

Parameters	μ-CT Data	Literature Range	Reference
Trabecular thickness [μm]	215	90–230	[19]
Trabecular spacing [μm]	691	680–750	[14,19–21]
Porosity (%)	80.5	78–88	[17,19–21]

The data obtained from the μ-CT analysis confirmed a good agreement with the literature data, especially for the osteoporotic case. This was not strictly verified for the physiological model, which can be related to the resolution limits of the 3D printer as well as to the dynamics of the molten polymer cooling during the 3D printing process, which has already been observed [22,23]. Three-dimensional images obtained through the μ-CT analysis are shown in Figure 4.

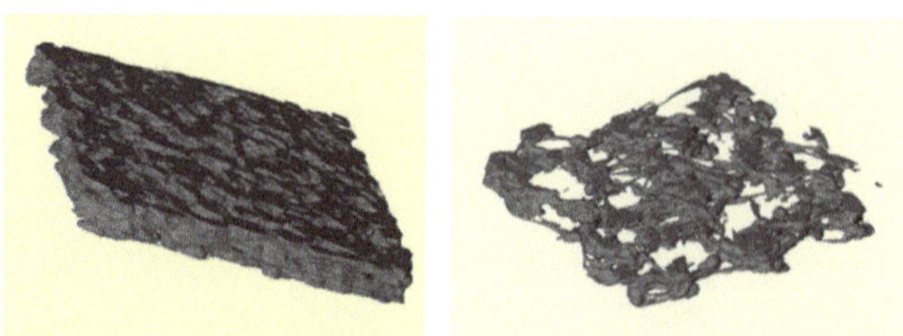

Figure 4. Three-dimensional images of the scaffolds obtained by μ-CT analysis: physiological P model (**left** panel) and osteoporotic O model (**right** panel).

3.3. Cell Growth and Metabolic Activity Study in RCCS Conditions

Simulated microgravity was investigated to evaluate the interaction of osteoblast-like cells with the physiological and osteoporotic-like PLA scaffolds (P and O models). Two osteoblast-like cell lines, SAOS-2 and U2OS, were cultured in dynamic conditions with both PLA scaffolds using the RCCS bioreactor to investigate their capability of colonizing and growing within the constructs by means of BrdU and WST-1 colorimetric assays, in comparison to those grown in 2D static conditions. These cell lines, derived from human osteosarcomas with different differentiation levels, have been used in several studies for bone tissue engineering. SAOS-2 cells have an osteoblast phenotype with characteristics similar to human primary osteoblasts and present a high level of ALP activity and mineralization ability [24,25]. U2OS cells instead exhibit properties of mesenchymal cells [26] and have a very low level of ALP and mineralization capacity [27]. The SAOS-2 and U2OS cells cultured for 4 days in dynamic conditions showed a positive interaction with polymer substrates, which were able to support both cell growth and metabolic activity processes (Figures 5 and 6).

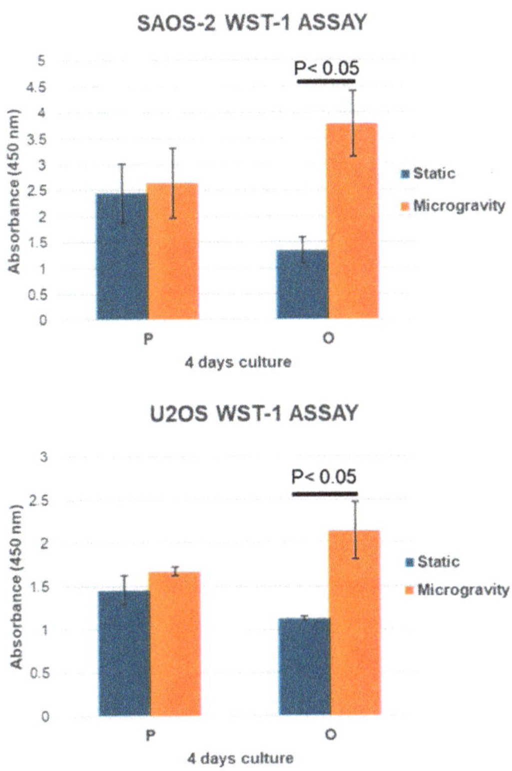

Figure 5. Analysis of cell metabolic activity on P and O scaffold models measured with the WST-1 assay at day 4 for both cell lines. $n = 3$, mean $\pm$ SD.

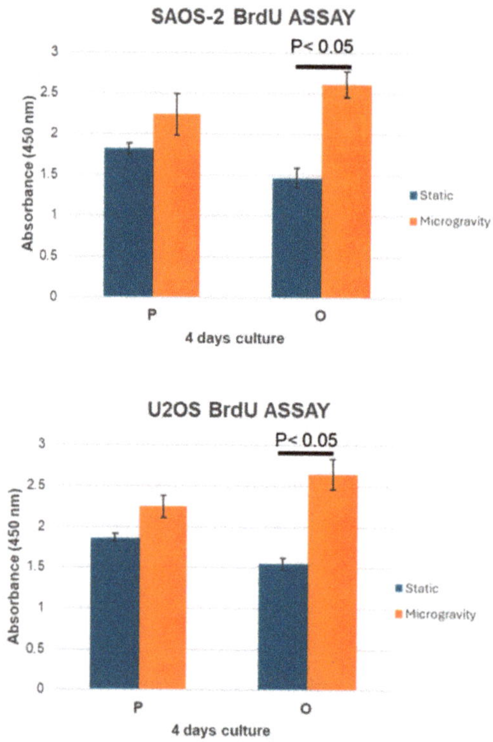

Figure 6. Analysis of cell growth on P and O scaffold models measured with the BrdU assay at day 4 for both cell lines. $n = 3$, mean ± SD.

The P scaffolds were characterized by a slight increase in cell colonization and metabolic activity in RCCS conditions, compared to the static one, while a higher growth and metabolic response were observed for the O model, in which this difference between static and dynamic cultures, obtained by BrdU and WST-1 tests, were statistically significant ($p < 0.05$). The BrdU and WST-1 assay results in a static culture revealed a decreasing trend from the less porous model to the more porous model. This trend could be due to the scaffold morphology: the P model, with an effective porosity of about 44.7%, prevents cells from settling to the bottom, allowing cells to better adhere to the scaffold surface and within the internal trabecular network, resulting in a higher cell growth after 4 days, consistent with a report by Zhou et al. [22]. Instead, the O model, due to its osteoporotic-like structure (about 80.53% of porosity), induced a lower cell interaction with the printed microarchitecture, settling at the bottom through the large pore network with a lower cell adhesion, as proven by the BrdU and WST-1 assays. In microgravity culture conditions, scaffolds and cells continuously interact, and the high porosity of the O scaffolds, by increasing the available surface area, improves the cells' attachment. For this reason, the osteoblast-like cells (SAOS-2 and U2OS) seeded on high-porosity scaffolds (O) and cultured in dynamic conditions showed a higher cell colonization compared not only to static conditions but also to low porosity substrates (P).

3.4. Cell Adhesion Study in RCCS Conditions

The adhesion and migration ability of cells into the scaffolds after culture in the RCCS bioreactor were studied using scanning electron microscopy (SEM) both for SAOS-2 cells, which are routinely used for investigations of osteoblast–biomaterial interactions, and for U2OS. SEM analysis allows for a detailed visualization of the scaffold's structure and the spatial distribution of cells within it. The images reported in Figure 7 confirmed a random porous microarchitecture. SAOS-2 and U2OS cells typically have polygonal, spindle-shaped, and fibroblast-like morphology (Figure S1) and are strongly adherent to the scaffold's surfaces, demonstrating their ability to colonize and penetrate the scaffold. This behavior was more pronounced in samples cultured under dynamic (microgravity) conditions compared to static ones. In particular, for both cells, SEM images revealed a noticeably higher cell density and more extensive surface coverage under dynamic conditions, suggesting improved adhesion and migration into the scaffold structure. In contrast, under static conditions, the cells appeared less uniformly distributed and in a lower density, indicating that the dynamic culture environment provided by the RCCS bioreactor enhances cell–scaffold interactions. These comparative observations, consistently confirmed for both SAOS-2 and U2OS cells, highlight the positive effect of a dynamic culture in the RCCS bioreactor, which creates a more favorable 3D environment for cellular interaction with the scaffold, supporting improved colonization.

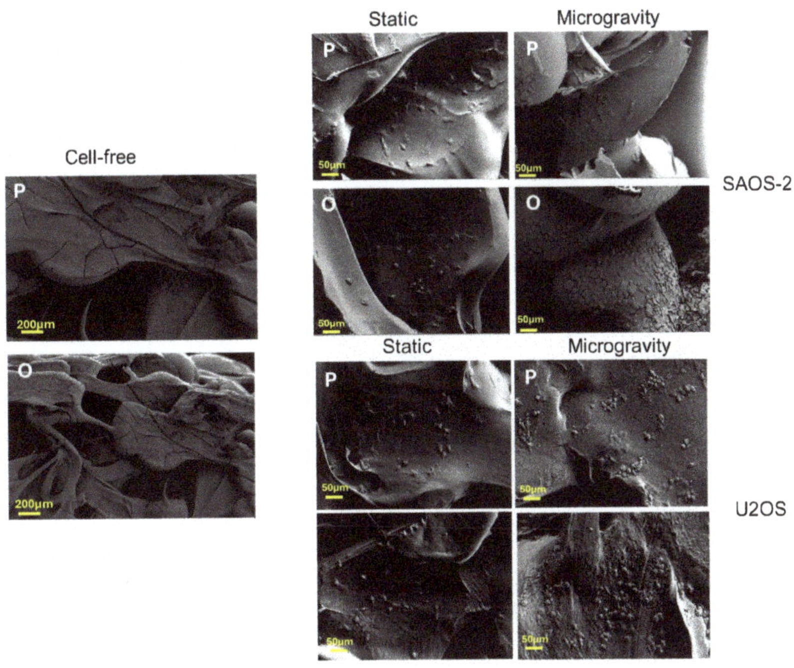

Figure 7. SEM analysis of cell-free scaffold and SAOS-2 and U2OS cells grown on P and O scaffold models for 4 days.

3.5. *TNF-α Secretion Analysis*

The results of cell growth, metabolic activity, and cell adhesion confirmed the biocompatibility of the analyzed scaffolds and the dynamic bioreactor. To further support these findings, and as a preliminary step to evaluate the potential effects of the combination of scaffolds and RCCS culture conditions, the pro-inflammatory TNF-α levels secreted by SAOS-2 and U2OS cells, grown on P and O scaffolds, were evaluated using an ELISA assay. TNF-α was selected as an initial inflammatory marker due to its well-established role in osteoimmune signaling. A substantial body of evidence identifies TNF-α as a key pro-inflammatory cytokine in osteoimmune interactions, promoting osteoclastogenesis, suppressing osteoblast activity, and facilitating bone resorption under inflammatory conditions [28,29]. Figure 8 shows the TNF-α concentration, measured as the absorbance level at 405 nm for all conditions studied at 4 days of culture.

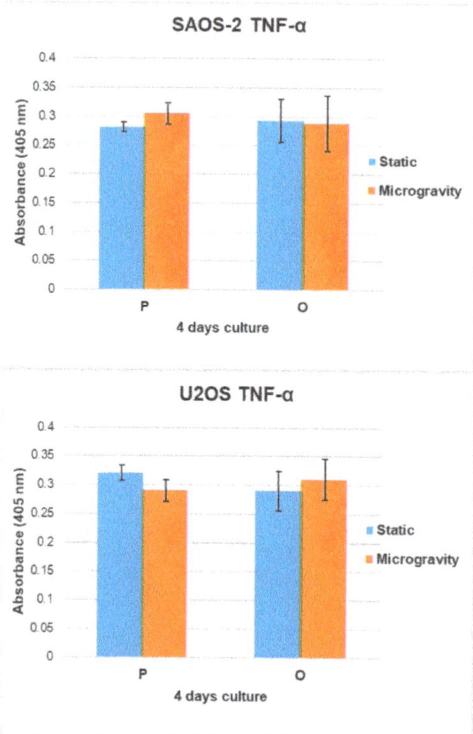

Figure 8. SAOS-2 and U2OS TNF-α concentration analysis grown on P and O scaffold models. $n = 3$, mean ± SD.

Evaluating the secretion levels of TNF-α, no experimental group produced detectable amounts of this proinflammatory factor: the concentration in the samples was undetectable compared to the measured absorbance and the standard curve of TNF-α concentration. This outcome states that the scaffolds do not induce an inflammatory response after 96 h of culture, further confirming the biocompatibility of the substrates and the dynamic platform.

4. Discussion

4.1. Bioreactor System and Scaffold Design

Three-dimensional cell culture systems represent a major advancement in biomedical research, providing an improved in vitro model to study cellular behavior in a more physiologically relevant microenvironment. In addition, their integration with dynamic bioreactors has opened new opportunities in tissue engineering and regenerative medicine [30]. In this study, we investigated the interaction between osteoblast-like cells and 3D-printed scaffolds that mimic physiological and osteoporotic bone structures under simulated microgravity conditions. Bone remodeling is significantly affected by microgravity, inducing a faster resorption process than ossification [31]. Gravity and mechanical stimuli, e.g., tensile and compressive stresses, fluid-exerted shear stress, and hydrostatic pressure, concur to sustain the maintenance of healthy tissues and cells [32]; these input conditions need to be investigated in detail to develop specific countermeasures. In this regard, additive manufacturing may be a suitable technology to prepare tissue replicas to foster bone research in

3D conditions. The possibility to modulate the architecture of the bone microenvironment, in terms of pore size, porosity, and trabecular arrangement, can be promptly realized to test the biological output of different bone-related cells [33]. To promote an advancement in the field and perform in vitro analyses resembling bone physiopathology, PLA bone scaffolds with increasing porosity from a low physiological value to an osteoporotic range were designed and 3D printed using fused deposition modeling. The morphometric characteristics of the P model, measured using μ-CT, were slightly different from the literature data with respect, in this case, to the physiological range of the human proximal ulna [18]; these differences can be ascribed to the resolution limit of the 3D printer. Regarding the design values selected for the O model, these were set to create a biomimetic trabecular architecture whose structural parameters fall within the osteoporotic range, referred to as the femoral head [14,17,19–21], which is a skeletal site dramatically prone to osteoporosis both under normal Earth gravity conditions and in weightlessness.

4.2. Cellular Responses and Biocompatibility

Regarding biological assessment, the results showed that both the osteoblast-like cell lines used positively interacted with all PLA substrates [34], demonstrating that the fabricated scaffolds and the dynamic cell culture system studied were biocompatible and able to support cell growth, adhesion, and metabolic activity without an inflammatory response, as proven by BrdU, WST-1, and ELISA assays.

SAOS-2 and U2OS cells, derived from human osteosarcomas with different levels of differentiation, represent well-established models for bone tissue engineering investigations. They can be used for studying the osteogenesis process and how it could be affected by different treatments. SAOS-2 cells have the capability to produce a mineralized matrix and express osteocyte marker genes, as well as the potential to mimic the interaction between primary human osteoblasts and biomaterials [13]. U2OS has a lower osteoblastic differentiation level and exhibits mesenchymal cell properties [26]. For this reason, both cell lines were chosen in this work as we aimed to confirm our results using two distinct cell lines with different differentiation levels, ensuring the robustness and reproducibility of the findings across different cellular contexts. We aimed to confirm our results using two distinct cell lines with different differentiation levels, ensuring the robustness and reproducibility of the findings across different cellular contexts. This comparative approach provided deeper insights into how scaffold properties and bioreactor conditions influence cellular behavior.

The cells were cultured for a short period of four days across the different experimental conditions, as the primary goal was to examine the early phase of cell–scaffold interactions. Specifically, this study aimed to assess how scaffold microarchitecture and dynamic fluid conditions influence initial cellular responses such as adhesion, viability, and metabolic activity. These preliminary findings were instrumental in validating the proposed in vitro platform, which combines 3D-printed scaffolds with a rotary dynamic culture system. Establishing this integrated setup represents a foundational step for future long-term studies focused on osteogenic differentiation and bone tissue remodeling in dynamic environments that more closely replicate physiological conditions.

The results obtained using BrdU and WST-1 assays highlighted that scaffold morphology, mainly porosity, affects cell adhesion and growth when comparing static and dynamic cell culture conditions. This effect is greater in dynamic conditions because the 3D culture environment and the fluid motion support cells to interact with a larger surface area of the substrate, thus amplifying the effect of the different scaffold porosity on cell adhesion and viability [35,36]. Particularly, the best cell colonization and metabolic response occurs for cells cultured within the O model in the RCCS bioreactor, suggesting that the combination of this culture condition and the most porous scaffold (O model) provides the best cell–scaffold interaction, probably due to a suitable fluid flow condition established within the substrate framework. This finding underscores the importance of dynamic culture conditions, which allow cells to interact with a larger surface area, improving attachment and viability.

The absence of detectable TNF-α secretion by cells cultured under both static and dynamic conditions on the scaffolds confirms the system's biocompatibility, as already demonstrated by the results on cell growth, adhesion, and metabolic activity, and further suggests a lack of inflammatory response [33,36]. TNF-α is a key mediator of early osteoimmune activation; however, it represents only one element of the broader inflammatory cascade. While the current findings provide encouraging evidence of scaffold and bioreactor biocompatibility, the absence of TNF-α alone offers only a preliminary indication of the system's inflammatory behavior and should be interpreted within this limited context. For a more comprehensive assessment of immunocompatibility in dynamic 3D culture systems, future studies should include additional markers such as IL-6 and IL-1β, and possibly co-culture models.

4.3. Implications for Bone Tissue Engineering Research

The combination of additive manufacturing and rotary cell culture systems enabled the creation of biomimetic platforms with precise control over scaffold microarchitecture, offering an innovative system to study bone remodeling and for tissue engineering applications. The advantages of 3D cell culture models are particularly evident when studying bone tissue engineering strategies. Traditional 2D cell cultures fail to replicate the complex structural and mechanical properties of bone, limiting their relevance in preclinical research. The RCCS bioreactor system, by mimicking a low-shear microgravity environment, provides a more biomimetic approach for investigating bone cell behavior, particularly in the context of osteoporosis and microgravity-induced bone loss. The enhanced nutrient diffusion and dynamic fluid conditions within the RCCS further contribute to a more physiologically relevant culture environment. At the same time, it should also be underlined that the dynamic conditions established within a rotating bioreactor lead to a non-uniform flow regimen, which might affect cell response in terms of shear stress and nutrition/waste transfer that is locally experienced (e.g., center vs. periphery of the vessel). However, the collected results, supported by additional studies, can have important implications for both space research and terrestrial biomedical applications. The ability to simulate bone loss conditions in microgravity offers a valuable tool for studying the mechanisms of osteoporosis and developing countermeasures. Additionally, the use of high-porosity scaffolds in a dynamic 3D culture system may provide novel insights into optimizing scaffold design for bone tissue engineering. Future research should focus on refining scaffold materials and architecture, as well as investigating the role of interstitial fluid dynamics in regulating osteogenic activity. Clearly, a refinement of the biological protocol is also necessary to evaluate the cell response to longer conditioning periods in the RCCS and to specifically address microgravity-induced alterations in bone remodeling by means of specific assays, including osteogenic marker expression analysis. By leveraging advanced 3D cell culture systems and dynamic bioreactors, this research may pave the way to bridge the gap between in vitro models and in vivo applications. The integration of biomimetic scaffolds with simulated microgravity conditions presents a promising strategy for improving bone tissue engineering approaches, ultimately contributing to the development of more effective therapies for osteoporosis and other degenerative bone disorders.

5. Conclusions

Data from the biological assays of this work show that 3D-printed PLA scaffolds with an osteoporotic bone-like structure, rather than worsening cell growth, improve cell response when cultured in a RCCS bioreactor to simulate weightlessness. This outcome is probably related to the fluid dynamics established within the scaffold, which contributes to enhancing the cell response. A 3D-printed osteoporotic-like scaffold may be regarded as a potential model to assess possible countermeasures to stimulate and improve cell activity, simulating bone disorders when exposed to dynamic culture conditions. A RCCS bioreactor can represent a suitable in vitro set-up for this aim thanks to the 3D culture capability, fluid dynamics, nutrient transfer, and gas exchange that allow us to perform more effective and specific studies with respect to the conventional static approach. These results need to be further refined, as microgravity-simulating bioreactors and osteoporotic bone models could be used to better analyze the effects of fluid flow in the osteo-activity regulation as a means to design a potential protocol to treat microgravity and/or age-related osteoporosis. Biological data also suggest that it is probably feasible to achieve a fluid condition suitable for osteogenic activities even in bone tissues with higher porosity, supporting future research projects that would be focused on the role of interstitial fluid flow in osteoporotic bone.

Supplementary Materials: The following supporting information can be downloaded at: https://www.mdpi.com/article/10.3390/jfb16080271/s1, Figure S1: Higher-magnification SEM images of SAOS-2 and U2OS cells grown on P and O scaffold models for 4 days. The cells have the typical polygonal, spindle-shaped, and fibroblast-like morphology and are strongly adherent to the scaffolds.

Author Contributions: Conceptualization, M.L., C.D.G. and A.L.; methodology, G.G., E.Z. and L.P.; software, E.S.; validation, E.Z., G.G. and E.S.; formal analysis, E.Z. and G.G.; investigation, E.Z., G.G., L.P., R.P. and A.C.; resources, M.L. and A.L.; data curation, M.L. and G.G.; writing—original draft preparation, M.L.; writing—review and editing, M.L., A.L., E.Z., E.S., C.D.G. and C.A.; supervision, M.L., A.L. and C.D.G.; project administration, M.L.; funding acquisition, M.L. and A.L. All authors have read and agreed to the published version of the manuscript.

Funding: This research was funded by MUR DSB.AD006.371.010/Invecchiamento attivo e in salute (FOE 2022) IFT CUP B53C22010140001. This research was co-financed by the European Union—Next Generation EU, within the PNRR project "Rome Technopole—Innovation Ecosystem" (ECS00000024, CUP B83C22002890005).

Institutional Review Board Statement: Not applicable.

Informed Consent Statement: Not applicable.

Data Availability Statement: The original contributions presented in this study are included in the article; further inquiries can be directed to the corresponding authors.

Acknowledgments: The authors would like to thank Francesca Villani, Matilde Paggiolu, and Pamela Papa for their administrative and technical support.

Conflicts of Interest: Eleonora Zenobi is employed in E. Amaldi Foundation and Hypatia Research Consortium; Elisa Scatena is also employed in E. Amaldi Foundation and Hypatia Research Consortium; Luca Panizza is employed in Violatech srl as Product Specialist; Carlotta Achille is employed in E. Amaldi Foundation. The remaining authors declare that the study was conducted in the absence of any commercial or financial relationships that could be construed as potential conflicts of interest.

References

1. Habanjar, O.; Diab-Assaf, M.; Caldefie-Chezet, F.; Delort, L. 3D Cell Culture Systems: Tumor Application, Advantages, and Disadvantages. *Int. J. Mol. Sci.* **2021**, *22*, 12200. [CrossRef] [PubMed]
2. Wang, Z.; Sun, Y.; Li, C. Advances in 3D printing technology for preparing bone tissue engineering scaffolds from biodegradable materials. *Front. Bioeng. Biotechnol.* **2024**, *12*, 1483547. [CrossRef] [PubMed]
3. Winarso, R.; Anggoro, P.W.; Ismail, R.; Jamari, J.; Bayuseno, A.P. Application of fused deposition modeling (FDM) on bone scaffold manufacturing process: A review. *Heliyon* **2022**, *8*, e11701. [CrossRef] [PubMed]
4. Hidaka, M.; Su, G.N.; Chen, J.K.; Mukaisho, K.; Hattori, T.; Yamamoto, G. Transplantation of engineered bone tissue using a rotary three-dimensional culture system. *In Vitro Cell. Dev. Biol. Anim.* **2007**, *43*, 49–58. [CrossRef] [PubMed]
5. Boskey, A. *Mineralization, Structure and Function of Bone*; Academic Press: San Diego, CA, USA, 2006; pp. 201–212.
6. Weiner, S.; Wagner, H.D. The material bone: Structure mechanical function relations. *Annu. Rev. Mater. Sci.* **1998**, *28*, 271–298. [CrossRef]
7. Salgado, A.J.; Coutinho, O.P.; Reis, R.L. Bone tissue engineering: State of the art and future trends. *Macromol. Biosci.* **2004**, *4*, 743–765. [CrossRef] [PubMed]
8. Seeman, E. The structural and biomechanical basis of the gain and loss of bone strength in women and men. *Endocrinol. Metab. Clin. N. Am.* **2003**, *32*, 25–38. [CrossRef] [PubMed]
9. Hammond, T.G.; Hammond, J.M. Optimized suspension culture: The rotating-wall vessel. *Am. J. Physiol. Renal. Physiol.* **2001**, *281*, F12–F25. [CrossRef] [PubMed]
10. Tang, Y.; Xu, Y.; Xiao, Z.; Zhao, Y.; Li, J.; Han, S.; Chen, L.; Dai, B.; Wang, L.; Chen, B.; et al. The combination of three-dimensional and rotary cell culture system promotes the proliferation and maintains the differentiation potential of rat BMSCs. *Sci. Rep.* **2017**, *7*, 192. [CrossRef] [PubMed]
11. Foroughi, A.H.; Valeri, C.; Jiang, D.; Ning, F.; Razavi, M.; Razavi, M.J. Understanding compressive viscoelastic properties of additively manufactured PLA for bone-mimetic scaffold design. *Med. Eng. Phys.* **2023**, *114*, 103972. [CrossRef] [PubMed]
12. Guo, W.; Bu, W.; Mao, Y.; Wang, E.; Yang, Y.; Liu, C.; Guo, F.; Mai, H.; You, H.; Long, Y. Magnesium Hydroxide as a Versatile Nanofiller for 3D-Printed PLA Bone Scaffolds. *Polymers* **2024**, *16*, 198. [CrossRef] [PubMed]
13. Zenobi, E.; Merco, M.; Mochi, F.; Ruspi, J.; Pecci, R.; Marchese, R.; Convertino, A.; Lisi, A.; Del Gaudio, C.; Ledda, M. Tailoring the Microarchitectures of 3D Printed Bone-like Scaffolds for Tissue Engineering Applications. *Bioengineering* **2023**, *10*, 567. [CrossRef] [PubMed]
14. Molino, G.; Dalpozzi, A.; Ciapetti, G.; Lorusso, M.; Novara, C.; Cavallo, M.; Baldini, N.; Giorgis, F.; Fiorilli, S.; Vitale-Brovaron, C. Osteoporosis-related variations of trabecular bone properties of proximal human humeral heads at different scale lengths. *J. Mech. Behav. Biomed.* **2019**, *100*, 103373. [CrossRef] [PubMed]
15. Ciarelli, T.E.; Fyhrie, D.P.; Schaffler, M.B.; Goldstein, S.A. Variations in three-dimensional cancellous bone architecture of the proximal femur in female hip fractures and in controls. *J. Bone Miner. Res.* **2000**, *15*, 32–40. [CrossRef] [PubMed]
16. Sandino, C.; McErlain, D.D.; Schipilow, J.; Boyd, S.K. Mechanical stimuli of trabecular bone in osteoporosis: A numerical simulation by finite element analysis of microarchitecture. *J. Mech. Behav. Biomed. Mater.* **2017**, *66*, 19–27. [CrossRef] [PubMed]
17. Porrelli, D.; Abrami, M.; Pelizzo, P.; Formentin, C.; Ratti, C.; Turco, G.; Grassi, M.; Canton, G.; Grassi, G.; Murena, L. Trabecular bone porosity and pore size distribution in osteoporotic patients-A low field nuclear magnetic resonance and microcomputed tomography investigation. *J. Mech. Behav. Biomed. Mater.* **2022**, *125*, 104933. [CrossRef] [PubMed]
18. Viveen, J.; Perilli, E.; Zahrooni, S.; Jaarsma, R.L.; Doornberg, J.N.; Bain, G.I. Three-dimensional cortical and trabecular bone microstructure of the proximal ulna. *Arch. Orthop. Trauma. Surg.* **2023**, *143*, 213–223. [CrossRef] [PubMed]
19. Vale, A.C.; Pereira, M.F.C.; Maurício, A.; Amaral, P.; Rosa, L.G.; Lopes, A.; Rodrigues, A.; Caetano-Lopes, J.; Vidal, B.; Monteiro, J.; et al. Micro-computed tomography and compressive characterization of trabecular bone. *Colloid. Surf. A* **2013**, *438*, 199–205. [CrossRef]
20. Nikodem, A. Correlations between structural and mechanical properties of human trabecular femur bone. *Acta Bioeng. Biomech.* **2012**, *14*, 37–46. [CrossRef] [PubMed]
21. Ozan, F.; Pekedis, M.; Koyuncu, S.; Altay, T.; Yildiz, H.; Kayali, C. Micro-computed tomography and mechanical evaluation of trabecular bone structure in osteopenic and osteoporotic fractures. *J. Orthop. Surg.* **2017**, *25*, 2309499017692718. [CrossRef] [PubMed]
22. Zhou, X.; Zhou, G.; Junka, R.; Chang, N.; Anwar, A.; Wang, H.; Yu, X. Fabrication of polylactic acid (PLA)-based porous scaffold through the combination of traditional bio-fabrication and 3D printing technology for bone regeneration. *Colloids Surf. B Biointerfaces* **2021**, *197*, 111420. [CrossRef] [PubMed]
23. Gremare, A.; Guduric, V.; Bareille, R.; Heroguez, V.; Latour, S.; L'Heureux, N.; Fricain, J.C.; Catros, S.; Le Nihouannen, D. Characterization of printed PLA scaffolds for bone tissue engineering. *J. Biomed. Mater. Res. A* **2018**, *106*, 887–894. [CrossRef] [PubMed]

24. Prideaux, M.; Wijenayaka, A.R.; Kumarasinghe, D.D.; Ormsby, R.T.; Evdokiou, A.; Findlay, D.M.; Atkins, G.J. SaOS2 Osteosarcoma Cells as an In Vitro Model for Studying the Transition of Human Osteoblasts to Osteocytes. *Calcif. Tissue Int.* **2014**, *95*, 183–193. [CrossRef] [PubMed]
25. Czekanska, E.M.; Stoddart, M.J.; Richards, R.G.; Hayes, J.S. In Search of an Osteoblast Cell Model for Research. *Eur. Cells Mater.* **2012**, *24*, 1–17. [CrossRef] [PubMed]
26. Mornet, E.; Stura, E.; Lia-Baldini, A.S.; Stigbrand, T.; Ménez, A.; Le Du, M.H. Structural evidence for a functional role of human tissue nonspecific alkaline phosphatase in bone mineralization. *J. Biol. Chem.* **2001**, *276*, 31171–31178. [CrossRef] [PubMed]
27. Orimo, H.; Shimada, T. Effects of phosphates on the expression of tissue-nonspecific alkaline phosphatase gene and phosphate-regulating genes in short-term cultures of human osteosarcoma cell lines. *Mol. Cell. Biochem.* **2006**, *282*, 101–108. [CrossRef] [PubMed]
28. Takayanagi, H. Osteoimmunology: Shared mechanisms and crosstalk between the immune and bone systems. *Nat. Rev. Immunol.* **2007**, *7*, 292–304. [CrossRef] [PubMed]
29. Schett, G. Effects of inflammatory and anti-inflammatory cytokines on the bone. *Eur. J. Clin. Investig.* **2011**, *41*, 1361–1366. [CrossRef] [PubMed]
30. Aishwarya, P.; Agrawal, G.; Sally, J.; Ravi, M. Dynamic three-dimensional cell-culture systems for enhanced applications. *Curr. Sci.* **2022**, *122*, 149–160. [CrossRef]
31. Zerath, E. Effects of microgravity on bone and calcium homeostasis. *Adv. Space Res.* **1998**, *21*, 1049–1058. [CrossRef] [PubMed]
32. LeBlanc, A.; Matsumoto, T.; Jones, J.; Shapiro, J.; Lang, T.; Shackelford, L.; Smith, S.M.; Evans, H.; Spector, E.; Ploutz-Snyder, R.; et al. Bisphosphonates as a supplement to exercise to protect bone during long-duration spaceflight. *Osteoporosis Int.* **2013**, *24*, 2105–2114. [CrossRef] [PubMed]
33. Leong, K.F.; Cheah, C.M.; Chua, C.K. Solid freeform fabrication of three-dimensional scaffolds for engineering replacement tissues and organs. *Biomaterials* **2003**, *24*, 2363–2378. [CrossRef] [PubMed]
34. da Silva, D.; Kaduri, M.; Poley, M.; Adir, O.; Krinsky, N.; Shainsky-Roitman, J.; Schroeder, A. Biocompatibility, biodegradation and excretion of polylactic acid (PLA) in medical implants and theranostic systems. *Chem. Eng. J.* **2018**, *340*, 9–14. [CrossRef] [PubMed]
35. Ingram, M.; Techy, G.B.; Saroufeem, R.; Yazan, O.; Narayan, K.S.; Goodwin, T.J.; Spaulding, G.F. Three-dimensional growth patterns of various human tumor cell lines in simulated microgravity of a NASA bioreactor. *In Vitro Cell. Dev.-An.* **1997**, *33*, 459–466. [CrossRef] [PubMed]
36. Jin, F.; Zhang, Y.; Xuan, K.; He, D.; Deng, T.; Tang, L.; Lu, W.; Duan, Y. Establishment of three-dimensional tissue-engineered bone constructs under microgravity-simulated conditions. *Artif. Organs* **2010**, *34*, 118–125. [CrossRef] [PubMed]

MDPI AG
Grosspeteranlage 5
4052 Basel
Switzerland
Tel.: +41 61 683 77 34

Journal of Functional Biomaterials Editorial Office
E-mail: jfb@mdpi.com
www.mdpi.com/journal/jfb

www.ingramcontent.com/pod-product-compliance
Lightning Source LLC
Chambersburg PA
CBHW061548210926
51850CB00009B/2032